W9-CBK-776

Tetracyclines in Biology, Chemistry and Medicine

Edited by M. Nelson, W. Hillen and R.A. Greenwald

Birkhäuser Verlag
Basel · Boston · Berlin

Editors:

Dr. Mark Nelson
Paratek Pharmaceuticals
75 Kneeland St
Boston MA 02111
USA

Prof. Dr. Wolfgang Hillen
Lehrstuhl für Mikrobiologie
Friedrich-Alexander Universität
Erlangen-Nürnberg
Staudstr. 5
D-91058 Erlangen
Germany

Prof. Dr. Robert A. Greenwald
Long Island Jewish Medical Center
Division of Rheumatology
Suite 107
410 Lakeville Rd
New Hyde Park, NY 11040
USA

Library of Congress Cataloging-in-Publication Data
Tetracyclines in biology, chemistry and medicine / edited by M. Nelson, W. Hillen, and R.A. Greenwald.
p. cm
Includes bibliographical references and index
ISBN 3764362820 (alk. paper) -- ISBN 0-8176-6282-0 (alk. paper)
1. Tetracyclines. I. Greenwald, Robert A., 1943- II. Hillen, Wolfgang. III. Nelson, Mark L.

RM666.T33 T48 2001
615'.329--dc21

Deutsche Bibliothek Cataloging-in-Publication Data
Tetracyclines in biology, chemistry and medicine / ed. by M. Nelson .... - Basel ; Boston ; Berlin : Birkhäuser, 2001
ISBN 3-7643-6282-0

ISBN 3-7643-6282-0 Birkhäuser Verlag, Basel - Boston - Berlin

Birkhäuser is a member of the BertelsmannSpringer Publishing Group
Printed on acid-free paper produced from chlorine-free pulp. TFC ∞
Cover design: in the background, crystals of the tetracycline repressor protein, TetR; in the foreground, chemical structure of the tetracycline-Mg^{2+} complex and schematic recognition pattern of amino acid side chains of TetR in the binding tunnel. Small pictures: overall structure of the TetR homodimer, and MOLCAD rendering of doxycycline.

Printed in Germany
ISBN 3-7643-6282-0

9 8 7 6 5 4 3 2 1

http://www.birkhauser.ch

Table of contents

Section III Use of tetracyclines as non-antimicrobial medicinal agents (Editor: Robert A. Greenwald)

List of contributors

Steven B. Abramson, Department of Rheumatology, Hospital for Joint Diseases, New York, NY 10003/Department of Medicine, Kaplan Cancer Center, New York University Medical Center, New York 10016, USA

Ashok R. Amin, Department of Rheumatology, Hospital for Joint Diseases, 301 E. 17th St., Room 1600, New York, NY 10003, USA; e-mail: amina01@popmail.med.nyu.edu

George J. Armelagos, Emory University, Atlanta, USA

Mukundan G. Attur, Department of Rheumatology, Hospital for Joint Diseases, NYU School of Medicine, New York, NY 10003, USA

Danielle M. Baker, Department of Oral Biology and Pathology, School of Dental Medicine, State University of New York at Stony Brook, Stony Brook, New York 11794-8702, USA

Christian Berens, Lehrstuhl für Mikrobiologie, FAU Erlangen-Nürnberg, Staudtstr. 5, D-91058 Erlangen, Germany; e-mail: cberens@biologic.uni-erlangen.de

Hermann Bujard, Zentrum für Molekulare Biologie der Universität Heidelberg (ZMBH), Im Neuenheimer Feld 282, D-69120 Heidelberg, Germany

Kristy Collins, Duke University, Durham, UK

Jennifer Cook, Florida State University, Tallahassee, USA

Mandar N. Dave, Department of Rheumatology, Hospital for Joint Diseases, New York, NY 10003, USA

Paul Emery, Rheumatology and Rehabilitation Research Unit, University of Leeds, 36 Clarendon Road, Leeds LS2 9NZ, UK

Christiane Fenske, Institut für Chemie und Biochemie, Ernst-Moritz-Arndt Universität Greifswald, Soldmannstr. 16, D-17487 Greifswald, Germany; e-mail: fenske@uni-greifswald.de

Lorne M. Golub, Department of Oral Biology and Pathology, School of Dental Medicine, State University of New York at Stony Brook, Stony Brook NY 11794, USA

Manfred Gossen, Max-Delbrück-Centrum, Robert-Rössle-Str. 10, D-13125 Berlin-Buch, Germany

Robert Greenwald, Division of Rheumatology, Long Island Jewish Medical Center, Suite 107, 410 Lakeville Road, New Hyde Park, NY 11042, USA; e-mail: rgreen@lij.edu

Roeland Hanemaaijer, Gaubius Laboratory TNO-PG, P.O. Box 2215, 2301 CE Leiden, The Netherlands; e-mail: R.Hanemaaijer@pg.tno.nl

Karen A. Hasty, Department of Orthopaedic Surgery/Campbell Clinic, University of Tennessee Health Science Center, VA Medical Center, 1030 Jefferson Ave., Memphis, TN 38104, USA; e-mail: khasty@utmem.edu

Winfried Hinrichs, Institut für Chemie und Biochemie, Ernst-Moritz-Arndt Universität Greifswald, Soldmannstr. 16, D-17487 Greifswald, Germany; e-mail: hinrichs@uni-greifswald.de

Kristi Kolbacher, State University of New York, Binghamton, USA

Yrjö T. Konttinen, Dept. of Oral Medicine, Surgical Hospital and Institute of Dentistry, University of Helsinki, Helsinki, Finland

Maria Krafeld-Daugherty, Emory University, Atlanta, USA

Jeffrey E. Kudlow, Department of Medicine/Endocrinology, University of Alabama at Birmingham, 1808 7th Avenue South, Rm 756, Birmingham, AL 35294-0012, USA; e-mail: kudlow@uab.edu

Stuart B. Levy, Tufts University School of Medicine, The Center for Adaptation Genetics and Drug Resistance, 136 Harrison Ave, Boston, MA 02111, USA

Jan Lindeman, Gaubius Laboratory TNO-PG, P.O. Box 2215, 2301 CE Leiden, The Netherlands/Dept. of Vascular Surgery, Leiden University Medical Centre, Leiden, The Netherlands

Helena Marzo-Ortega, Rheumatology and Rehabilitation Research Unit, University of Leeds, Leeds, UK

Nirupama Mohandas, Department of Rheumatology, Hospital for Joint Diseases, NYU School of Medicine, New York, NY 10003, USA

Mark L. Nelson, Paratek Pharmaceuticals, Inc., 75 Kneeland Street 5th Floor, Boston, MA 02111, USA; e-mail: mnelson@paratekpharm.com

Indravadan R. Patel, Aventis Pharmaceuticals, Bridgewater, New Jersey 08807, USA

Elisabeth Pook, Institut für Mikrobiologie, Biochemie und Genetik, Friedrich-Alexander-Universität Erlangen-Nürnberg, Staudtstr. 5, D-91058 Erlangen, Germany

Maria Emanuel Ryan, Department of Oral Biology and Pathology, School of Dental Medicine, State University of New York at Stony Brook, Stony Brook, New York 11794-8702, USA

Tuula Salo, Dept. of Diagnostics and Oral Medicine, University of Oulu, Oulu, Finland

Siegfried Schneider, Institut für Physikalische und Theoretische Chemie, Friedrich-Alexander-Universität Erlangen-Nürnberg, Egerlandstr. 3, D-91058 Erlangen, Germany; e-mail: schneider@chemie.uni-erlangen.de

Gerald N. Smith, Department of Medicine, Indiana University School of Medicine, 1110 W. Michigan L0545, Indianapolis, IN 46202, USA; e-mail: gnsmith@iupui.edu

Timo Sorsa, Faculty of Medicine, University of Helsinki, Helsinki, Finland

Roland M. Strauss, Dermatology Department, Royal Hallamshire Hospital, Sheffield, UK

Natascha van Lent, Gaubius Laboratory TNO-PG, P.O. Box 2215, 2301 CE Leiden, The Netherlands

Brad R. Zerler, CollaGenex Pharmaceuticals, Inc., 41 University Drive, Newtown, PA 18940, USA; e-mail: bzerler@collagenex.com

Foreword

The tetracyclines have an illustrious history as therapeutic agents which dates back over half a century. Initially discovered as an antibiotic in 1947, the four-ringed molecule has captured the fancy of chemists and biologists over the ensuing decades. Of further interest, as described in the chapter by George Armelagos, tetracyclines were already part of earlier cultures, 1500–1700 years ago, as revealed in traces of drug found in Sudanese Nubian mummies.

The diversity of chapters which this book presents to the reader should illustrate the many disciplines which have examined and seen benefits from these fascinating natural molecules. From antibacterial to anti-inflammatory to anti-autoimmunity to gene regulation, tetracyclines have been modified and redesigned for various novel properties. Some have called this molecule a biologist's dream because of its versatility, but others have seen it as a chemist's nightmare because of the synthetic chemistry challenges and "chameleon-like" properties (see the chapter by S. Schneider).

My laboratory entered the tetracycline field in the mid-1970s when we sought to understand the mechanism of bacterial resistance to this family of antibiotics. Bacteria resistant to this commonly used antibiotic had emerged but the molecular basis for resistance was not known. We began our studies in *Escherichia coli* and uncovered what was then the first antibiotic resistance mechanism based on energy dependent drug efflux. For years the tetracycline efflux systems represented an unusual mechanism of resistance and their modulation is covered by my colleague, Mark Nelson, who has spent many years in my laboratory studying tetracycline chemistry as well as its history, biology and significance in medicine. While efflux is one of the major ways how bacteria and other microorganisms avoid the action of chemotherapeutic agents, our work together has shed new light on antibiotic resistance phenotypes and tetracyclines capable of reversing resistance.

This book provides readers a chance to sample various areas of tetracycline interest reported by experts in the fields of biology and chemistry. Paul Emery describes the benefits provided by tetracyclines in treating autoimmunity – generally linked to their effect on metalloproteases. In fact, special attention has been given to tetracyclines' effect on mammalian metalloproteases and cartilage (K. Hasty, R. Hanemaijer, A. Amin). Maria Ryan describes their use in counteracting gingivitis and promoting dental health. In all these areas, the antibacterial activity of the chemical structure is not needed for the activity. Chapters dealing with the separation of biologic from antibacterial activities (L. Golub and R. Greenwald) involve activities of tetracyclines against diverse diseases from periodontitis to arthritis to cancer.

The chemical interaction of these molecules with metals (S. Schneider) provides ways of monitoring structural permutations via spectroscopy and explores the chemical properties of tetracyclines. Their interactions with RNA suggest other potential uses in controlling intrinsic and externally acquired diseases (C. Berens). The photosensitivity of the molecule (B. Zerler) provides a reason for sun sensitivity of patients taking the drug, but more importantly, suggests ways of using these agents to probe biologic questions. In yet another unique example, tetracyclines are part of a defined regulatory system where they control gene expression (W. Hinrichs, J.E. Kudlow, M. Gossen and H. Bujard).

It is hoped that the book will serve to stimulate others to look at tetracyclines for their broad potential use in the many areas described and in yet other areas. They are ideal substances with which to work. Available in bulk quantities from fermentation, they can be modified to fit human and animal health needs.

The volume also presents to the reader an extensive bibliography beginning with the early work on tetracyclines through the present. The tetracyclines have experienced a multifaceted history and an exciting present. We can look forward to an equally fascinating future.

Stuart B. Levy M.D.

Section I
Chemistry, biology and microbiologic action of tetracyclines

(Editor: Mark L. Nelson)

Tetracyclines in Biology, Chemistry and Medicine
ed. by M. Nelson, W. Hillen and R.A. Greenwald

The chemistry and cellular biology of the tetracyclines

Mark L. Nelson

Paratek Pharmaceuticals, Inc., 75 Kneeland Street, 5th Floor, Boston, MA 02111, USA

Chance favors the prepared mind.

Louis Pasteur,
Quoted in *Dictionary of Scientific Biography*,
Ed. Charles C. Gillespie

Introduction to microbial antagonists

Microbes produce a vast array of toxic chemicals as secondary metabolites in an effort to establish territory and fend off both prokaryotes and eukaryotes alike. These chemicals act as part of a defensive strategy to enable host survival in a hostile microbial environment. When microbes and man initiate "antibiosis", a term created by Vuillemin in the late 1800s to describe when "one creature (the antibiote) destroys the life of another to preserve its own"[1], the antibiote produces and uses chemical toxins that diffuse radially outward, changing the physiological state of the offending organism by a variety of often lethal molecular mechanisms. In the long term, the antibiote antagonizes the growth and survival of intruding cells.

The principle of microbial antagonism, where one species dominates over another by using chemical signals, is an example of chemical Darwinism, where one organism thrives and survives as the fittest to reproduce in an ecological niche. These well-suited microbes have evolved genetically based chemical defense systems through the process of natural selection to produce chemical toxins to enable them to survive.

Most antibiotics, including the penicillins, aminoglycosides and especially the tetracyclines, the subject of this chapter, possess the chemical capability of modifying both prokaryotic and eukaryotic cellular processes.

Beginnings of the antibiotic era

Even before the discovery of microorganisms and the Germ Theory of Disease, the concept of microbial antagonism emerged in the 1800s, where folkloric medicine prescribed fungal concoctions to paint wounds and cuts in an effort to reduce fever [2]. In 1928, Alexander Fleming, a microbiologist at St. Mary's Hospital in London, demonstrated through both serendipity and empirical methods the concept of fungal antagonism, when a culture of *Staphylococcus aureus* became contaminated with the fungal mold *Penicillium notatum.* Fleming found that colonies of *Staphylococcus* were lysed and destroyed around and distal to the edge of the fungi colonies, producing a zone of growth inhibition. He recognized that the mold must have produced a chemical substance that diffused through the agar and inhibited the growth of the bacteria, causing him to exclaim about the "remarkable power of [agent] diffusion into the agar and of inhibiting growth" [3]. Fleming published this observation shortly thereafter [4] and then abruptly redirected his research efforts after he found the mysterious chemical from the mold hard to produce, unstable and seemingly medically unpromising, although he had now named the substance penicillin.

A decade later, biochemists at Oxford University in England, Howard Florey and Ernst Chain, with chemist Norman Heatley, found his report and began producing penicillin in large enough quantities for further study and use against infectious diseases, and met with large success. It not only inhibited the growth of bacterial pathogens *in vitro*, it also stopped life-threatening infections *in vivo*, at least in mice, a finding that Fleming had also studied [5]. Several years later Heatley and chemist Andrew J. Moyer increased the production of penicillin dramatically, by improving fermentation methods and by scaling up the process, growing the mold first in bedpans and then in large beer vats [6].

They also developed chemical and biological methods for the isolation, standardization and characterization of penicillin, sparking the interest of and investment into further research by large pharmaceutical companies in the United States. It was these combined efforts that led to further understanding of its biological potential and use in public health. Soon penicillin became the first chemical ever to be called a "wonder drug" in both scientific and written history, treating bacterial infections both acute and chronic.

This initiated a rush throughout the 1930s and '40s by scientists to discover other antibiosis and wonder drugs, financed by large pharmaceutical companies. Finding antagonist compounds to common infections and diseases even became a major focus of research for numerous pharmaceutical houses, where the benefits and profits from the discovery of such a drug far outweighed the costs of research and development.

In 1944, the antibiotic streptomycin was discovered from the soil bacterium *Streptomyces griseus* by the eminent soil microbiologist Selman Waksman [7]. This compound was one of the first found to be active against the tuberculosis bacillus, *Mycobacterium tuberculosis*, a bacterium for which there was no cure, aside from the isolation and rest assured at the growing number of sanatoria found throughout the U.S. and the rest of the world. It was Waksman who first coined the term "antibiotic", taken to mean "destructive to life", and further developed streptomycin as another major wonder drug isolated from a soil-producing microbe [8]. Streptomycin was also mass-produced and used against other numerous infectious diseases, just as its earlier rival, penicillin.

The discovery of the tetracyclines

In 1947, Benjamin Minge Duggar (1872–1956), botanist emeritus from the University of Wisconsin (Fig. 1), was hired at the age of 76 by the Lederle Laboratories Division of American Cyanamid, to bioprospect for soil microorganisms producing new antibiotic substances. Out of the many samples he obtained, one of them, from a timothy hay field near Columbia, Missouri, labeled sample A377, produced an unusual yellow-colored microorganism from the genus *Streptomyces*. He aptly named the species *aureofaciens* (gold

Figure 1. Benjamin Minge Duggar (1872–1956) discovered the tetracyclines.

surface), and reported his finding first in 1948 in the *Annals of the New York Academy of Sciences* [9], and then in 1949, as the first patent ever in the field of tetracycline research entitled *Aureomycin and Preparation of Same* (US patent 2,482,055) [10]. He reported that the strain produced in high yield a chemical of unknown identity and structure, and named it *Aureomycin*, in reference to its yellow color and isolation from Waksman's *Streptomyces* species. Furthermore, aureomycin was effective against many bacteria and had remarkable activity against rickettsias and viral pathogens.

But chemically aureomycin did not resemble the material Waksman reported 4 years earlier, streptomycin; it had a unique chemical structure that for the next several years proved hard to solve.

In 1950, the company Chas. Pfizer and Co. described a compound of similar character and named their substance *Terramycin*, obtaining patent protection for its fermentation and production [11]. Now two large companies had entered the race to determine the chemical identities of these substances.

As the fermentation and purification processes for producing tetracyclines improved, the compounds were subjected to rigorous chemical analysis by leading academic research groups of that time, including those of the late chemist-legend and Nobel Laureate Robert Burns Woodward, of Harvard University. Woodward and his group raced to be the first to prove the chemical structure and character of either of the new substances, aureomycin or terramycin. In the 1950s, Woodward at Harvard, whose laboratory chemically characterized other complex chemical structures such as strychnine and vitamin B12, was well staffed and equipped to study both aureomycin I and terramycin II (Fig. 2) and he collaborated with the Pfizer chemists to solve these chemical structures. The chemical structure of terramycin began to be described in 1953 and was given the generic name oxytetracycline II [12]. This work spanned over four years since the discovery of aureomycin and was done in an era that predated modern instrumental techniques, such as mass-spectrometry and NMR.

During this era of organic chemistry, complex natural products had their chemical structure determined only after laborious and sometimes inconclusive chemical degradation methods, where molecules are destroyed or modified by other reagents producing products from which their structure can be deduced. The use of instrumental methods, such as infrared and ultraviolet-visible spectroscopy, were in their infancy and their use was restricted to only a few privileged laboratories.

Woodward and his Pfizer colleagues, led by chemist Lloyd Conover, were able to correctly deduce the correct structure of both aureomycin I and terramycin II by definite, but "albeit small" differences in the ultraviolet spectrum of the two, compared to the aureomycin product obtained by catalytic hydrogenation, deschloroaureomycin III (Fig. 2) [13]. Deschloroaureomycin lacked a chlorine atom and was chemically similar in every way to the compound Tetracyn, a compound that was discovered in fermentation broths in 1954 as a biosynthetic product of *Streptomyces aureofaciens* by researchers at

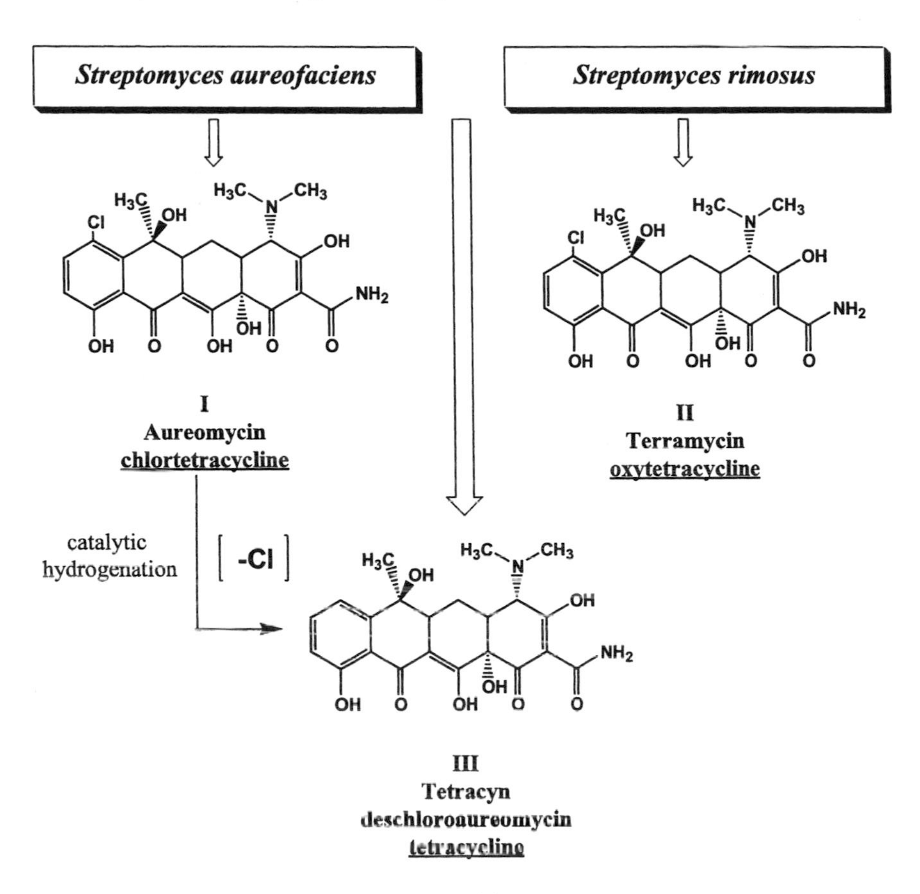

Figure 2. Industrial producers *S. aureofaciens* and *S. rimosus* and the primary tetracyclines they biosynthesize. Tradename; generic chemical name I aureomycin, chlortetracycline, II terramycin, oxytetracycline;. tradename, original name, generic chemical name III tetracyn, deschloroaureomycin, tetracycline. The chemical conversion of chlortetracycline I to tetracycline III used hydrogenolysis reagents H_2 and palladium on carbon [14].

Pfizer. It was Conover at Pfizer, however, who first took aureomycin and chemically modified it by catalytic hydrogenation to produce the more stable and usable antibiotic tetracycline III, the generic structure for which the tetracycline family of molecules is named [14].

During the time that the chemistry and the structure determination of the tetracyclines were slowly evolving, the compounds were rapidly gaining wonder-drug status for possessing the ability to inhibit the growth of a broad range of bacteria. Not only was the growth of Gram-positive bacteria (those that stain violet with the dye crystal violet) such as *Staphylococcus aureus* affected, but also Gram-negative (those that counterstain red with Safranin O) bacteria, such as *E. coli*, *Klebsiella* and the causative agent of epidemic cholera, *Vibrio cholera,* were susceptible to the action of tetracyclines. This unique property of the tetracyclines to affect the growth and physiology of a large spectrum of bacteria allowed for the first time the term *broad spectrum* to be used to describe their activity. The early tetracyclines are represented by the initial members of this class of compounds: chlortetracycline I, oxytetracycline II and tetracycline III (Fig. 2).

Research on the tetracycline family of compounds in this era of basic discovery, the 1950s and 1960s, primarily focused on: the isolation of novel tetracyclines from microorganisms and their fermentation broths, tetracycline chemical characterization, the semisynthesis of novel Tcs and unnatural Tcs (the chemically modified tetracyclines or CMTs, see later in this chapter) and the total synthesis of compounds within the tetracycline family. This global research effort was fueled by the hopes of increasing both the potency of these antibacterial agents and broadening their spectrum of activity against a larger variety of microorganisms.

Biosynthesis of tetracyclines

The biochemistry and regulation of biosynthesis of tetracyclines by *Streptomyces* species were studied primarily by McCormick at Lederle Laboratories in the 1960s and further elaborated by Hostalek and others, who studied the biosynthetic pathways using blocked-pathway mutants and the formation of tetracycline metabolites [15, 16]. By the end of the decade, the biosynthetic pathway and the enzymatic steps of functional group introduction into the tetracycline nucleus had been fairly well elucidated.

Tetracyclines are a diverse family of compounds produced primarily by the sequential enzymatic synthesis and biotransformation of polyketide precursors. The most studied bacteria are within the order *Actinomycetales,* where *Streptomyces aureofaciens* and *Streptomyces rimosus* species, as well as numerous other soil *Actinomycetes*, have been found to produce structural variants of the tetracyclines. For industrial processes and commercial fermentation, *Streptomyces aureofaciens* is typically used to produce chlortetracycline and tetracycline, while *Streptomyces rimosus* produces both tetracycline and oxytetracycline, with the latter produced in high yield (Fig. 2).

Incorporation experiments using radioactive malonate and acetate macromolecular precursors indicate that the building blocks of tetracycline antibiotics are oligoketides, forming the backbone of the tetracycline nucleus (Fig. 3). The synthesis of the initial polyketide chain proceeds by the sequential addition of malonyl–CoA units I to the malonamoylCoA starter unit to form the backbone of the tetracycline ring II [17]. Once a nonaketide is formed, the tetracycline nucleus is produced by a series of enzyme-mediated foldings, ring closings and stereochemistry transformations to produce a naphthacene ring system with multiple pendant functional groups.

The enzymes responsible for the synthesis of the naphthacene ring system are designated as type II synthetases and are derived genetically from specific

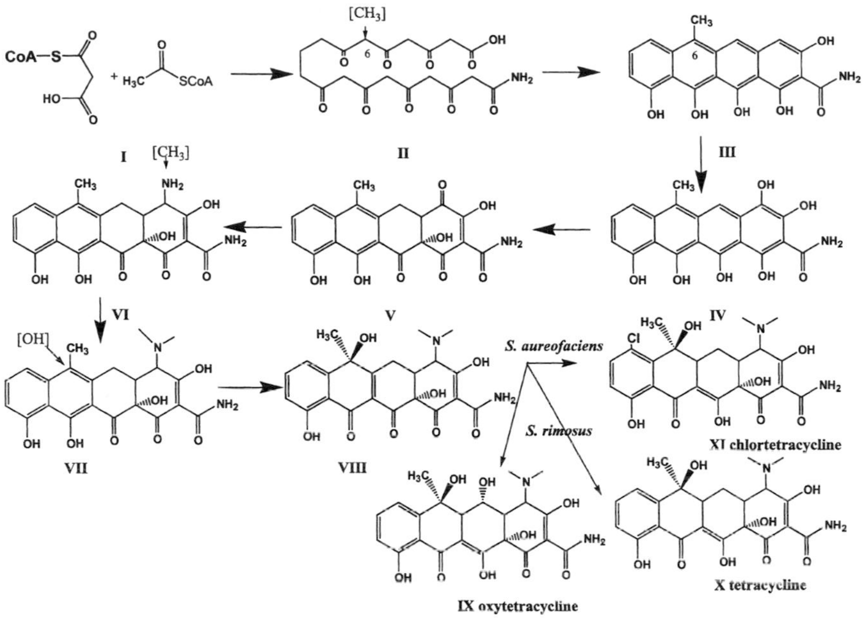

Figure 3. Biosynthesis of oxytetracycline, tetracycline and chlortetracycline by *Streptomyces* species.

gene clusters that function concertedly to produce the nascent β-diketone substructure that is eventually folded into the tetracycline naphthacene ring. In the folding steps, ketoreductase and aromatase enzymes are critical to the formation of the tetracycline nucleus as an intermediate, where methylation at position 6 results in the production of 6-methylpretetramid (III), the first biochemically distinct and isolatable naphthacene ring system obtained in this pathway [18]. The exact details of the enzymatic steps related to the folding and condensation forming the linearly arranged ABCD rings and their aromatization and methylation to the naphthacene nucleus remain to be further elucidated.

6-methylpretetramid III, as a pivotal intermediate, is subjected to a series of enzymatic reactions producing the final products. The C4 position is hydroxylated to the intermediate 4-hydroxy-6-methylpretetramid IV, which then is converted by other hydroxylases and ketoreductases, introducing a hydroxy group into the 12a position of the nucleus while reducing the 4-hydroxyl group to a ketone group, producing 4-oxo anhydrotetracycline V as an intermediate [19]. Amination of position 4 to the 4-amino derivative occurs, forming 4-amino anhydrotetracycline VI, which is then sequentially methylated by methylase enzymes, forming 4-dimethylamino anhydrotetracycline VII. Hydroxylation at carbon 6 by hydroxylase enzymes produces a hydroxyl group at C6 along with an unsaturated bond between the 5a and 11a carbons and is designated 5a(11a)-dehydrotetracycline VIII.

Table 1. Other biologically active tetracyclines and natural polyketides from the order Actinomycetales

Compound	Structure	Source
Demecycline I *antibacterial agent*	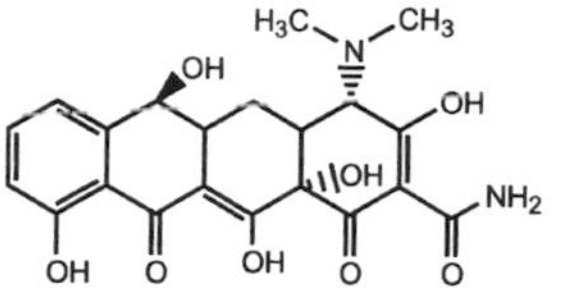	*Streptomyces aureofaciens*
Demeclocycline II *antibacterial agent*	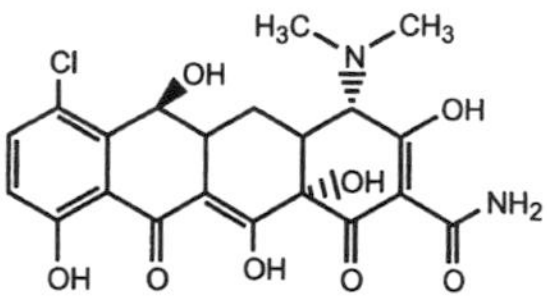	*Streptomyces aureofaciens*
Dactylocycline A III *antibacterial agent*	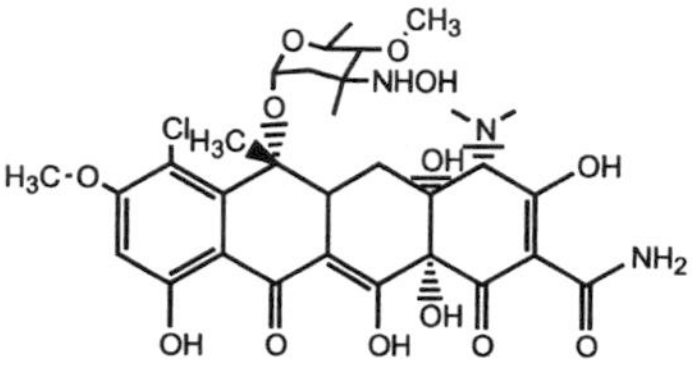	*Dactylosporangium*
Dactylocycline B IV *antibacterial agent*	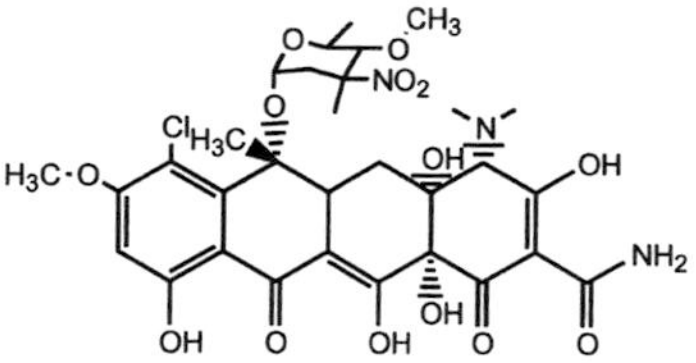	*Dactylosporangium*
2-Acetyl-2-decarboxamido-7-chlortetracycline V *antibacterial agent*	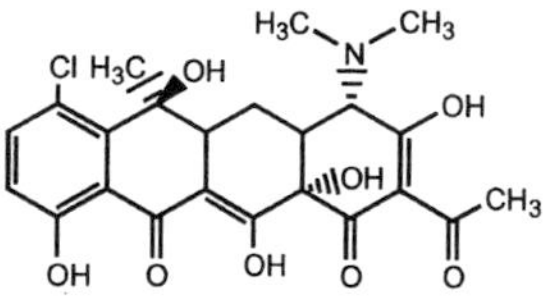	*Streptomyces aureofaciens*
Terramycin X VI *antibacterial agent*	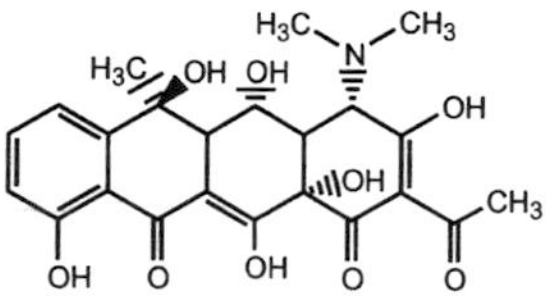	*Streptomyces psammoticus*
Chelocardin VII *antibacterial agent*	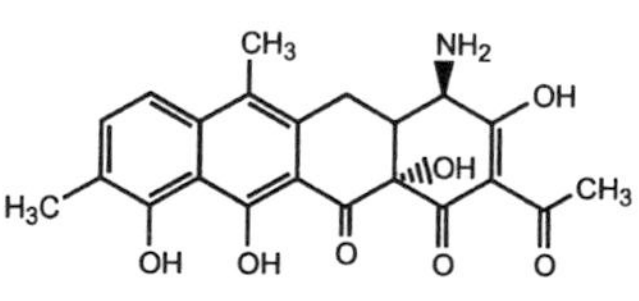	*Streptomyces* species

(continued on next page)

Table 1. (continued)

Compound	Source
Daunorubicin VIII *anticancer agent*	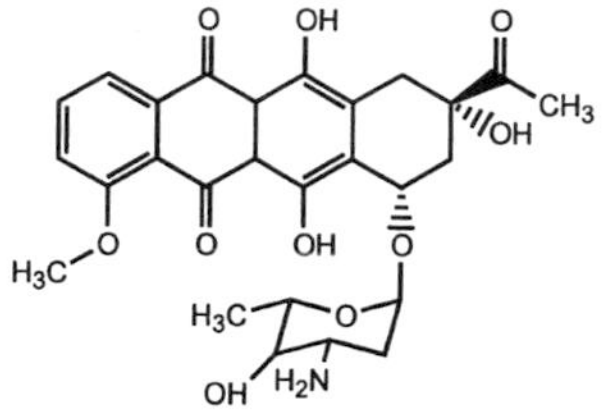 *Streptomyces peucetius*
SF2575 IX *antitumor agent*	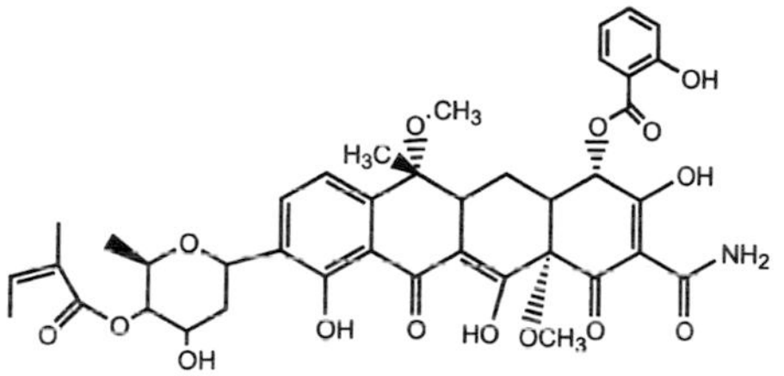 *Streptomyces* sp. SF2575
TAN 1518 A X and B XI *antitumor agent*	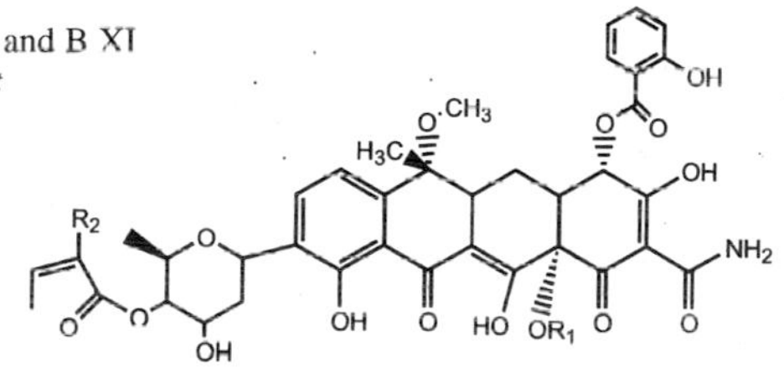 *Streptomyces* sp. AL-16012

TAN 1518	R1	R2
A	H	Me
B	Me	Et

With *S. rimosus* as the antibiotic producer, the substrate is finally hydroxylated at position 5 to yield the final product, oxytetracycline IX. In *S. aureofaciens*, this hydroxylation does not occur. Instead, 5a(11a)-dehydrotetracycline can undergo transformation to two different products, either directly to tetracycline X or the halogenated derivative, 7-chlortetracycline XI, via haloperoxidase enzymes [20].

Other natural tetracyclines

Natural tetracyclines can be obtained by other antibiotic producers of the *Actinomycetales* order, whose members number over 3,000 distinct species in over 40 genera. The species are grouped taxonomically by physiology and ribosomal RNA comparisons and biosynthesize secondary polyketide metabolites in a large chemical diversity of natural products [21]. Polyketide compounds also have a broad range of biological activities against both prokaryotes and eukaryotes. A partial list of lesser-studied natural tetracyclines produced by *Streptomyces* soil bacteria, as well as other genera in the order *Actinomycetales*, is shown in Table 1.

Uncommon tetracyclines, demecycline I and demeclocycline II, derivatives of tetracycline and 7-chlortetracycline lacking the 6-α-methyl group, are readily obtained from fermentation broths and are also characterized as structurally modified tetracyclines and valuable intermediates for the synthesis of other tetracycline compounds [22].

Dactylocyclines A (III) and B (IV) are two fermentation tetracyclines produced by the Actinomycete *Dactylosporangium* sp. SC14051, a soil bacterium discovered in a New Jersey wetland [23, 24]. Dactylocyclines are chemically unique, where the position 4a ring juncture carbon possesses a 4-OH group, while glycosylated at C6 with an aminosugar in an unusual stereochemical orientation, opposite that of the 6-OH of typical tetracyclines. Both compounds were found to have biological activity against antibiotic-resistant Gram-positive bacteria.

2-acetyl-2-decarboxamido-7-chlortetracycline V and terramycin X VI are tetracyclines possessing similar functionality to the common tetracyclines with one chemical distinction: the presence of a 2-acetyl group instead of the 2-carboxamido functional group [25]. Both are produced by several different *Streptomyces* species and have biological activity against bacteria.

Chelocardin VII is yet another tetracycline found to possess a 2-acetyl group and it possesses a position 4-amino group that is also stereochemically opposite from that found in typical tetracyclines [26]. It also possesses a 9-methyl functional group and an aromatized C-ring, adding to its uniqueness as an antibacterial agent. These changes convey activity primarily against Gram-positive bacterial pathogens and decreases Gram-negative antibacterial activity.

Streptomyces species also produce a vast array of polyketide-derived natural products that have been found to be bioactive against eukaryotic cells. The anticancer agent daunorubicin VIII and its congeners doxorubicin and other anthracyclines are produced by several different *Streptomyces* species and possess as a substructure a naphthacene ring system [27]. As anticancer agents they are quite effective, forming reactive species that intercalate with DNA, acting as DNA synthesis inhibitors in actively dividing cells. There is also a pleiotropic component or mechanism of action to these compounds, due presumably to their ability to form free radical species and non-specifically modify macromolecular targets.

The antitumor antibiotic SF2575 IX was isolated from *Streptomyces* sp. SF2575 and has been deposited at the Fermentation Research Institute, Agency of Industrial Science and Technology, Japan, under the name *Streptomyces* sp. SF2575 [28, 29]. This tetracycline was cytotoxic against several different cancer cell lines with an undetermined mechanism of action. Other antibiotics with similar structures have also been isolated from fermentation and described, e.g., the TAN series of anticancer agents. These tetracyclines were isolated from *Streptomyces* species AL-16012, producing three different structural variants designated TAN A (X), TAN B (XI) and TAN X, the latter of which had the same chemical composition and structure as SF-2575 IX [30].

Chemical structure of the tetracyclines

The tetracycline nucleus possesses four linearly fused rings, labeled the ABCD rings by IUPAC [31] convention, and is in the naphthacene ring family of organic polycyclic hydrocarbons. Naphthacene rings I (Fig. 4) are numbered starting at C1 on the A-ring moving counterclockwise, where all skeletal and exocyclic carbon atoms receive a number designation. Bridgehead tertiary carbon atoms receive both a number and letter designation.

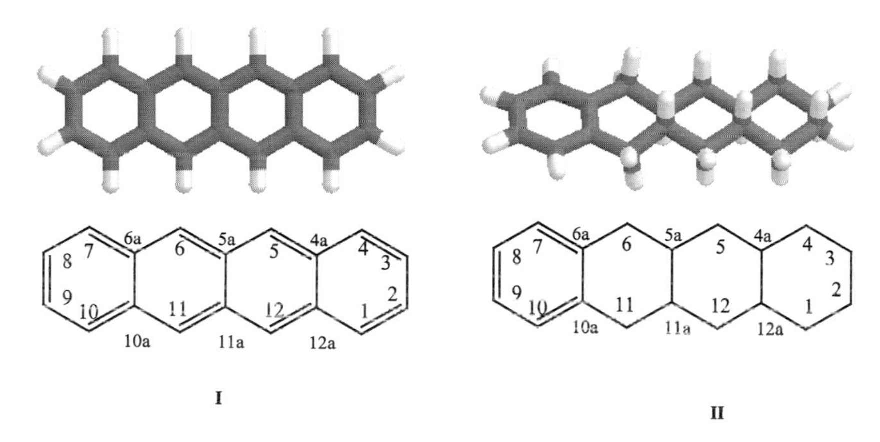

Figure 4. 3-dimensional structure of naphthacene I and dodecahydronaphthacene II and the Fischer projection structure representation with structural locants of the ABCD naphthacene ring system. Drawn using Alchemy 2000 software with PM3 and MM3 geometry optimization. Tripos, Inc., St. Louis, MO.

The tetracycline nucleus is perhydrogenated, where only the D-ring retains aromatization. The basic tetracycline nucleus, possessing all of its hydrogens, is named by convention 1,2,3,4,4a,5,5a,6,6a,11,11a,12,12a-dodecahydronaphthacene II and is essentially a flat, planar molecule with little molecular dynamicism, possessing both alicyclic alkane (the ABC rings) and aromatic character (D-ring).

Major changes in the structural and molecular character of the tetracycline nucleus occur by the introduction of the six oxygen functional groups into the lower region of the molecule and to the C3 carbon, along with a dimethylamine group at C4 producing the compound sancycline I (Fig. 5). Sancycline is the simplest tetracycline with the most basic chemical structure needed for activity in prokaryotes, while eukaryotes and their macromolecular targets can be affected by the 4-dedimethylamino derivatives of tetracyclines II, i.e., those lacking the C4 dimethylamine group.

Introduction of the C10 phenol, the C11-C12 keto-enol substructure, the tertiary hydroxyl group at C12a and the C1 ketone in conjunction with enolate ion formation at C3, adds dynamic structural attributes to the molecule, changing

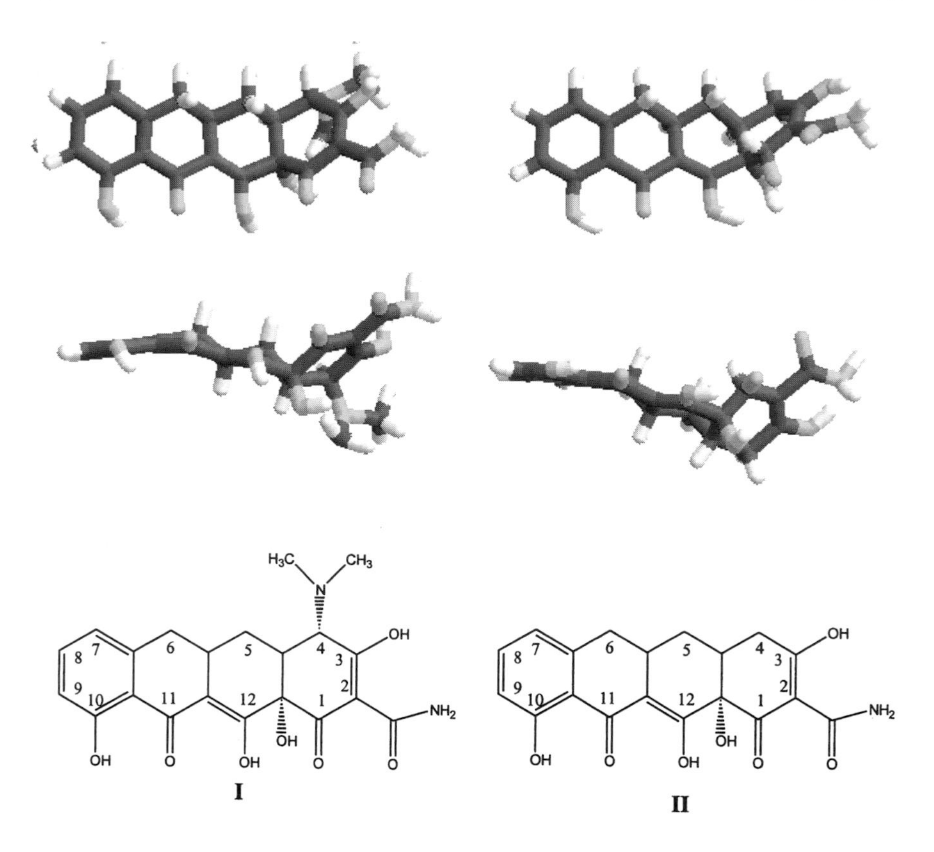

Figure 5. Geometry minimized 3-dimensional structure of sancycline I and 4-dedimethylaminosancycline II. The top structures are the front view and the middle view from the pharmacophore region. The chemical structures are presented in the bottom view.

the naphthacene ring to the structure shown, where the A-ring is bent upwards at an angle about the C4a-C12a bond. Sancycline I and 4-dedimethylsancycline II display a similar three-dimensional shape to that which was obtained from oxytetracycline through x-ray crystallographic methods [32].

The presence of the highly dense oxygen functional group system has an impact on the physicochemical and three-dimensional properties of the naphthacene ring system and is the major region responsible for the pharmacological properties of the tetracycline family. Along the lower periphery of the ring system is the pharmacophore region, the area of three-dimensional spatial arrangement of atoms that convey binding affinity to biological macromolecules and receptors and it is the area most sensitive to chemical modification.

Tetracyclines, in particular oxytetracycline, have been shown to exist as an equilibrium of two different forms, based on solvent environment, although this view is probably simplistic. Hydrophilic phases favor the formation of the zwitterionic "A" form, whereby proton transfer between the 3-OH group and

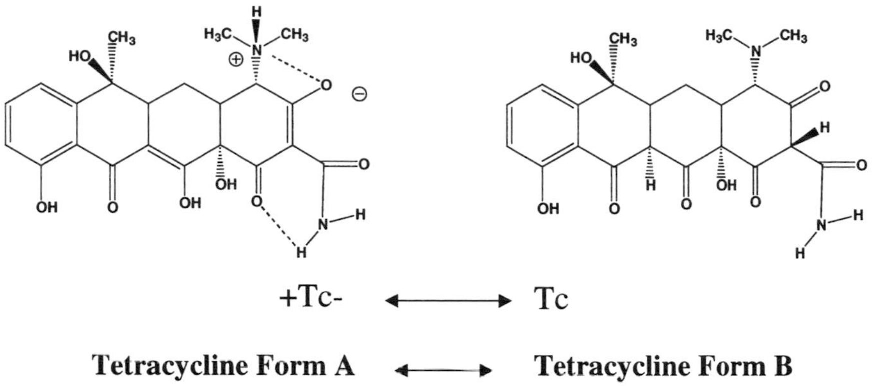

+Tc- ⟷ Tc

Tetracycline Form A ⟷ Tetracycline Form B

Figure 6. X-ray crystallography structures of two bioactive forms of tetracycline, the A zwitterions form and the B unionized form. Adapted from Stezowski [32].

the neighboring C4 dimethylamine occurs to form a twisted charged species I (Fig. 6) [33]. Conversely, an uncharged lipophilic form occurs, the "B" form, which causes the overall structure to extend and become more planar II about the C4a-C12a-axis. It has been suggested that both forms are required for bioactivity, at least in bacteria, where the charged form is required for binding at the ribosome, and the uncharged form is responsible for crossing biological membranes.

Both derivatives show the effect of each individual exocyclic oxygen and/or C4 nitrogen atom and C2 carboxamide on the overall shape and therefore subsequent function of the tetracyclines. Modifications of the C1 and C11 ketone, C3 or C11a-C12 enolization, or the C10 phenol drastically change the structure to a more linear flattened or extended structure, similar to the B form. The β-diketone structure between rings B and C and the process of tautomerization [34] are extremely important for maintaining biological activity, whereby blocking of the 11a position causes severe twisting about the C5a-C11a ring juncture, changing the structure from its normally angled shape. This molecular deformation occurs with C11a halogenated tetracycline derivatives, which inhibits the formation of the C11a-C12 enolization and is inactive against both prokaryotic and eukaryotic targets.

Comparisons and observations of the effect of its functional groups on overall structure suggest that the C4 dedimethylamino tetracycline A form has a similar structure to the C4 dimethylamine A form. The superimposability of both tetracyclines indicates that perhaps the compounds would bind similarly to receptors and show equipotency against cellular targets and biological endpoints. Both compounds, sancycline and the chemically modified tetracycline C4 dedimethylamino derivative, CMT-3 (also known as COL-3), have been studied both chemically and biologically, and data can be readily compared in biological models.

Nomenclature

Tetracyclines are named according to IUPAC convention and the substituents found at the various positions along the naphthacene ring. Thus, sancycline, the simplest tetracycline both chemically and biologically, is named by convention I (Fig. 7):

4-dimethylamino-1,4,4a,5a,6,11,12a-octahydro-3,10,12,12a-tetrahydroxy-1,11-dioxo-2-naphthacenecarboxamide.

4α-sancycline **4β-sancycline**

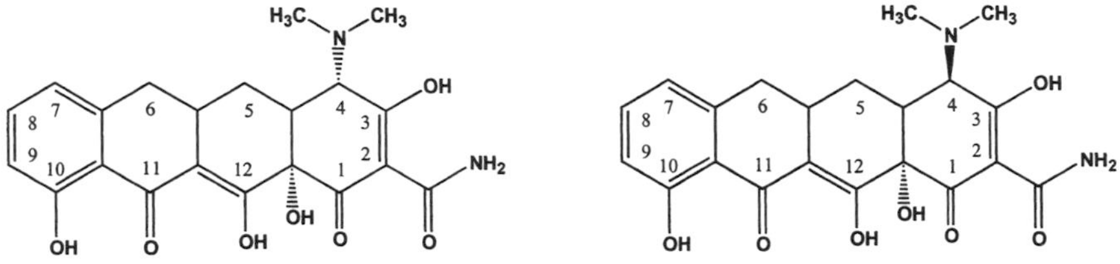

Figure 7. Chemical structures and ring numbering system of 4α-and 4β-sancycline.

Other IUPAC tetracycline names for several clinically significant tetracyclines are shown below, where the changes based on the nomenclature of sancycline are bolded.

Chlortetracycline
7-chloro-4-dimethylamino-1,4,4a,5,5a,6,11,12a-octahydro-3,6,10,12,12a-pentahydroxy-6-methyl-1,11-dioxo-2-naphthacenecarboxamide

Oxytetracycline
4-dimethylamino-1,4,4a,5,5a,6,11,12a-octahydro-3, **5**,6,10,12,12a-hexa**hydroxy**-6-methyl-1,11-dioxo-2-naphthacenecarboxamide

Tetracycline
4-dimethylamino-1,4,4a,5,5a,6,11,12a-octahydro-3, **6**,10,12,12a-hexa**hydroxy**-6-methyl-1,11-dioxo-2-naphthacenecarboxamide

Minocycline
4,**7**-bis-(**dimethylamino**)-1,4,4a,5,5a,6,11,12a-octahydro-3, 10,12,12a-tetrahydroxy-1,11-dioxo-2-naphthacenecarboxamide

Stereochemistry located at the chiral centers within the tetracycline nucleus is denoted within the chemical name, where the orientation is added as S or R, depending upon its three-dimensional spatial arrangement. The designation of the alpha (α) I and beta (β) II orientation depicts the spatial arrangement of two possible diastereomers of the C4-dimethylamino group, the α derivative occupying space below the plane of the naphthacene ring, and the β orientation occupying substituent space above the plane of the ring. The IUPAC designation for minocycline now shows the stereochemistry, where these designations are bolded:

Minocycline
[4**S**-(4α,12aα)-4,7-bis(dimethylamino)-1,4,4a,5,5a,6,11,12a-octahydro-3,10,12,12a-tetrahydroxy-1,11-dioxo-2-naphthacenecarboxamide

There are more basic ways to name tetracyclines and their derivatives, and one will find these interchangeably within the scientific literature. By taking the core structure of tetracycline and then removing and/or adding functional groups, the derivation of simple chemical names for the tetracyclines is possible. For example, oxytetracycline, a generic name, can be described distinctively as 5-hydroxytetracycline. Demeclocycline is thus named 6-demethyl-7-chlortetracycline, and demecycline is denoted as 6-demethyltetracycline. Sancycline is thus named 6-demethyl-6-deoxy tetracycline and so forth. This becomes important when describing the functional groups attached at the various positions forming more complex derivatives.

Another nomenclature convention has emerged recently. The term "chemically modified tetracyclines" or CMTs, refers to tetracyclines that at some position along the naphthacene ring have been synthetically altered and this term delineates semisynthetic derivatives of tetracyclines. CollaGenex Pharmaceuticals, Inc., of Newtown, PA, USA, has described certain CMTs with the prefix COL, designating them as compounds used in their research programs as potential therapeutics in several different areas, some of which will be discussed in the following text. The CMTs, COL designation and their chemical structure are shown in Figure 8, as a guide and reference for other chapters throughout this book.

Chemical properties of the tetracyclines

A tetracycline molecule has two major areas residing within the naphthacene ring system that can be designated the lower (inviolate) and upper (violate) peripheral regions. Both regions are based on their bacteriological activity upon being chemically modified (Fig. 9). The linear arrangement of the ABCD rings is crucial for bioactivity, and, as a general rule, chemical modification along the lower peripheral region decreases bioactivity, whereas the upper peripheral region can be usually chemically modified to produce other bioactive semi-synthetic tetracyclines. The lower peripheral region, however, is the major pharmacophore region sustaining bioactivity in both prokaryotic and eukaryotic systems, where chemical functional groups dictate binding and

CMT-1 COL-1

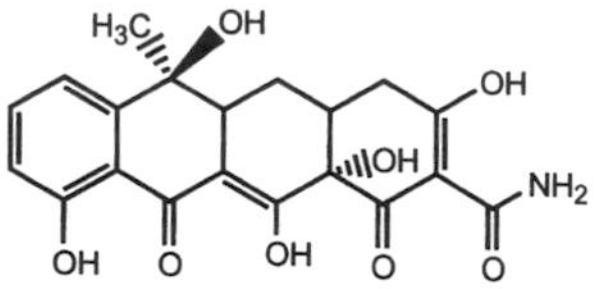

CMT-2 COL-2

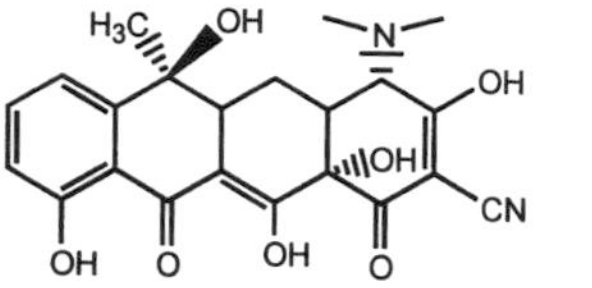

CMT-3 COL-3

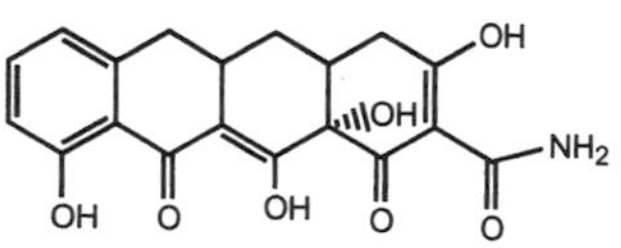

CMT-4 COL-4

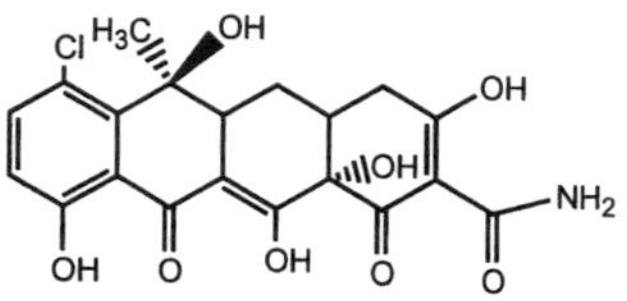

CMT-5 COL-5

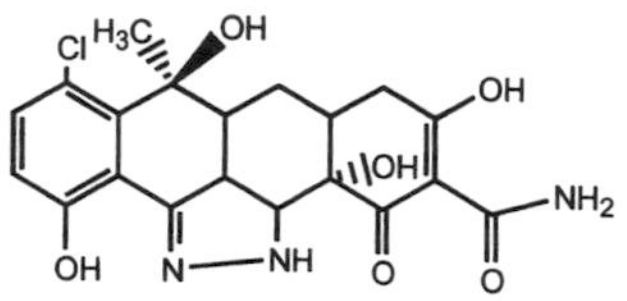

CMT-6 COL-6

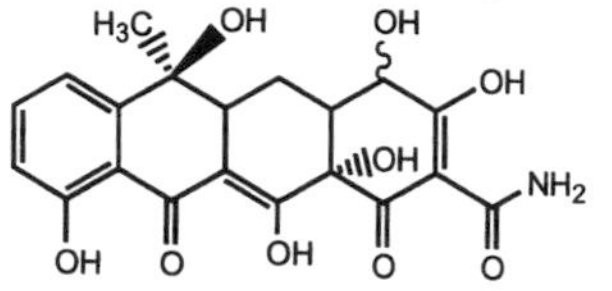

CMT-7 COL-7

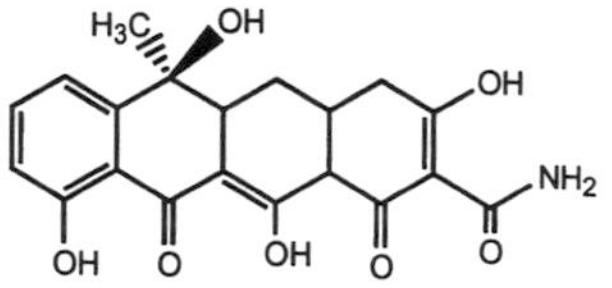

CMT-8 COL-8

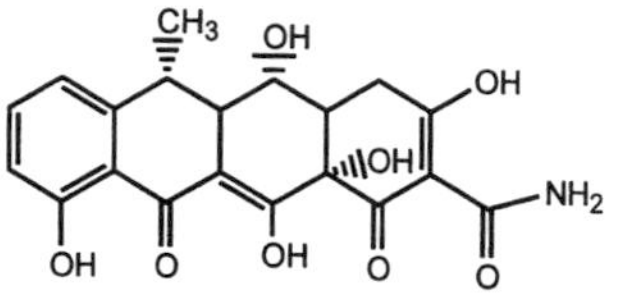

Figure 8. Chemical structures of the chemically-modified tetracyclines (CMTs) and their COL number designation as a reference for use with other chapters.

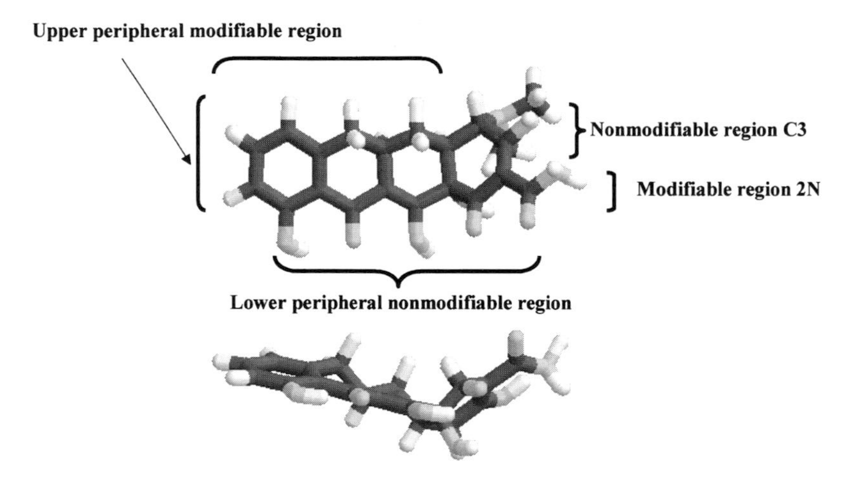

Figure 9. *Top* Geometry minimized 3-dimensional structure of sancycline viewed from the pharmacophore lower peripheral region.

activity. As a complex molecular entity, tetracycline can form different molecular substructures based on the individual dynamics and arrangement of the oxygen functional groups.

Individual attributes of the ABCD ring system contribute to the total tetracycline structure, while their correlation to antibacterial activity is a ready measure of potential activity against different targets, even eukaryotic cells.

Many other factors influence tetracycline binding to receptors, including pH and inorganic cation complexation, and all ultimately affect biological activity. The description of binding events of tetracyclines and metals and their activity profiles remains to be totally elucidated. Further exploration of the role of cations will be discussed in the chapter on Tetracyclines and Proton and Metal Binding.

A-ring chemistry

Within the A-ring of the tetracycline nucleus, five different functional groups are present simultaneously. C1 possesses a carbonyl group, C2 an exocyclic carboxamide group, C3 a keto-enol functional group, which will be elaborated on further, and a dimethylamino group at C4. There is also a tertiary hydroxy group at C12a at the ring AB juncture, separating the A-ring chromophore from the BCD chromophore substructure (Fig. 10).

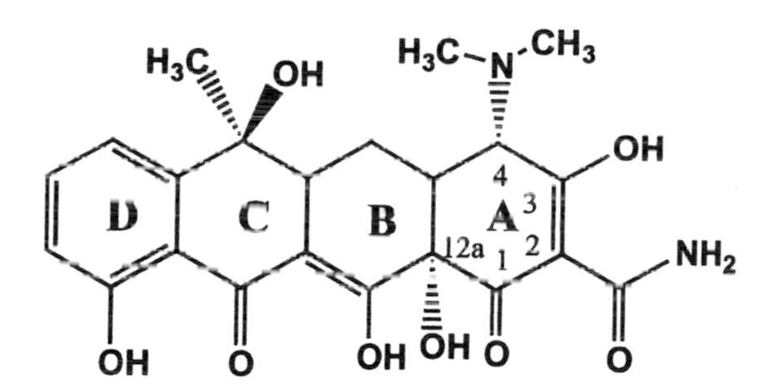

Figure 10. Structural attributes of the tetracycline A-ring.

Given the dense chemical functionality within the A-ring, one can expect and find complex chemical behavior and distinctive chemical dynamicism, which coincide with its ability to ionize and and produce mixtures of zwitterions or charged forms of tetracycline.

The A-ring possesses a tricarbonylmethane keto-enol system spanning C1 I, C2 II and C3 III (Fig. 11), producing an acidic substructure with a pKa of 2.8–3.3. At physiological pH this functional group is ionized, allowing formation of a tautomeric substructure that is crucial for biological activity (Fig. 11). Modification of any one of the carbonyl functional groups or inhibition of tautomerization results in a loss of bioactivity.

The C4 dimethylamine group functions as a base forming a conjugate acid with a weak pKa of approximately 9.4. 4-dedimethylamino derivatives, the

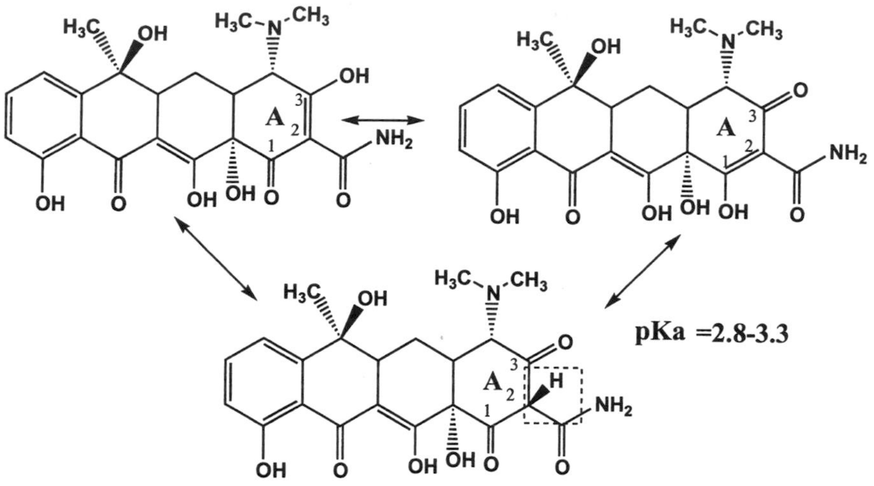

Figure 11. Tautomeric structures of the A-ring tricarbonyl group.

CMT or COL compounds, lack this basic functional group, although ring tautomerization from the C1-C2-C3 tricarbonyl should still occur.

The tetracycline chelocardin VII (Tab. 1), possessing a 2-position methyl carbonyl group, retains biological activity. Tetracyclinenitrile, produced from tetracyclines by carboxamide dehydration using p-toluenesulfonic acid or dicyclohexylcarbodiimide [13, 35], results in a loss of biological activity (Fig. 12). Inactivity may be due to diffusion through biological membranes, where nitrile derivatives are 400-fold less permeant in a liposome model, compared to tetracyclines with a 2-position carbonyl [36]. Chemical modification of C2 position derivatives indicates that the exocyclic carbonyl is crucial for activity, whereas the carboxamide N atom may be synthetically modified.

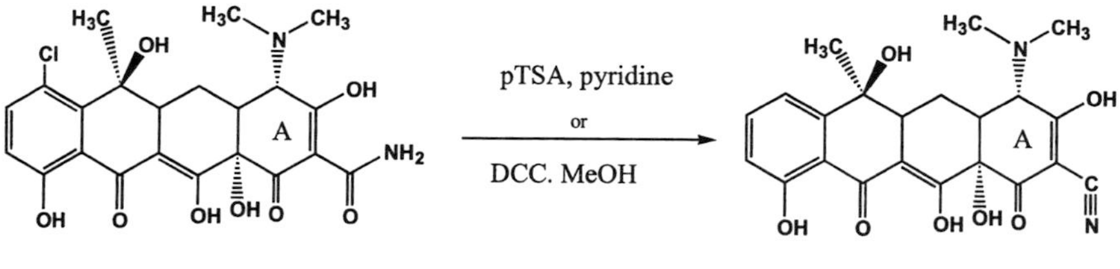

Figure 12. Synthesis of tetracycline C2 position nitrile derivatives.

C2 carboxamide chemistry

The C2 carboxamide nitrogen group is capable of reacting with chemical reagents to produce numerous other tetracycline derivatives. The Mannich reaction, a combination of a carboxamide, aldehyde and nitrogenous base, has been successfully used to covalently synthesize tetracycline prodrug forms I that increase the both the water solubility and the bioavailability of tetracyclines (Fig. 13) [37]. Rolitetracycline, the tradename for the clinically used prodrug form of tetracycline, is formed by the reaction of formaldehyde and pyrrolidine in t-butanol producing the condensation product N-(pyrrolidinylmethyl)-tetracycline II (Fig. 13) [38]. In an aqueous phase, Mannich base prodrugs hydrolyze back to the parent compound with a half-life that is dependent upon the nitrogenous base that is used. The half-life of tetracycline Mannich base prodrugs can vary dramatically, where rolitetracycline has a half-life of 45 min at pH 7.2 and 37 °C [39]. The morpholinomethyl derivative has a half-life of greater than 300 min under the same conditions.

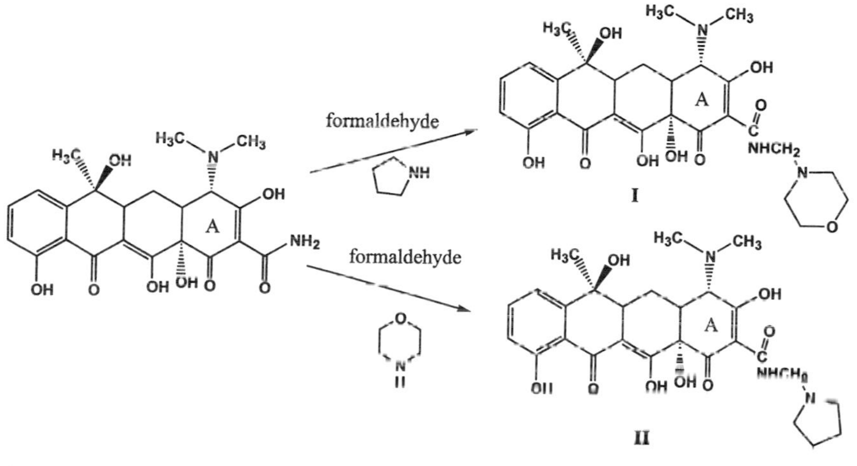

Figure 13. Synthesis of the tetracycline C2 position Mannich bases morpholinomethyltetracycline I and II rolitetracycline.

Many Mannich bases of tetracyclines are reported in the literature, with the most notable being those composed of simple nitrogenous bases and amino acids. The most studied Mannich bases of the tetracyclines have been the pyrrolidine I, morpholine II, lysine III and hydroxyethylpiperazine IV derivatives (Fig. 14).

The carboxamide functional group reacts with aldehydes and alcohols producing N-alkoxymethyl tetracyclines I. Although this class of compounds has not received as much attention as the Mannich products (Fig. 15) [40], they retained antibacterial activity against Gram-positive bacteria.

Tetracyclinenitriles react with tertiary alcohols and isobutylene functional groups to produce monoalkylated tetracyclines I. They retain exceptional activity against Gram-positive bacteria but lose activity against the Gram-negative bacteria (Fig. 16) [41]. The nitrile group may be reconverted back to the carboxamide II using BF_3 followed by hydrolysis [42].

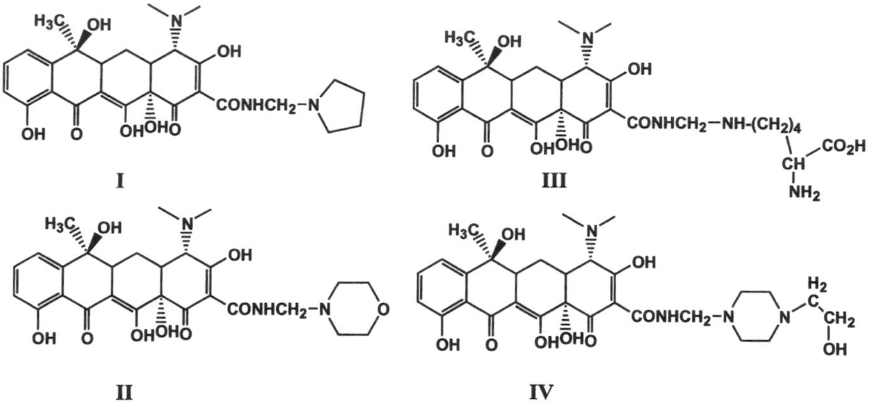

Figure 14. Structures of rolitetracycline I, N-morpholinomethyltetracycline II, N-lysinomethyltetracycline III, and N-hydroxyethylpiperazine tetracycline IV.

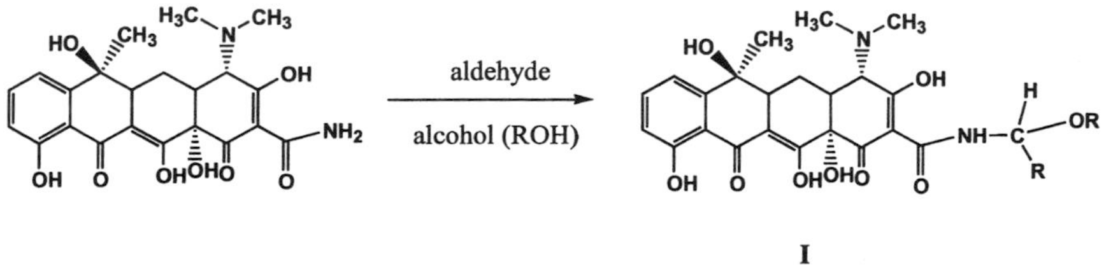

Figure 15. Synthesis of 2N-alkoxymethyltetracyclines.

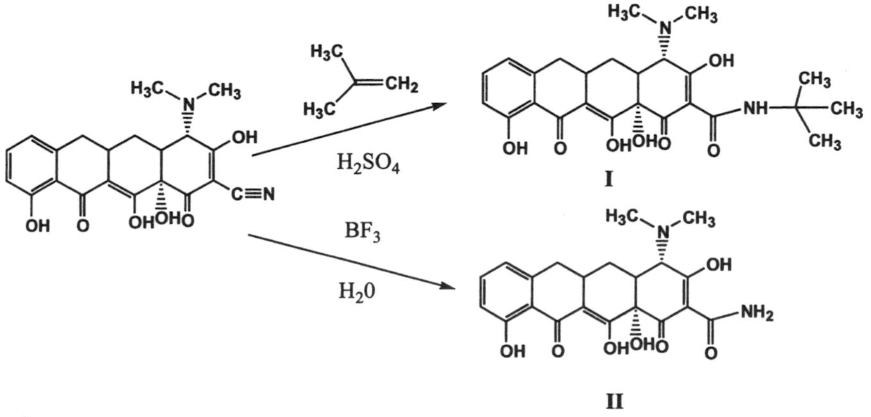

Figure 16. Synthesis of 2N-alkyl derivatives of tetracyclines and nitrile hydrolysis to the carboxamide.

C4 dimethylamino chemistry

The C4 dimethylamino group has been the most studied of all the functional groups on the tetracycline nucleus, after it was discovered that C4 stereochemical orientation is extremely important for antibacterial bioactivity. The C4 dimethylamino group can undergo a reversible process *in vitro* and *in vivo,* producing an unnatural C4 β-epimer product that was earlier called "quatramycin"(Fig. 17) [43, 44].

Tetracycline epimers show differences in antibacterial bioactivity, where the β-epimer possesses only 5% of the activity of the natural α-epimer against Gram-negative bacteria such as *Klebsiella*. Epimerization can be induced by various solvent systems within a pH range of 2–6, although epimerization can occur above pH 7.5 [45]. The kinetics of epimerization are influenced by numerous factors, including the position C2 and C5 substituents, the presence of chelating metals, solvent and buffer systems, and temperature [46]. While neighboring substituents and chelating metals tend to decrease the epimerization process, aqueous environments and acidic solutions increase the production of epimers. C4 epimers are also formed *in vivo*, where epimerization of tetracycline has been demonstrated and measured in muscle and organ tissue samples in animals [47].

4α-tetracycline and the 4-epimer product of tetracycline, 4β-epitetracycline, show considerable differences in antibacterial activity, where 4α-tetra-

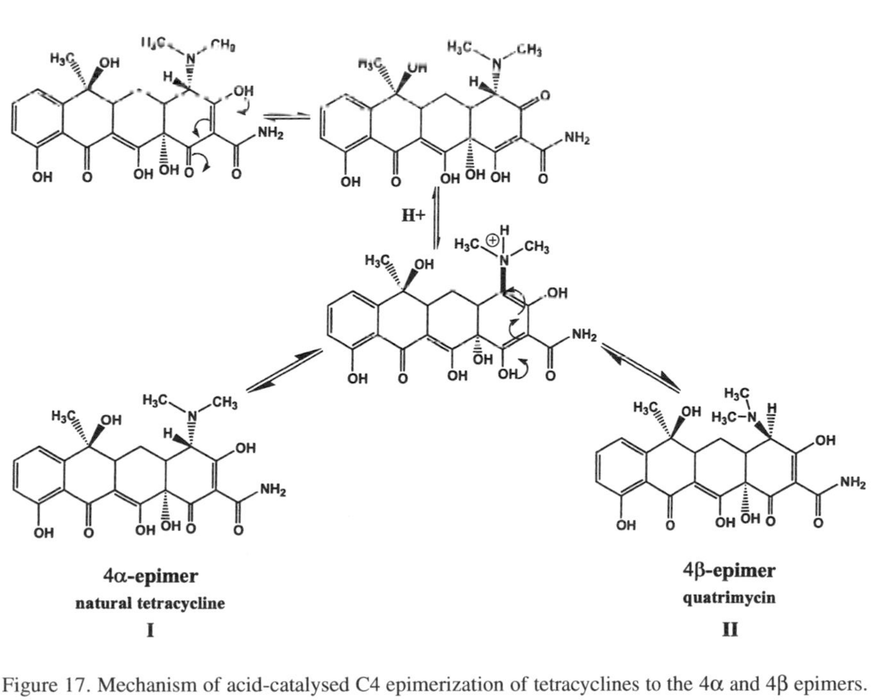

Figure 17. Mechanism of acid-catalysed C4 epimerization of tetracyclines to the 4α and 4β epimers.

cycline has a minimum inhibitory constant (MIC) value of 0.21 and 0.06 μg/mL against *Staphylococcus aureus* and *Streptococcus pyogenes*, respectively, whereas 4β-epiTc has MIC values at 6.0 and 3.0 μg/mL [43]. The 30- to 40-fold decrease in activity upon epimerization suggests that protein-synthesis inhibition is substantially decreased, where 4β-epimer has less agonist activity and is less effective in inhibiting ribosomal function.

Epimerization kinetics have been studied by UV-visible spectroscopy, where absorbance at 250–300 nm, the A-ring chromophore region, is typically monitored [45]. Changes in this region indicate that during epimerization, only the A-ring is changed from one form to another, whereas 4β-epitetracyclines exhibit a lower molar absorptivity (ε).

Nuclear magnetic resonance can also be used to monitor epimerization kinetics, where the C4-β proton in natural tetracyclines shifts downfield when isomerized to the C4-α epimer proton. Minocycline undergoes C4 epimerization in polar protic solvents, such as water, where epimerization is a function of pH, temperature and time, as well as in polar aprotic solvents, such as DMSO. The chemical shift of the natural C4 β-proton of minocycline is clearly evident in an NMR spectrum exhibiting a sharp singlet at δ4.27 ppm, as depicted in Figure 18a. Isomerization and proton exchange at C4 can be demonstrated by heating minocycline in dimethylsulfoxide-d_6 solution, with or without deuterium chloride (DCl) as a deuterium exchange catalyst, for 1 h at 70 °C.

After 1 h at 70 °C in DMSO-d_6 without DCl, the spectrum of minocycline (Fig. 18b) shows the disappearance of the C4-β proton at δ4.27 ppm and the appearance of the C4-α proton downfield at δ4.72 ppm, indicating epimerization at the C4 position.

One can also quantify the relative epimer ratios in a C4 epimer pair by careful integration of the proton spectra. Isomerization produces a downfield chemical shift of about 0.5 ppm, where now the C4-α proton epimer species represents approximately 65% of the sample, while the remainder is the natural isomer. The presence of a mixture of C4 isomers is also evident in the aromatic region of the spectra, where a doublet of doublets is indicative of two A-B spin-coupled systems, consistent with the presence of two compounds.

The NMR spectrum of minocycline (Fig. 18c) in DMSO at 70 °C with DCl shows the characteristic 4α and 4β protons at δ 4.27 and 4.72 ppm, respectively, although the peaks are somewhat minuscule. The DCl added to the DMSO has produced a deuterium exchange, whereby the C4 protons are now exchanged with deuterium, resulting in a loss of C4 proton signal.

Other C4 derivatives

The conversion of the C4 dimethylamine of tetracyclines to quaternary amines I is possible by reacting the starting free base of tetracyclines with alkyl iodides I (Fig. 19) [48]. Tetracyclines possessing a 6-OH group form numer-

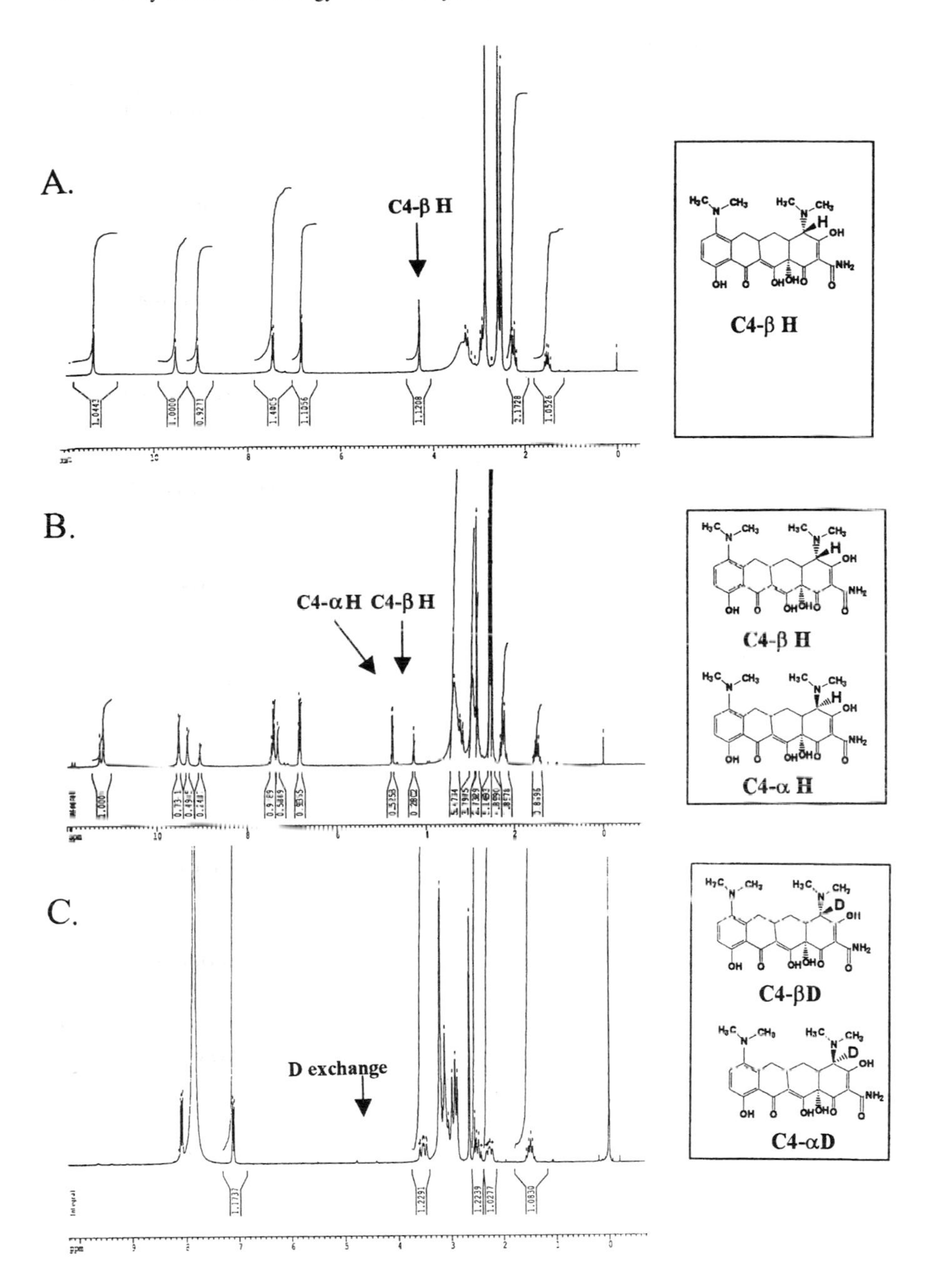

Figure 18. ^{1}H-NMR spectra of minocycline (A).

ous by-products, presumably due to the instability of tetracycline methiodides, with the formation of C-ring dehydration product 4-trimethylammoniumanhy-

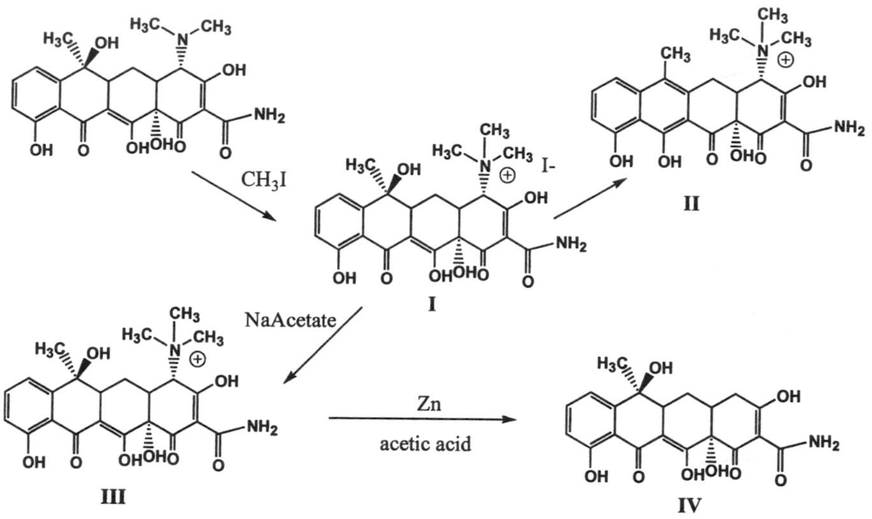

Figure 19. Synthesis of quaternary amine I, methyl betaine II, 4-trimethylammonium anhydro III and 4-dedimethylamino IV tetracyclines.

drotetracycline II. Methiodides of tetracyclines form methyl betaines III by treatment with sodium acetate at pH 4.5. The methyl betaine UV absorption spectrum is similar for the methiodide except for a decreased absorbance at 223–226 nm, indicating a loss of iodide ion.

Quaternary derivatives of tetracyclines typically abolish all antibacterial activity compared to the parent materials from which they are prepared. However, they are readily synthetically modifiable to the corresponding 4-dedimethylaminotetracycline family of compounds IV.

4-dedimethylamino tetracyclines, CMTs or COL compounds

Reductive degradation of the 4-dimethylamino group in tetracyclines can be attained by using chemical reagents such as zinc in glacial acetic acid (Fig. 20) and was one of the first reactions performed on chlortetracycline as part of its structure determination, forming 4-dedimethylamino chlortetracycline I [13]. More recent methods use reductive electrolysis of C4-trimethylammonium tetracyclines in methanol for the scale-up synthesis of 4-dedimethylamino doxycycline II, also known as CMT-8 or COL-8 [49].

The antibacterial activity of 4-dedimethylamino derivatives depends on whether the target pathogens are Gram-positive or Gram-negative bacteria, and whether or not they are used *in vivo*. 4-dedimethylamino-6-demethyl-6-deoxy tetracycline, also called CMT-3 or COL-3, is reported to have activity *in vitro* against *Staphylococcus aureus* and *Streptococcus pyogenes* with MICs

chlortetracycline → (Zn, acetic acid) → I

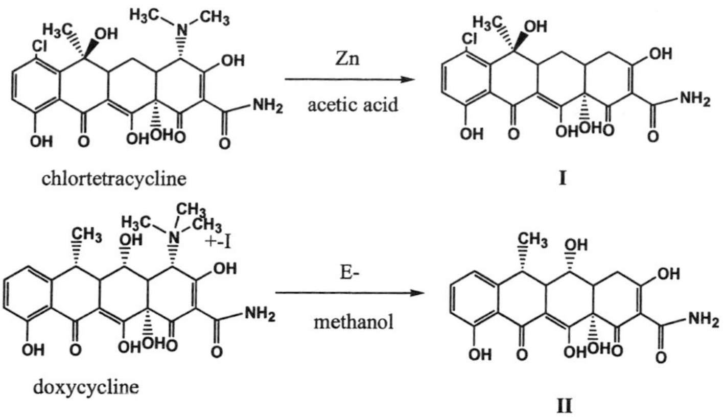

doxycycline → (E-, methanol) → II

Figure 20. Synthesis of 4-dedimethylamino tetracyclines using reductive degradation methods.

of 0.8 and 1.5 μg/mL [50]. However, these compounds were ineffective against Gram-negative bacteria and were devoid of activity *in vivo* against Gram-positive strains. The lack of activity against Gram-positive strains *in vivo* suggests that aqueous solubility and/or bioavailability relates to activity, although this question remains to be answered. The solubility of the 4-dedimethylamino tetracyclines in aqueous buffers or physiological systems may be considerably less due to their increased hydrophobicity as compared to the hydrophilic 4-dimethylamino derivatives and their inorganic or organic salts.

The C12a hydroxyl group at the ring A–ring-B juncture is extremely important for maintaining the overall three-dimensional shape of the tetracycline molecule, as well as for maintaining bioactivity. The selective removal of this group is possible using zinc dust and 15% ammonium hydroxide [51] (Fig. 21). The product, 12a-deoxytetracycline I, has only about 2% of the activity of the parent tetracycline when tested *in vitro* against *S. aureus*. 12a-deoxytetracycline is also readily converted to the degradation product dedimethylaminoterrarubein II, also known as 6-methyl pretetramide, a major biosynthetic precursor discussed earlier in the polyketide pathway of oxytetracycline.

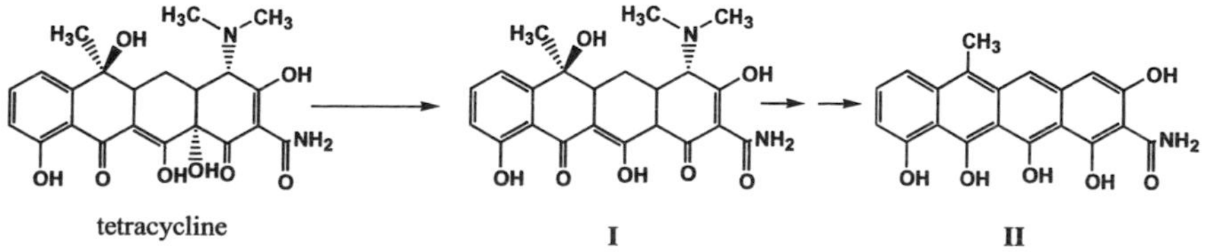

tetracycline → I → → II

Figure 21. Synthesis of 12a-deoxy tetracyclines and degradation products.

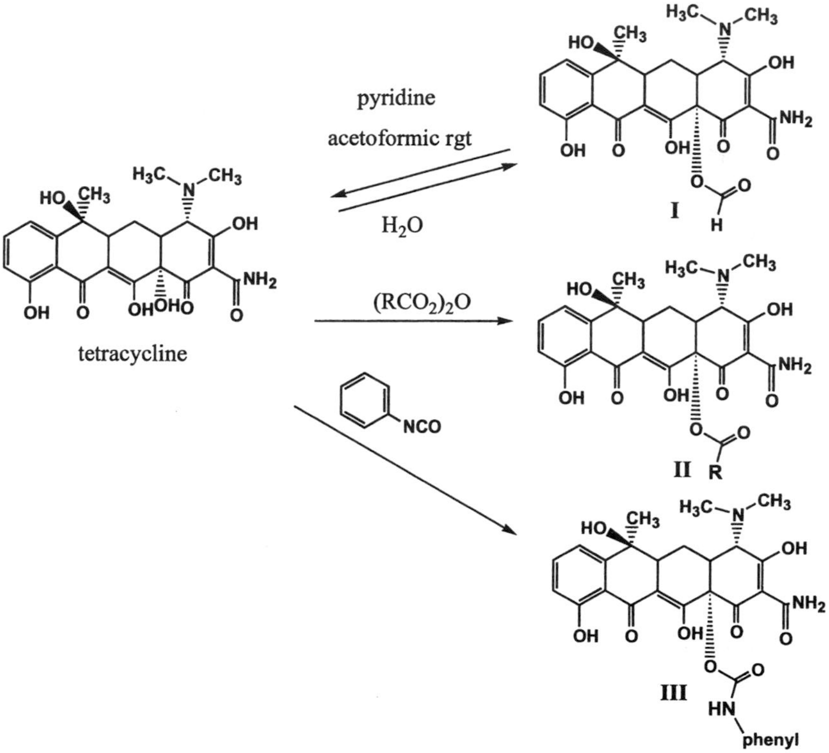

Figure 22. O^{12a}-transformations of tetracycline.

O^{12a}-formyl tetracyclines I (Fig. 22) are obtained by their reaction in an acetoformic acid reagent in cold pyridine or by anhydride reagents in acid [52]. O^{12a}-formyltetracycline, as a prodrug form of tetracycline, shows essentially the same bioactivity as the parent material, reverting back to parent in 5 min in aqueous buffers at pH 7.5. O^{12a} esters II have also been synthesized, retaining activity against Gram-positive bacteria but losing activity against Gram-negative bacteria. Esters greater than 2 carbons also lose their activity against Gram-positive bacteria [53].

The reactivity of the O^{12a} group with organic isocyanates has also been described, where O^{12a}-phenyl carbamoyl III derivatives were produced in a reaction with anhydrous tetracyclines. The compounds were claimed to retain antibacterial activity [54].

B-ring chemistry

There are relatively few position 5 modifications found within the literature; however, the 5-OH group of doxycycline may be chemically modified using

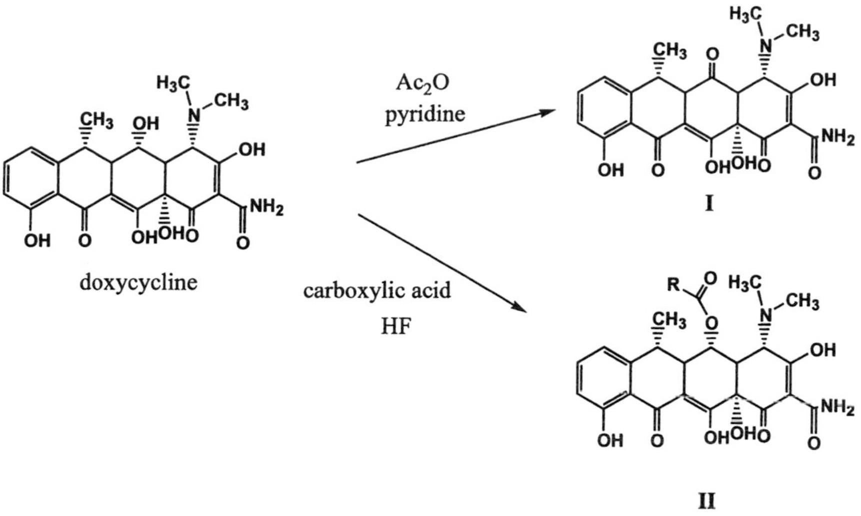

Figure 23. C5-OH transformations of doxycycline.

DMSO and acetic anhydride to produce the 5-deoxy-5-oxo doxycycline I in high yield (Fig. 23) [35]. 5-deoxy-5-oxo-methacycline was purportedly synthesized using the same reagents, using the IR spectrum to confirm the structural assignment of the C5 carbonyl at 1740 cm^{-1}.

Acid-stable tetracyclines, those lacking a 6-OH group, form esters at the C5-OH position via the reaction of carboxylic acids in anhydrous hydrogen fluoride or methanesulfonic acid as both a solvent and catalyst II (Fig. 23) [55]. C5-esters generally have increased activity against Gram-positive bacteria and decreased activity against Gram-negative bacteria, although a limited number have been studied.

The lower peripheral part of the B-ring possesses a keto-enol substructure in conjunction with the C-ring, spanning carbons C11, C11a and C12. By allowing keto-enolization to occur, the third pKa macroscopic acidity constant may be derived. This region has a pKa value of 7.5, while the β-diketone substructure is sensitive to chemical modification. Disruption of the system results in a loss of bioactivity against both prokaryotic and eukaryotic targets.

The CMT-5 is an example of a tetracycline modified at this position, produced by the reaction of hydrazine with the β-dicarbonyl tautomer between rings B and C [56] (Fig. 24). Tetracycline hydrazine derivatives, 11,12-pyrazolotetracyclines I, have no bioactivity, showing the importance of the C11-C12 β-diketone structure in maintaining structure and function while allowing the formation of chelates with inorganic cations.

Although hydrazine derivatives cannot form cation chelates at the ligand-receptor interface and are inactive, they serve to suggest the role of cations in the metal-binding pharmacophore C11-C12 region and the role of cations in

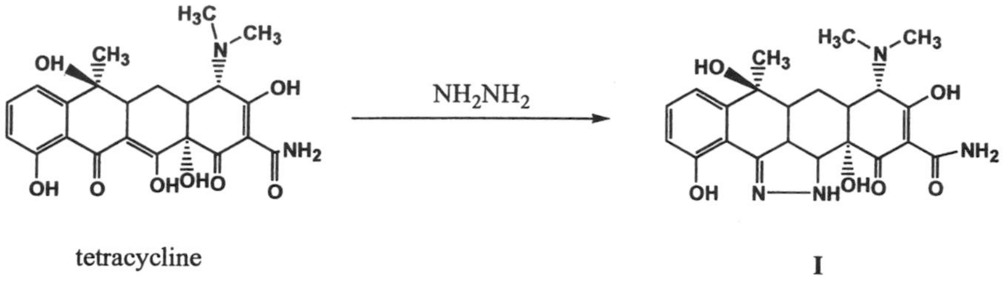

NH_2NH_2

tetracycline I

Figure 24. Synthesis of 11, 12-pyrazolotetracyclines.

biological modulation. Binding of cations in this region appears to be a major molecular interaction at the tetracycline pharmacophore-receptor interface.

C-ring chemistry: synthesis of anhydrotetracyclines, methacycline and doxycycline

Oxytetracycline serves as the starting material for several different tetracycline derivatives, depending upon the reaction pathways used (Fig. 25). Due to the relative instability of the C6 benzylic hydroxyl group, strong acids dehydrate the C-ring through the 6-OH group, while concertedly aromatizing the C-ring, forming 5-hydroxyanhydrotetracycline I [57]. In a similar reaction, tetracycline undergoes C-ring dehydration to form anhydrotetracycline II, chlortetracycline to form anhydrochlortetracycline III, while methacycline forms 5-hydroxyanhydrotetracycline I.

As degradation by-products, anhydrotetracyclines are found in most chemical samples of tetracyclines which possess a reactive 6 position functional group, unless extra care is taken to ensure that these mixtures are not present [58]. The biological assay of 6-OH tetracyclines may be compromised by anhydrotetracycline contaminants, showing variable results and/or toxicity.

Anhydrotetracyclines have several deleterious biological effects. *In vivo* they are phototoxic and hepatotoxic and they may cause the development of anemias [59–61]. They are thought to show their atypical activity as being similar to membrane perturbation agents [62], by non-selectively changing cytoplasmic cell membrane shape, cell morphology and releasing internal cellular constituents, such as β-galactosidase or lactate dehydrogenase, from the cell, signaling lethal cellular damage. They do not, however, cause lysis of *E. coli* spheroplasts, in contrast to a true membrane perturbation agent. Atypical tetracyclines may be acting at the level of membrane energization, where electrochemical changes can initiate autolytic processes and cell lysis.

Anhydrotetracyclines are also high-affinity ligands for the Tet repressor protein, a natural bacterial expression system in bacteria that senses tetracycline antibiotics and up regulates antibiotic efflux resistance mechanisms, which will be addressed in detail later in this and other chapters. As a side note,

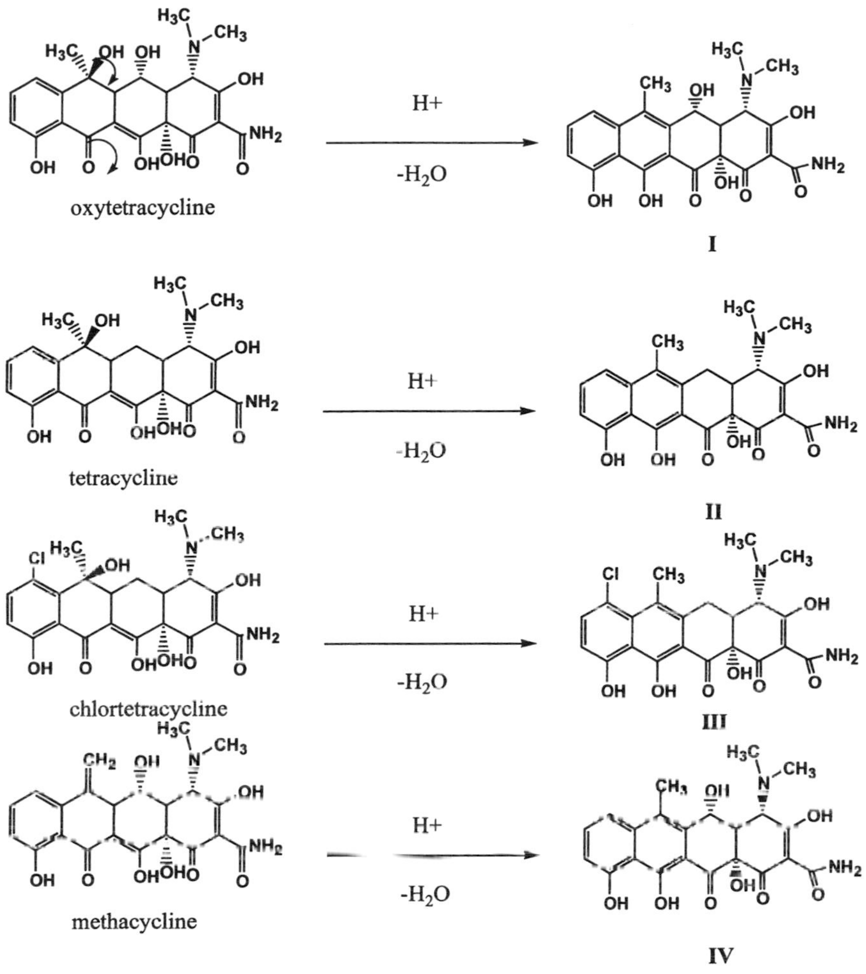

oxytetracycline → (H+, $-H_2O$) → I

tetracycline → (H+, $-H_2O$) → II

chlortetracycline → (H+, $-H_2O$) → III

methacycline → (H+, $-H_2O$) → IV

Figure 25. Dehydration mechanism common to 6-OH tetracyclines and methacycline forming anhydrotetracyclines.

the Tet repressor protein is produced naturally by bacteria that are trying to avoid the toxic effects of anhydrotetracyclines, and this defense mechanism probably evolved specifically to sense this and other toxic tetracyclines.

Synthesis of methacycline

The formation of anhydrotetracyclines can be avoided during semi-synthesis procedures, where protection of the 11a position as the 11a-chloro derivative I (Fig. 26) is readily afforded by N-chlorosuccinimide [63]. Chemically introducing an 11a-Cl halogen group stabilizes the oxytetracycline C-ring, allow-

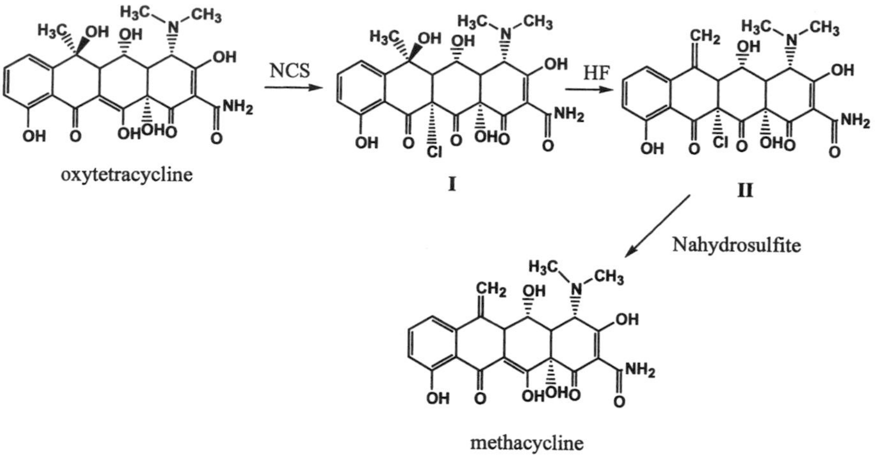

Figure 26. Synthesis of methacycline from oxytetracycline.

ing it to be reacted with acids, bases and other chemical reagents under harsh chemical conditions. Treatment with anhydrous hydrogen fluoride forces an exocyclic dehydration to occur to produce 6-methylene-11a-Cl-oxytetracycline II. The 11a-Cl group is readily removed by reaction with sodium hydrosulfite to produce methacycline III, a clinically significant tetracycline that possesses activity against a broad spectrum of bacteria [64].

Synthesis of doxycycline

The exocyclic double bond of 6-methylene tetracyclines can be further reacted with mercaptans via a radical reaction producing the anti-Markovnikov sulfur adduct of methacycline (Fig. 27) [64]. Benzylmercaptan reacted with methacycline produces 6α-benzylthiomethylene-5-OH tetracycline I, which can then be reduced with Raney nickel catalysts under hydrogen to produce 6α-methyl-6-deoxy-5-hydroxytetracycline [65], now known as doxycycline II, a tetracycline that is still currently widely used against a broad spectrum of bacteria.

Alternatively, methacycline may be directly stereospecifically reduced under hydrogen using rhodium metal complexes, producing doxycycline in high yields by a method proven useful enough for its industrial-scale synthesis [66].

13-benzylmercapto-α-6-deoxytetracycline sulfide is readily oxidized to the S-oxide I using oxidizing reagents such as peroxides or Oxone, a mixture of permanganate salts (Fig. 28) [64, 67]. The S-oxide is readily converted into 7,13-epithio-α-6-deoxytetracycline by treatment in strong acid II [64]. The 7, 13-epithio compound was found to have superior activity compared to the 13-benzyl derivative, particularly against Gram-positive bacteria.

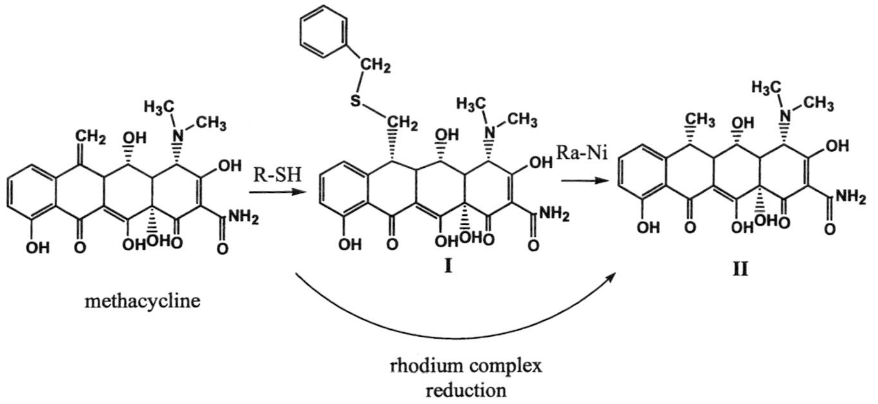

Figure 27. Synthesis of doxycycline from methacycline.

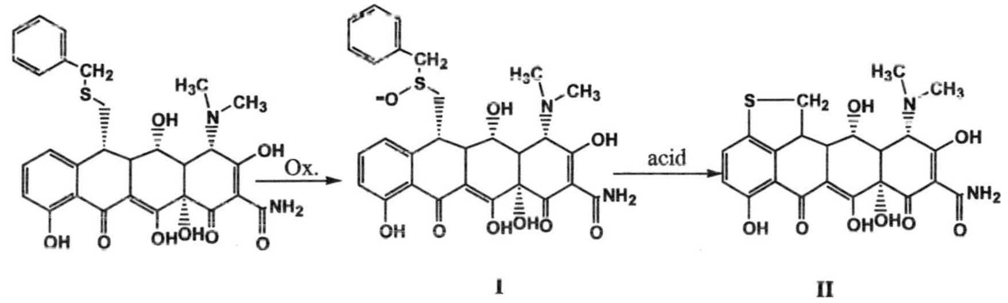

Figure 28. Synthesis of 7,13 epithio-5-OH tetracycline.

The synthesis of C13 mercaptan adducts of tetracyclines has been studied in our laboratory, where we have found them to be active against antibiotic-resistant bacteria possessing drug efflux and ribosomal protection mechanisms. We have also studied their effect against tetracycline efflux proteins and have found them to act as inhibitors. This subject will be discussed shortly.

Other C6 position derivatives

Modification of demecycline with N-chlorosuccinimide produces the 11a-Cl derivative I, which in liquid HF results in the displacement of the 6-OH with fluorine to yield two possible stereoisomers of the 6-fluorotetracyclines, α-isomer II and β-isomer III (Fig. 29) [68]. Demecycline can also be directly acetylated using HBr and acetic acid to produce 6-α-acetoxy-6-demethyl tetracycline IV. Both the fluorine and acetoxy β-isomers are more active against *S. aureus,* while the α-isomer shows significant decreases in activity.

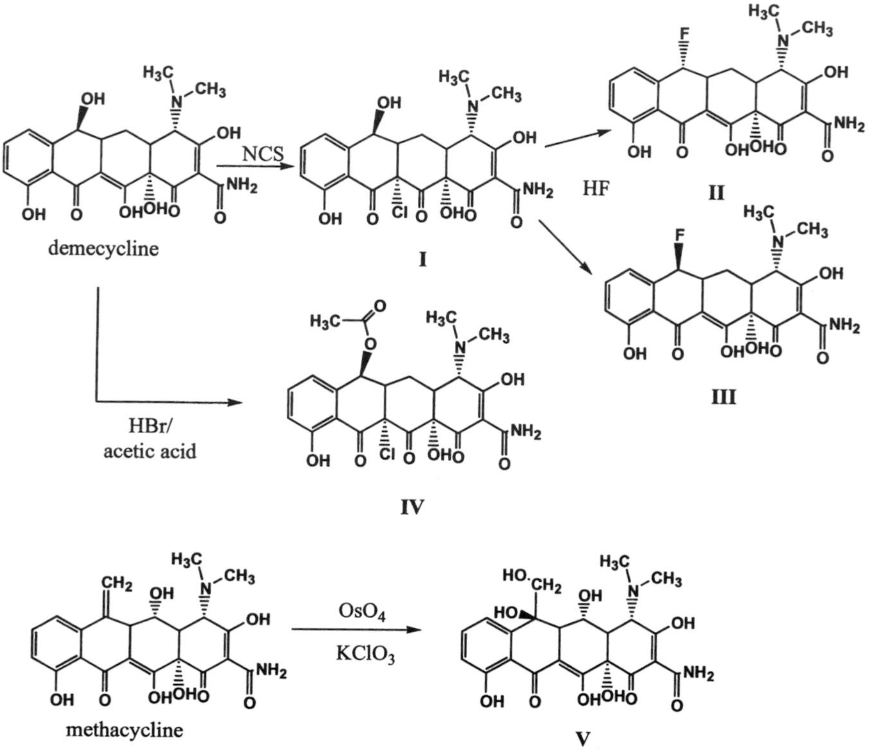

Figure 29. Synthesis of C6 position chemical derivatives.

Methacycline may be *cis*-dihydroxylated between the C6 and C13 exocyclic carbons. Oxidation with osmium tetroxide, OsO_4, or with oxidizing salts such as potassium perchlorate, produces 6, 13-dihydroxytetracyclines V (Fig. 29), although this compound was found to be inactive against both Gram-positive and Gram-negative bacteria [35].

D-ring chemistry: synthesis of minocycline and GAR-936

The aromatic D-ring, composed of positions C7, C8, C9 and a phenol at C10, has been a useful chemical platform for the semi-synthesis of numerous other tetracycline derivatives (Fig. 30). Earlier tetracyclines, those possessing a 6-OH group, were chemically unstable, and the lability of this position limited synthetic progress in D-ring chemistry. Semi-synthesis of 6-deoxytetracyclines and 6-deoxy-6-demethyltetracycline, sancycline and other chemically stable tetracyclines, led to the development of new D-ring derivatives possessing both chemical diversity and, in some cases, potent antibacterial activity.

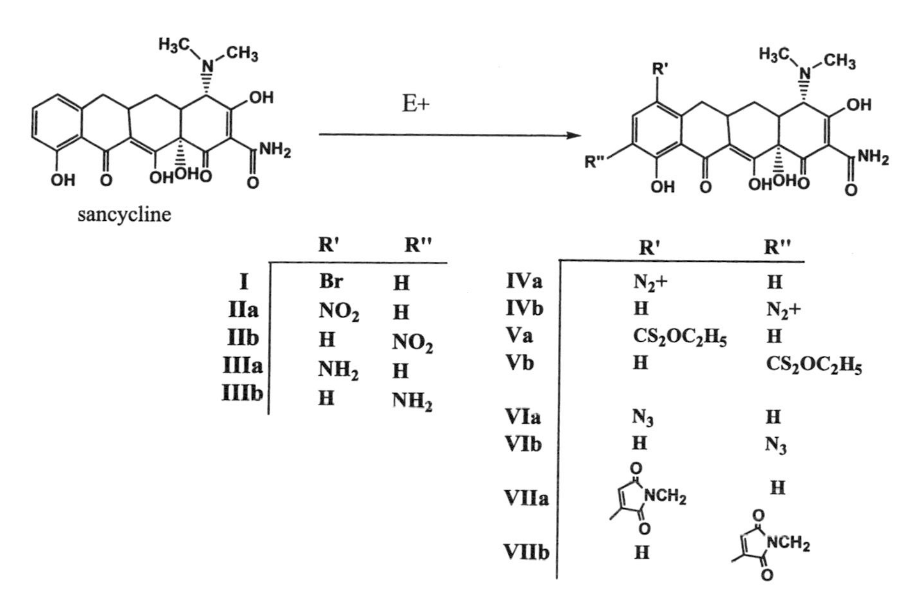

Figure 30. Synthesis of C7 and C9 derivatives of sancycline.

The D-ring undergoes chemical modification primarily via electrophilic aromatic substitution, where electrophilic reagents, E+, react to produce position 7 and 9 derivatives of 6-deoxy-6-demethyltetracyclines I (Fig. 30) [69]. The types of substitutions possible include halogens I, nitro groups IIa, IIb, amino groups IIIa, IIIb, diazonium functional groups IVa, IVb, and their reaction products, such as the ethyl xanthates Va, Vb and azides VIa, VIb. Reaction with N-hydroxyphthalimides in strong acid also is possible, producing corresponding 7 and 9 position analogues VIIa, VIIb.

The reactivity of the D-ring is based on several factors, the major one being the electron-donating characteristics of the phenol group activating the ortho C9 and para C7 positions to electrophilic substitution. At the same time, C8 is deactivated by both the C10 phenol and the C11 carbonyl, leading to a loss of reactivity. Electron density calculations performed in our laboratory confirm the reactivity at these positions, where C7 and C9 possess greater electron density than the C8 position (Fig. 31). A single point semi-empirical molecular orbital calculation using the AM1 Hamiltonian was conducted to determine the relative electrostatic point charges at the 7, 8 and 9 positions. Shown is the MOLCAD rendering of the Coulson charges.

Synthesis of minocycline

Nitration of sancycline results in the electrophilic substitution at either the C7 or C9 position (Fig. 32). Separation of the 7 or 9 nitro regioisomers and reduc-

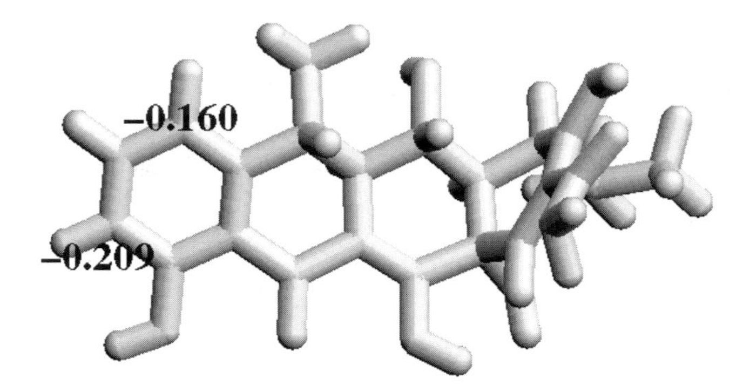

Figure 31. MOLCAD rendering of doxycycline (Gasteiger-Huckel Charges) and the electron density at C7. C8 and C9. Modeling was done using Silicon Graphics hardware and version 6.6 Sybyl software, Tripos, St. Louis, MO.

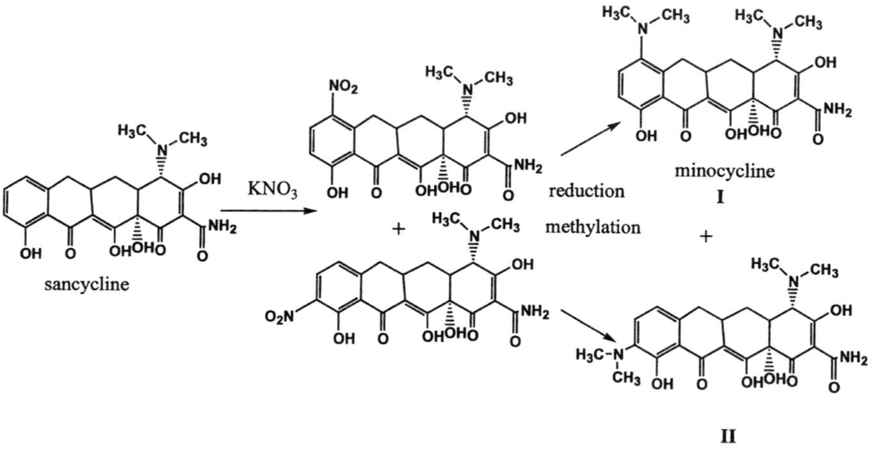

Figure 32. Synthesis pathway for 7 and 9 aminominocycline derivatives and minocycline.

tive hydrogenation and methylation using 10% Pd/C at atmospheric pressure produce 7-dimethylamino-6-demethyl-6-deoxytetracycline II or 9-dimethylamino-6-demethyl-6-deoxytetracycline III [70]. The 7-dimethylamino derivative, named minocycline, has superior activity against a broad spectrum of bacteria and is one of the most potent of all the tetracyclines used today against tetracycline-susceptible bacteria. The initial synthesis route for this compound had major problems, the separation of the regioisomers at positions C7 and C9 is a formidable task, where industrial-scale separations by chromatography are not economically feasible.

Two different reaction pathways are now used for the industrial-scale synthesis of minocycline, and American Cyanamid Lederle Laboratories have been instrumental in the development of both synthetic strategies.

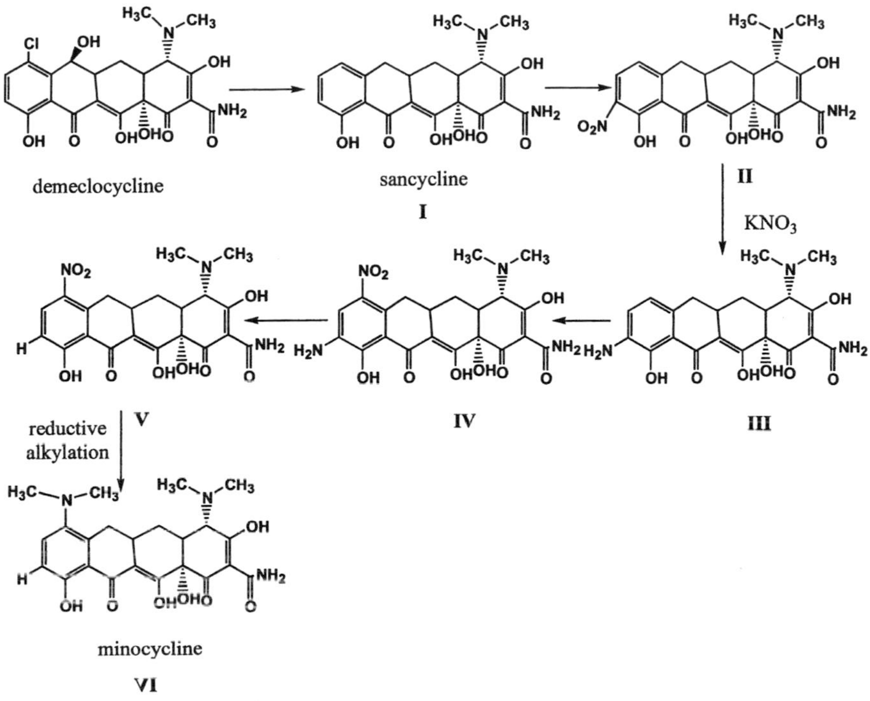

Figure 33. Multistep synthesis of minocycline via diazonium intermediates.

The first synthesis scheme uses the fermentation product demeclocycline as the starting material, which is readily reduced by catalytic hydrogenation to sancycline I (Fig. 33) [70]. In anhydrous HF and under the correct nitration conditions, 9-nitro sancycline II is produced. Catalytic reduction of the nitro group to the amine produces 9-amino sancycline III. Further nitration produces 7-nitro-9-amino sancycline IV. The amino functionality is removed via the diazonium salt intermediate to produce 7-nitro sancycline V. The nitro group is reduced and simultaneously reductively alkylated using formaldehyde and catalytic reducing conditions to produce minocycline VI.

Another reaction scheme utilizes the reactivity of the 7 position of sancycline with diazo compounds (Fig. 34) [71]. Sancycline in acid reacted with dibenzyl azodiformate in tetrahydrofuran produces 7-N, N'-dicarbobenzyloxy hydrazine Tc I. This intermediate is then catalytically reduced to 7-aminosancycline II. The amino compound is finally reductively alkylated using formaldehyde and catalytic hydrogenation conditions to minocycline III. The regiospecific reaction at position 7 with amine-forming reagents is highly fortuitous, producing minocycline in a shorter synthetic sequence and in high yield.

By either method, minocycline was the last tetracycline to be approved for use against bacterial infections in the United States. It still is produced today

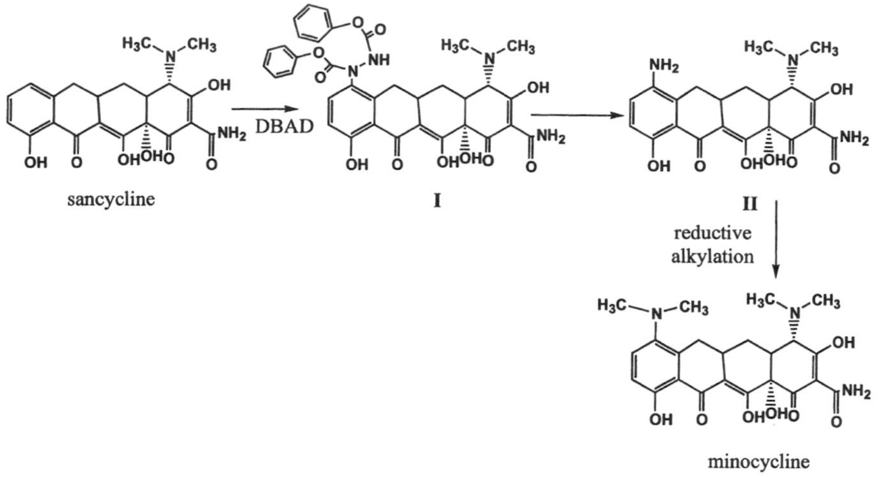

Figure 34. Regiospecific synthesis of minocycline.

and marketed under a variety of tradenames, the most prominent of which is Minocin, from American Home Products, Inc., (AHP) Pearl River, New York, formerly Lederle Laboratories.

Today AHP has continued research into new tetracyclines by producing a family structurally related to sancycline and minocycline, the glycylcyclines. These promising compounds have been found to be effective against antibiotic-resistant bacteria, both *in vitro* and *in vivo*, and are the first new tetracyclines to be developed in almost 30 years.

Synthesis of the glycylcyclines

Glycylcyclines are synthesized using either minocycline or sancycline as the starting material (Fig. 35), where nitration of the D-ring by reagents produces 9-nitro minocycline, or 7 and 9 nitro regioisomers of sancycline [72]. The D-ring nitro derivatives are reduced to the amino functional group and are pivotal intermediates for further modification. The 9-amino derivatives of minocycline I and sancycline II are treated with N, N-dimethylglycinyl chloride in an aprotic solvent to afford DMG-Mino III and DMG-DMDOT IV.

While the dimethylaminoglycyl derivatives were potent as antibacterials, another derivative from this series, 9-t-butylglycylamido minocycline, also known as TBG-MINO or GAR-936 V [73] (Fig. 35), has demonstrated superior potency as an antibacterial compared to the dimethylamine derivatives. Other glycylcycline-like compounds have been described since this work began, namely, 9-glycylamino doxycycline derivatives and amino acid derivatives of doxycycline IV, but there was no improvement in spectrum or potency of activity [74].

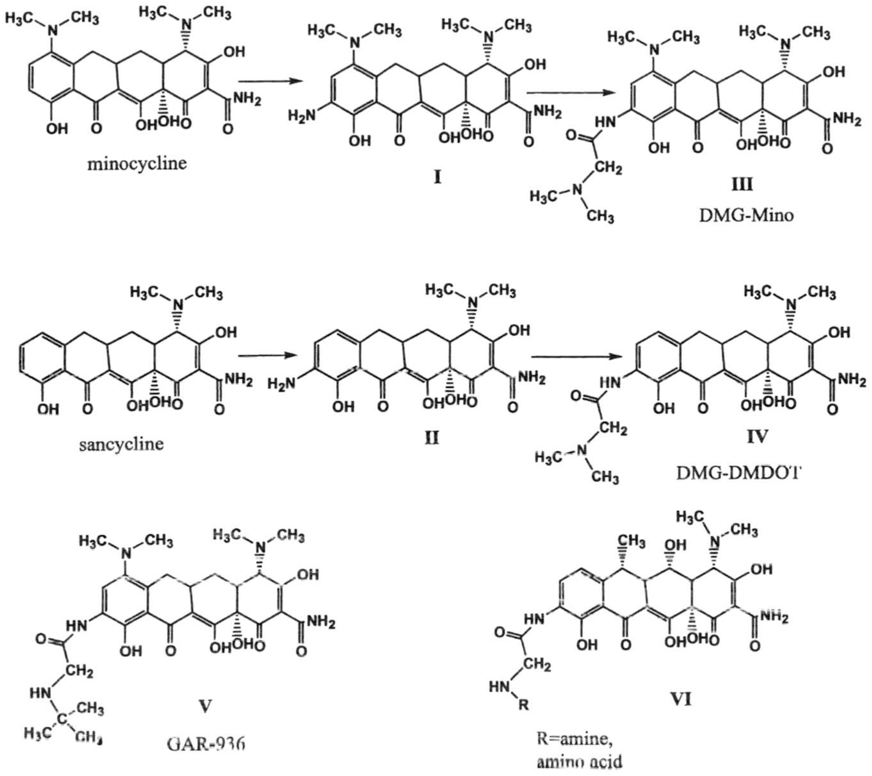

Figure 35. Synthesis of the glycylcyclines and DMG-Mino and DMDOT.

The attachment of N-alkyl glycylamido derivatives at position 9 generally increases activity against a broad spectrum of tetracycline-susceptible and resistant Gram-positive and Gram-negative bacteria, including strains possessing efflux and ribosomal protection resistance determinants. GAR-936 is currently undergoing development by AHP as a new antibacterial agent.

Cellular mechanisms of action of the tetracyclines

Studies of the cellular mechanisms of action of the tetracyclines against both prokaryotes and eukaryotes show that they have a diverse range of biological activities. The major mechanism of action of the typical tetracyclines against bacteria is protein-synthesis inhibition, at least upon initial examination. Both typical and atypical tetracyclines also have pharmacological effects against eukaryotic cells across multiple cell types; their molecular mechanisms of action are just beginning to be understood. A discussion of the effects of tetracyclines against both bacteria and mammalian cells demonstrate the chemically “promiscuous” [75] nature of the tetracycline molecule, where it can interact at a variety of receptors, both prokaryotic and eukaryotic in origin, and modulate cell processes.

Inhibition of protein synthesis

Soon after their discovery, tetracyclines were initially described as having inhibitory action in bacteria similar to chloramphenicol, an antibiotic and known protein-synthesis inhibitor. There were also reports that tetracyclines produce metabolic disturbances in cells and could modulate oxidative phosphorylation, electron transport, nucleic acid biosynthesis and bacterial cell wall synthesis [76–79]. Tetracyclines were also found to inhibit metalloflavoenzymes such as NADH and cytochrome oxidoreductase [80].

Attempts to localize the primary area of the mechanism of action of tetracyclines used *S. aureus* cells and radiolabelled macromolecular precursors such as ^{14}C-glutamine and ^{14}C-lysine and concentrations of Tc between 0.1 and 1.0 µg/mL. Protein synthesis decreased within 5 min, indicative of easy penetration of tetracycline into the bacterial cell [81].

Within the first 30 min after drug contact, the mode of action of tetracycline is primarily bacteriostatic, inhibiting bacterial growth without killing, while during this time macromolecular processes such as cell wall, DNA and RNA synthesis are found to be unaffected. Thus, the primary effect of tetracyclines was first described as inhibition of protein synthesis [82].

While some thought that the primary targets of tetracyclines were secondary processes such as protein and enzyme-dependent DNA and RNA synthesis, these findings were refuted by studies showing that tetracyclines mostly inhibit protein-synthesis activity in cell-free protein-translation assays [83, 84]. Cell-free systems have been used to establish a quantitative measure of the inhibition of protein synthesis using tritium-labeled tetracycline (^{3}H-Tc) as a probe. Tetracyclines were also found to decrease polyuridine-directed phenylalanine incorporation into cell-free *E. coli* systems [85].

The macromolecular synthesis of protein, as measured by incorporation of radioactive precursors, is more sensitive than other biochemical pathways to tetracyclines, while total cellular growth is inhibited at concentrations that inhibit 90% of the rate of protein synthesis [86]. Similar inhibition patterns are seen for both chlortetracycline and oxytetracycline, while still other studies show inhibition of other pathways by these compounds such as cell division, protein and nucleic acid synthesis in *E. coli* [87]. It was also found that tetracyclines caused inhibition of rRNA synthesis in *E. coli* [88] and down regulation of precursor uptake mechanisms [89] along with protein-synthesis inhibition.

Inhibiting protein synthesis may have an effect on membrane and phospholipid synthesis and turnover in *E. coli* [90], as shown by electron microscopy where distinct morphological abnormalities occur in the presence of tetracyclines [91]. However, cellular abnormalities due to tetracyclines may also be due to changes in the synthesis of envelope proteins [92].

Tetracyclines were found to bind firmly to isolated ribosomes using ^{3}H-Tc, binding with the 30S and 50S ribosomes of *E. coli* [93]. Other methods for studying the binding of tetracyclines to macromolecules include observation of the binding of ^{3}H-Tc tetracyclines by a two-phase partitioning system [94]. This demonstrated affinity of ^{3}H-Tc to the bacterial 30S ribosomal subunit, where tetracyclines also inhibited the formation of peptide amide bond linkages on growing polypeptide chains.

Ribosomal binding

When bacterial ribosomes encounter tetracycline in concentrations needed to inhibit bacterial growth, they present many different molecular receptors for tetracycline and tetracycline-metal complexes, and therefore a multitude of tetracycline-RNA-protein binding interactions are possible.

In prokaryotes, *E. coli* ribosomes are composed of 2 particles, designated the 30S and 50S subunits. Each subunit is composed of numerous proteins and RNA molecules. The 30S unit is further composed of a 16S RNA molecule and 21 S proteins, in order of largest to smallest. The 50S subunit is composed of two RNA molecules, 5S and 23S particles, and 32 proteins designated as L proteins. A functioning ribosome, with particles and S and L proteins, is an intricate architecture of molecular machinery used specifically to biosynthesize proteins. Tetracyclines bind to a variety of macromolecular systems in both prokaryotic and eukaryotic cells, but binding to ribosomes has been the most studied.

Among the major tetracyclines, tetracycline binds to *E. coli* ribosomes with a high degree of specificity [83, 93, 95] as does oxytetracycline [96]. Most tetracyclines bind to bacterial ribosomes to some degree, as studied by the aqueous two-phase partitioning system [92]. Bacterial ribosomes show binding of tetracyclines to both the 30S and 50S subunits, with a delineation of a high-affinity site [97] on the head region in the 30S ribosome and multiple low-affinity sites on both the 30S and 50S subunits. The low-affinity sites accompany the S proteins, namely, the S5, S7, S13, S14 and S18 proteins [93, 95], while binding to the 50S subunit L proteins has never been described. Recent proteolysis studies show most of the proteins affected by tetracyclines are located on the 30S head region and the 50S interface side, where one of the tetracycline-binding regions is near the ribosomal peptidyl transferase center [98]. Photoaffinity labeling experiments also suggest the S7 protein as a reasonable affinity receptor for tetracyclines, although other S proteins can be labeled as well. Mutant ribosomes lacking S proteins show decreased binding of tetracyclines as compared to the native subunits.

Spectroscopic study of the interaction of tetracyclines with ribosomal subunits from *E. coli* are possible using fluorometry and fluorescence anisotropy, where, using demeclocycline as a probe, both high-affinity and low-affinity binding sites can be observed [99]. Demeclocycline binds to not only 30S, 50S and 70S ribosomal particles with high affinity, but also reveals secondary binding to many other unspecific binding sites [100] at higher compound concentrations.

Photoaffinity labeling of ribosomes

E. coli ribosomes have been photoaffinity-labeled with a high degree of specificity using ^{3}H-Tc to probe the binding sites within the ribosome [101]. In these experiments, S proteins were demonstrated to be labeled, with S7 demonstrating high-affinity labeling, while other S proteins, S4, S13 and S14, incorporated less ^{3}H-Tc [102].

The success of these studies relies on the inherent photolability of tetracycline and its ability to form radical species to covalently modify and map ribosomal binding sites and ribosomal components. ^{3}H-Tc photoincorporates in the presence of ribosomes and other substructures, but free radical quenching agents must be used to increase specificity of the labeling reaction [103]. It was also noted that competing side reactions do occur [104], where photolysis of tetracyclines and RNA results in a 1:1 ratio of labeling of RNA and the S7 ribosomal protein.

Molecular mechanism of action of the tetracyclines

While the site of action was determined to be the ribosome and its components, the mechanism of typical tetracyclines was found to be inhibition of codon-specific binding of aminoacyl-tRNA to ribosomes and the blocking of protein synthesis [105, 106]. Tetracyclines were also found to inhibit formylMET-tRNA binding to ribosomes [107]. Supportive evidence further pointed to the tRNA binding site, where inhibition by tetracycline occurred via an allosteric effect inhibiting appropriate binding of the amino acid-tRNA at the acceptor site or "A"-site [108].

During protein synthesis, mRNA, with an appropriate start codon, initiates peptide formation by signaling the ribosome to self-assemble. The codons of the RNA encode for each amino acid in the chain. The ribosomal complex has two sites, the A-site (acceptor) and the P-site (peptidyltransferase). The A-site is a receptor to accept tRNA-amino acid complexes from the cytoplasm, binding at the ribosome complex, providing one specific amino acid for each codon triplet. The amino acid is then catalytically transferred from the A-site to the growing polypeptide chain on the P-site by peptidyltransferase enzymes, forming a peptide bond in the process. The A-site is free again to accept the next tRNA-amino acid complex to complete the growing protein chain.

When translation is completed, by the final reading of the mRNA, the polypeptide chain is released along with the RNA. Once the protein is released and post-translational modifications and protein folding occur, it is used by the bacterium for survival and growth. However, tetracycline binding to ribosomes decreases the ability of tRNA to bind at the acceptor A-site of ribosomes [109]. Once this occurs, the P-site lacks the tRNA-amino acid substrate for its reaction, and the aminoacyl transfer reaction is inhibited and protein synthesis stops.

It is believed that tetracyclines act at the A-site by non-competitive or allosteric effects, inhibiting binding and backing up tRNA-amino acid complexes within the cytosol. While tetracyclines are shown to be specific for the A-site, it is also probable that during binding they can inhibit the function of release factors that act in chain termination with stop codons, and chain release factors [110], while some effect may also reside at the polypeptide exit site on the *E. coli* ribosome [111].

Other mechanisms of action of the tetracyclines: the typical and atypical tetracyclines

It has been postulated that tetracyclines can inhibit cellular growth in bacteria by other mechanisms of action, whereby tetracyclines change bacterial membrane integrity and mechanical properties, eventually causing macromolecular dysfunction, cell lysis and death.

The typical Tcs, those that act as classic protein-synthesis inhibitors, such as tetracycline, doxycycline, minocycline and chlortetracycline, exhibit bacteriostatic activity, at least initially in bacteria. Other tetracyclines have been found to be bacteriocidal, killing bacteria in an atypical mechanism. Atypical tetracyclines are believed to act by disruption of cellular membranes, inhibiting all cellular processes and macromolecular synthesis pathways. Molecules that change cytoplasmic membrane shape and change cell morphology also trigger the release of cytosolic constituents such as β-galactosidase or lactate dehydrogenase from the cell [62, 112, 113].

Using cell-free protein translation systems, different compounds of the tetracycline family showed different abilities to inhibit protein synthesis. Typical tetracyclines, tetracycline, chlortetracycline, minocycline and doxycycline, showed primarily protein-synthesis inhibition. Other tetracyclines, anhydrotetracycline, 4-epi-anhydrotetracycline and 6-thiatetracycline, were found to be intermediate inhibitors of protein synthesis, while two compounds, sancycline and chelocardin, also appeared to inhibit protein-synthesis ability [114]. The intermediate protein-synthesis inhibitors also decreased radiolabeled precursor incorporation into protein, DNA and RNA, suggesting membrane perturbation and loss of precursor uptake capability.

It is also feasible that activity may also be a consequence of drug retention in the cell membrane, inhibiting macromolecular synthesis via inhibition of precursor uptake. One argument for this scenario is that only the more lipophilic tetracyclines seem to have this ability, whereas hydrophilic tetracyclines tend to be only protein-synthesis inhibitors. Another argument is that the equilibrium between the hydrophilic form (the A form) and the lipophilic non-ionized form of Tc (the B form) is changed, where the more lipophilic B form tetracyclines remain in the membrane hydrophobic environment, causing inhibition of intrinsic proteins.

But clearly some tetracyclines do not fall into either of the two categories. Sancycline is capable of exerting a dual behavior, inhibiting protein synthesis in the typical manner, yet it can also decrease the incorporation of precursors into macromolecular synthesis of DNA, RNA and even peptidoglycan.

Several atypical tetracyclines can interfere with membrane function [112]; 6-thiatetracycline, anhydrotetracycline and 4-epi-anhydrochlortetracycline initiate the immediate release of β-galactosidase, whereas chelocardin and anhydrochlortetracycline caused release after a short delay period. Tetracycline, chlortetracycline and minocycline did not cause an increase in release of cytosolic contents.

There is also an implication of intrinsic membrane proteins as targets for the atypical tetracyclines. Atypical tetracyclines do not directly solubilize cell membranes, as shown by the cell membrane spheroplast assays. This leaves other distinct possibilities: atypical tetracyclines may cause membrane de-energization and autolysis, or they may inhibit crucial enzymes responsible for cell wall biosynthesis.

Since atypical tetracyclines spend more residency time in the cytoplasmic membrane due to lipophilicity, they may be affecting enzymes and ion-mediated enzymatic reactions crucial to the cell wall synthesis and integrity.

Eukaryotic effects of tetracyclines

Tetracyclines also have been found to bind to a variety of different eukaryotic proteins as well as to affect different processes and ribosomal function in eukaryotic cells. Differences between bacterial and mammalian ribosomes are clear, where bacteria have a 70S ribosome comprised of 30S and 50S subunits, while the mammalian 80S ribosome is composed of 40S and 60S subunits. Even with their differences in chemical composition, tetracyclines have been found to affect mammalian ribosomal function.

Chlortetracycline is found to have an effect on leucine incorporation in rat liver ribosomes [115] while changing the protein-synthesis abilities of mammalian cells [116, 117]. Tetracycline also is covalently incorporated into rat liver ribosomes, both the 40S and 60S subunits, as monitored at 254 nm by UV spectroscopy [118].

Tetracyclines can also inhibit protein synthesis in mammalian cells [119], while affecting the protein synthesis in human lymphocytes treated with Concanavalin A. In these studies, it was found that using clinically achievable doses of tetracyclines significantly affected production of migration inhibition factor, a product of protein biosynthesis [120].

Tetracyclines also have profound activity against mammalian cell mitochondria. They inhibit mitochondrial protein synthesis [121] and can depolarize membrane potential and decrease respiratory control [122]. The inhibitory effects of tetracyclines and the effect of calcium suggest a Ca-tetracycline complex is responsible for the modulation of energy metabolism. Prolonged inhibition of protein synthesis in mitochondria also leads to proliferation arrest and the accumulation of growth-arrested cells in the G1-phase of the cell cycle, indicating that tetracyclines can inhibit cell cycles in a specific manner [123]. This might explain certain aspects of tetracyclines as immunosuppresive agents, where tetracyclines may impair lymphocyte function through inhibition of their mitochondria [124]. Tetracyclines are thought to impair mitochondrial biogenesis by inhibition of mitochondrial protein synthesis, thereby decreasing blast cell transformation.

Tetracyclines have activity as antiparasitic agents for perhaps the same reason, i.e., inhibitory effects against mitochondria. Oxytetracycline is a potent inhibitor of mitochondrial protein synthesis in *Theileria parva* and of lymphoblast mitochondrial protein synthesis [125].

As mitochondrial stress agents, tetracyclines may cause a myriad of effects. Tetracyclines also affect regulation of differentiation-specific gene expression, where it impaired myotube formation and induction of muscle creatine kinase activity which was observed in differentiated myocytes. Transcription of muscle-specific proteins, creatine kinase and troponin-I were also significantly suppressed in a dose-dependent manner [126].

Bacteriological uses of tetracyclines

The tetracycline family of antibiotics is currently in the stages of a renaissance, whereby the chemistry and biology are now being re-evaluated as the need for new antibiotics increases and the number of effective antibiotics against drug-resistant pathogens decreases.

Tetracyclines possess multifactorial and diverse, and in some cases unknown, modes of action against pathogenic bacteria, giving hope that chemotherapy using tetracycline analogues will help in the fight against drug-resistant bacteria. The use of tetracyclines in human medicine is limited to those that have been isolated from fermentation broths or by chemical modification through semi-synthesis to produce chemically stable and effective compounds. A list of medically important tetracyclines is described in Table 2, along with their primary and secondary uses against bacterial pathogens [127].

Natural products oxytetracycline, tetracycline and demeclocycline are used to a lesser extent, while the semi-synthetic tetracyclines, minocycline and doxycycline, are some of the most widely used antibiotics in medicine, animal health and agriculture, although antibiotic resistance has curtailed the use of some of these agents.

Table 2. The clinically used natural and semi-synthetic tetracyclines

natural

Oxytetracycline

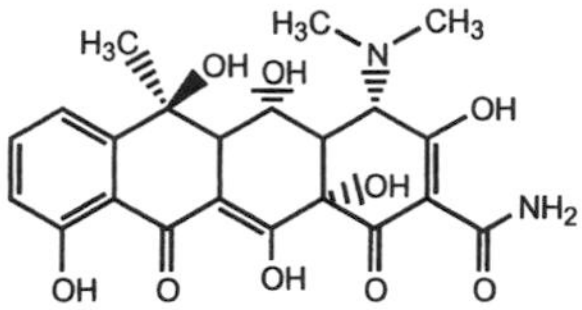

Demeclocycline 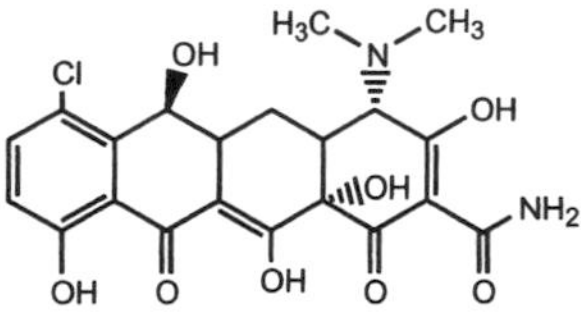 Tetracycline

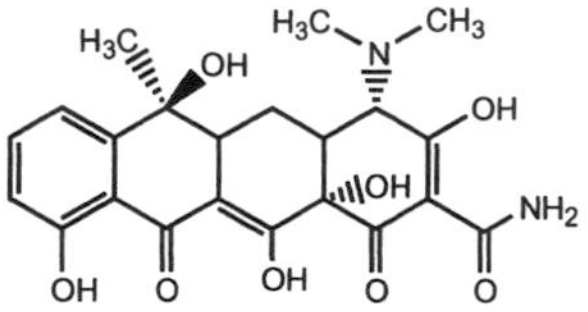

semisynthetic

Minocycline Doxycycline

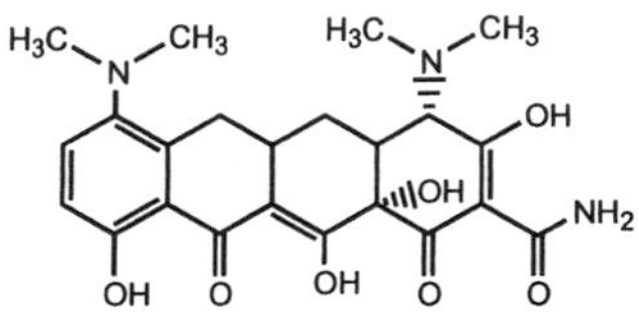

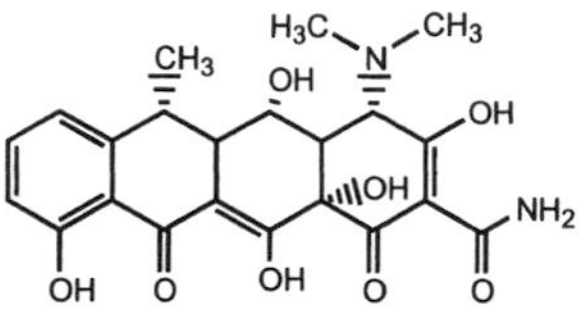

Primary indications	Secondary indications[1]
Rickettsia	*Streptococcus* species[2]
Mycoplasma pneumoniae	
Psittacosis and Ornithosis	Amoebiasis[3]
Borellia recurrentis and other species	
Hemophilius ducreyi	Severe acne[4]
Yersinia pestis	
Bacteroides species	*E. coli*[2]
Vibrio cholera	
Neisseria species	*Enterobacter aerogenes*[2]
Treponema species	
Clostridium species	*Shigella* species[2]
Bacillus anthracis	
Chlamydiaspecies	*Klebsiella* species[2]
Malaria prophylaxis	

[1] When other antibiotics as penicillin are contraindicated.
[2] When bacteriological testing indicates susceptibility.
[3] Adjunct to amoebicides.
[4] Adjunct to standard therapy

Tetracyclines as modulators of antibiotic resistance phenotypes

Bacteria become antibiotic-resistant to compounds when they acquire mechanisms to survive in the presence of toxic chemicals and antibiotics. There are three main mechanisms of resistance that bacteria may use: target receptor alteration, antibiotic alteration by enzymatic modification, and antibiotic efflux.

Numerous bacteria, once susceptible to many common antibiotics, are treated with more costly, toxic and sometimes ineffective antibiotics. Among these organisms are vancomycin-resistant *Enterococci*, *Pseudomonas aeruginosa* and *Acinetobacter baumanii*, for which no novel chemotherapy treatment exists, while other strains, such as methicillin-resistant *Staphylococcus aureus*, are treatable only with vancomycin or a diminishing list of last-resort antibiotics.

While consumer and health professional education is one approach to help stem the increasing problem of antibiotic resistance, other approaches to the problem include finding new antibiotics against bacterial pathogens and finding new targets within bacteria which can be utilized to inhibit their growth.

Another approach, currently being studied in our laboratory, is to take antibiotics that have lost their efficacy due to resistance and to modify their chemical structure, producing more potent analogues that may circumvent, inhibit, or bypass antibiotic resistance phenotypes. The tetracyclines are a model family of compounds where this is possible. Currently, our efforts at Paratek Pharmaceuticals, Inc., Boston, MA, are focused on modifying the chemical structure and nature of the tetracyclines through organic semi-synthesis, expanding their chemical and biological properties, while increasing their effectiveness against antibiotic-resistant bacteria.

Tetracycline bacterial resistance phenotypes

The cytoplasmic membrane of bacteria harbors many different types of proteins, both extrinsic as well as those that are intrinsic and membrane bound. Intrinsic efflux proteins in cell membranes are responsible for maintaining homeostasis within the cell by removing toxic compounds, metabolites, and antibiotics that enter the cell during chemotherapy. Intrinsic efflux proteins act as secondary transporters either specifically to remove only one class of toxins from intracellular compartments, or non-specifically, exhibiting broad specificity for compounds that are structurally unrelated.

In 1980, tetracycline efflux among Gram-negative bacteria was first described by Stuart B. Levy and Laura McMurry, at Tufts University School of Medicine, Boston, MA [128]. The mechanism was noted at first as a "keep out" mechanism that resulted in lower concentrations of ^{3}H-Tc within the cell (L. McMurry, personal communication). Much later, a second mechanism of tetracycline resistance was discovered by Vickers Burdett and involves the production of ribosomal protection proteins [129]. Both mechanisms of tetracycline resistance are found in over 20 resistance-determinant classes in both Gram-negative and Gram-positive bacteria; they are given a letter designation to describe the class of transferable genetic determinant responsible for the resistance phenotype (Tab. 3) [130].

Table 3. Tetracycline resistance determinant class with principal bacterial host

Efflux					
Gram-negative		Gram-positive		Ribosome protection	
A	*Pseudomonas*	K	*Staphylococcus*	M	*Enterococcus*
B	*Escherichia coli*		*Streptococcus*	O	*Campylobacter*
C	*Salmonella*	L	*Staphylococcus*	Q	*Bacteroides*
D	*Aeromonas*		*Streptococcus*	S	*Listeria*
E	*Aeromonas*	P_A	*Clostridium difficile*	T	*Streptococcus*
G	*Vibrio*			G	*Enterococcus*
H	*Pasteurella*			P_A	*Clostridium difficile*

The model of tetracycline transport in antibiotic resistant bacteria

Tetracycline efflux in bacteria is mediated by a family of inner membrane proteins, named the Tet proteins [131]. Tet proteins are highly conserved and homologous throughout this family, with slight variations of amino acid composition. They function as intrinsic membrane proteins to reduce the intracellular tetracycline concentration, ensuring survival of the bacterium.

Early studies examined the uptake of radiolabeled tetracycline into resistant and susceptible bacteria, identifying active efflux by de-energizing cells. Susceptible cells accumulated manifold more tetracycline when given an energy source than resistant cells possessing plasmid R222, which bears the tetracycline resistance determinant on transposon Tn*10* (Fig. 36, A). When the cells were energy-depleted, they took up the same amount of tetracycline. When the cells were treated with the oxidative uncoupling agent 2,4-dinitrophenol (DNP), the opposite occurred. Tetracycline accumulated in the resistant cells, whereas the susceptible cells showed no accumulation (Fig. 36, B). This pointed to an active transport mechanism in the resistant cell that extruded tetracyclines via energy-dependent processes. The Tet efflux mechanism has been found to be proton motive force-dependent; it functions as a tetracycline-cation antiport system with cations, extruding protons in exchange for a tetracycline-cation complex [131, 132].

The Tet protein is easily expressed in *E. coli*, and its protein structure has been determined by classical protein determination methods, where the primary protein structure reveals a two-domain protein separated by a large cytoplasmic loop, consisting of 12-membrane spanning segments [133]. Tet proteins belong to the multifacilitator superfamily of efflux proteins and have evolved as a multimer within the inner membrane to specifically pump tetracyclines out of bacterial cells.

Functioning transport proteins like Tet proteins can also be expressed in *E. coli* everted membrane vesicles, produced by French press lysis, which now bind tetracycline and transport it into membrane vesicles [128] instead of by

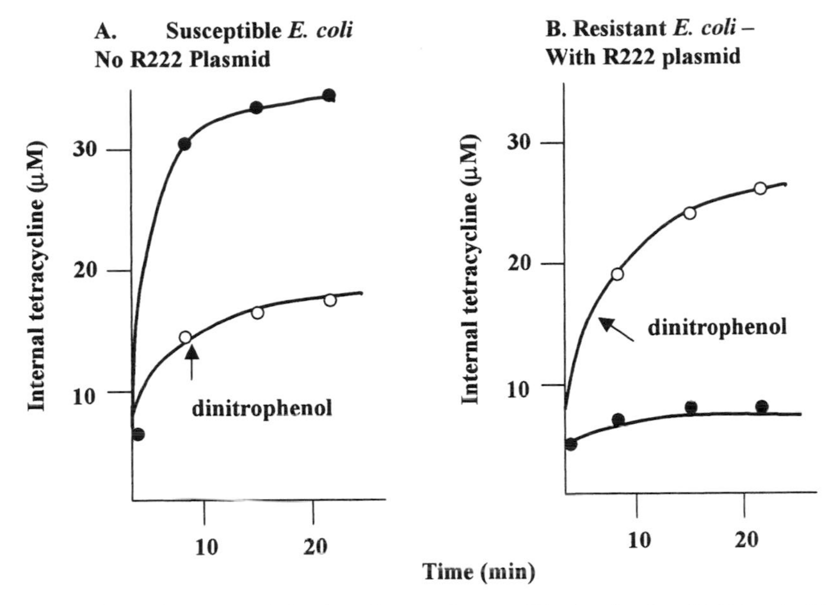

Figure 36. Tetracycline accumulation in resistant and susceptible *E. coli* cells with (A) and without (B) the plasmid R222 with determinant Tn10. Resistant cells accumulated radiolabeled tetracycline (^{3}H-Tc) in the presence of dinitrophenol (DNP), indicating that an active transport process was inhibited. Adapted from ref. 128.

efflux from the whole cell (Fig. 37). Transport dynamics of substrates can be followed for radioactive substrates such as calcium [134] as well as for radiolabeled tetracyclines.

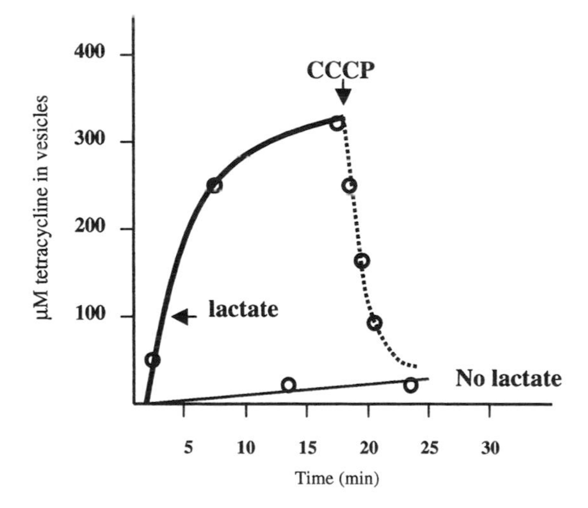

Figure 37. ^{3}H-Tc uptake by everted membrane vesicles possessing the Tet(B) protein. Vesicles were prepared from *E. coli* ML308-225 bearing plasmid R222. Lactate energized the vesicles, which allowed ^{3}H-Tc transport and accumulation.

Modulation of tetracycline efflux proteins

In order to determine the molecular requirements for inhibition of tetracycline efflux, we examined numerous tetracycline analogues for their ability to inhibit uptake of ^{3}H-Tc into everted inner membrane vesicles derived from a tetracycline-resistant *E. coli* D1-209 possessing the plasmid R222, which bears the class B determinant in Tn10. Class B is the most common Tc-resistant determinant found in many genera and has the highest transport activity of all the Tet proteins studied thus far.

After cell lysis under French pressure cell conditions, inner membranes of *E. coli* reorient themselves inside-out, producing Tet(B) proteins trapped in a lipid bilayer vesicle that is reversed – now pumping tetracyclines into the vesicle instead of pumping them out of the cell (Fig. 37). When everted membrane vesicles with Tet(B) are energized by Li-lactate, NADH, or ATP, they accumulate ^{3}H-Tc, which is readily measured by nitrocellulose membrane filtration and scintillation counting [128].

We hypothesized that when a competitor for or inhibitor of tetracycline transport is added to energized vesicles along with ^{3}H-Tc, the amount of inhibition can be measured as a decrease in accumulation of ^{3}H-Tc. Everted membrane vesicles therefore serve as an efficient assay for finding molecules capable of blocking Tet protein-mediated efflux and thus blocking tetracycline efflux. Inhibiting efflux proteins and increasing intracellular drug concentrations may constitute an effective approach to reversing tetracycline resistance, changing a resistant cell to a susceptible one.

We investigated tetracycline positional variants in relation to the tetracycline molecule and structural requirements for inhibition of efflux. The compounds initially tested were obtained from other laboratories, compound stores and pharmaceutical companies or were synthesized accordingly. The compounds chosen reflected partial ring derivatives of tetracycline as well as those that would define the minimal structural attributes needed to inhibit Tet(B) protein function. Furthermore, these assays would provide data for modifications of tetracycline that would provide for effective Tet(B) protein-blocking agents as synergists in combination with antibacterial tetracyclines.

All compounds were also examined for their ability to disrupt membrane energetics or pH gradient formation in vesicles from tetracycline-susceptible *E. coli* using an acridine orange assay [135]. If a compound were to act as a Tet(B) antagonist by membrane perturbation or disruption, the proton gradient collapses, and fluorescence changes signal vesicle disruption. In our studies none of the tetracycline compounds tested interfered with pH gradient formation or disrupted the membrane vesicles.

Based on over 50 compounds examined, we found that an intact ABCD naphthacene structure is needed for maximum inhibition of efflux (Fig. 38) [136]. Also required is the presence of the phenolic keto-enol structure along the lower peripheral region spanning C10, C11, C12 and C1. The function of these groups correlates well with activity, as divalent cations are required for

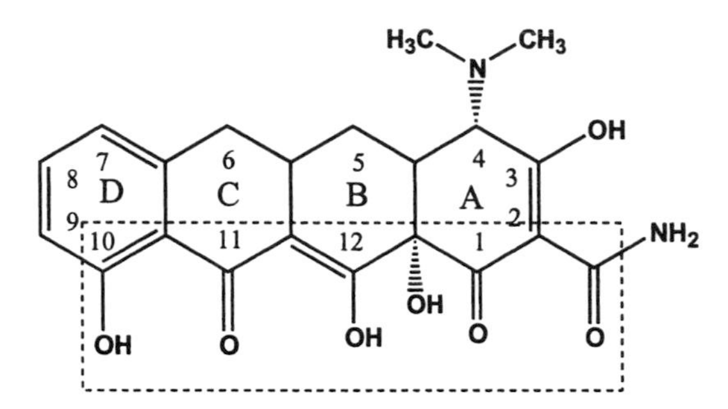

Figure 38. Minimal molecular requirements for inhibition of the Tet(B) efflux protein. Areas in boxes represent molecular substructures most sensitive to modification.

efflux. The C4 position dimethylamino group is not necessarily needed to inhibit activity, as dedimethyl tetracycline derivatives have marginal inhibitory activity alone.

Chemical modifications at positions 2, 4, 5, 6, 7, 8, and 9 produced compounds of variable activity, but indicated that synthetic derivatives may be more potent efflux inhibitors. We found that 13-alkylthio derivatives, produced by the synthesis scheme described previously in Figure 27, where methacycline is reacted with mercaptans, were among the most potent inhibitors of efflux (Fig. 39). In particular, 13-propylthio-5-OH tetracycline I and its halogenated derivative 13-[(3-chloropropyl)thio]-5-OH tetracycline II inhibited efflux with IC_{50} values of 0.7 and 0. 6, respectively. In contrast, the k_m of tetracycline is between 15 and 20 μM, while clinically used doxycycline has an IC_{50} value of 9.2 μM.

Studies of both derivatives against susceptible and resistant bacteria showed that they were effective alone against tetracycline-susceptible Gram-positive *S.*

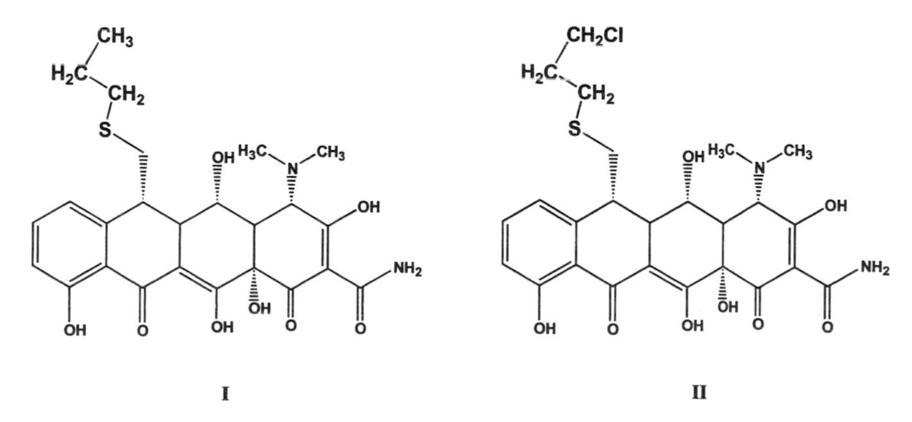

Figure 39. Chemical structures of efflux protein inhibitors 13-thiopropyl-5-OH tetracycline I and 13-[(3-chloropropyl)]-5-OH tetracycline II.

Table 4. Minimum inhibitory concentrations of doxycyclines and 13-(propylthio)-5-OH Tc analogues against tetracycline susceptible and resistant strains alone and in combination

		[a]MIC, µg/mL		
		Doxycycline	13-Propylthio I	13[(3-chloropropyl)]
susceptible				
E. coli	ML308-225	0.5	40	10
E. coli	D31m4	0.5	10	2.5
S. aureus	(RN450)	0.25	0.6	0.6
E. faecalis	(ATCC9790)	0.25	1.25	0.6
drug alone				
E. coli	ML308-225 Tet A	50	>50	50
E. coli	ML308-225 Tet B	>50	>50	>50
E. coli	D31m4 Tet B (LPS-)	12.5	12.5	6.25
S. aureus	(RN4250) Tet K	12.5	12.5	6.25
E. faecalis	(ATCC9790) Tet L	1.56	1.56	1.56
In combination: doxycycline/analogue				
E. coli	ML308-225 Tet A		6/3 synergy	3/3 synergy
E. coli	ML308-225 Tet B		NE	NE
E. coli	D31m4 Tet B (LPS-)		6/1.5 additive	3/1.5 synergy
S. aureus	(RN4250) Tet K		6/3 additive	3/1.5 synergy
E. faecalis	(ATCC9790) Tet L		1.5/0.8 additive	1.5/0.4 synergy

[a] Amount of compound to inhibit growth after 18 h at 37 °C. NE = no effect.

aureus and *E. faecalis* (Tab. 4). Even more encouraging was that they were also effective against *S. aureus* and *E. faecalis* that possessed Tet(K) and Tet(L) efflux proteins, and were more effective than doxycycline, especially against *E. faecalis*. Against susceptible *E. coli*, they were both ineffective, especially against tetracycline efflux strains possessing the Tet A and Tet B determinants, as was doxycycline.

The checkerboard MIC titration of the two derivatives with doxycycline indicated, however, that compound II acted in synergy with doxycycline against *E. coli* D1-299 (bearing Tet A), a weaker efflux system (Tab. 4). Furthermore, compound I acted in synergy with doxycycline against most of the efflux-mediated strains, both Gram-negative and Gram-positive.

We speculated that the efflux inhibitor I has an increased affinity for the Tet protein and displaces the normal substrate, doxycycline. We further postulated that once efflux inhibitors entered the cell and blocked efflux, therapeutic concentrations of doxycycline can enter and accumulate, leading to ribosome dysfunction and inhibition of growth.

This led to a more detailed study of the synthesis and evaluation of C13-alkylthio and 13-arylthio derivatives in the tetracycline efflux model. The 13-alkylthio series of compounds and the 13-phenylthio and 13-benzylthio compounds were next examined and clear trends in the structure-*versus*-activity emerged [67]. Position 13-substituted compounds exhibited activity that was dependent on the size, lipophilicity and polarity of the C13 substituent. Correlation of the bioassay results to Verloop-Hoogenstraten STERIMOL values, a measure of size of molecular substituents, suggested that the more potent inhibitors had steric length values of 4.4–6.2 angstroms, while optimal width was between 3.0 and 4.2 angstroms [137]. Lipophilicity also correlated well with bioactivity, where introduction of polar hydroxy groups or carboxylic acids, even within the same homologous series, caused a decrease in efflux inhibitory activity. Most of the active compounds fell within a range of pi values between 1.0 and 2.5.

We concluded that optimal efflux inhibition was attained when the size and thus lipophilicity of the substituent was constrained within limits of a hydrophobic pocket located within the domains of the Tet(B) protein, and that compounds not meeting or exceeding these parameters would be less active (Fig. 40).

From these studies one analogue, 13-cyclopentylthio-5-OH tetracycline (13-CPTC) (Fig. 41), was chosen for further study for its effects on Tet protein function [138]. 13-CPTC showed the most potency in inhibiting efflux with an IC_{50} value of 0.4–1.0 μM, and was examined for its effects on the Tet(B) protein.

The effect of 13-CPTC on accumulation of ^{3}H-Tc in everted vesicles indicated that it acted by competitive inhibition, as plotted using Michaelis-Menten kinetics and Lineweaver-Burke methods (Fig. 42). These findings showed that 13-CPTC had a greater affinity and was competing with tetracycline for the binding site on the Tet(B) protein. The increase in $1/k_m$ apparent upon addition of the inhibitor changes the distribution of available Tet protein

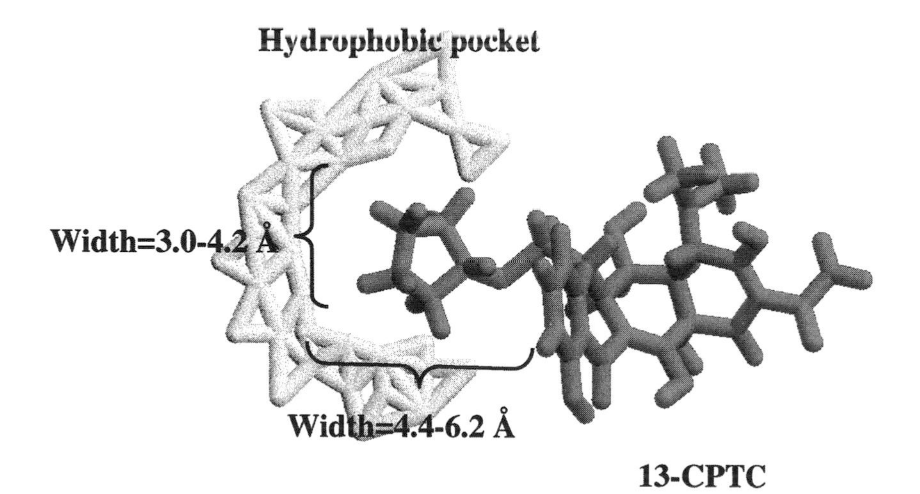

Figure 40. Hydrophobic pocket in the Tet protein outlined by 13-thio derivatives and their activity as efflux inhibitors. Using C13 derivatives, the constraints of length and width of the substituent optimal for activity was revealed.

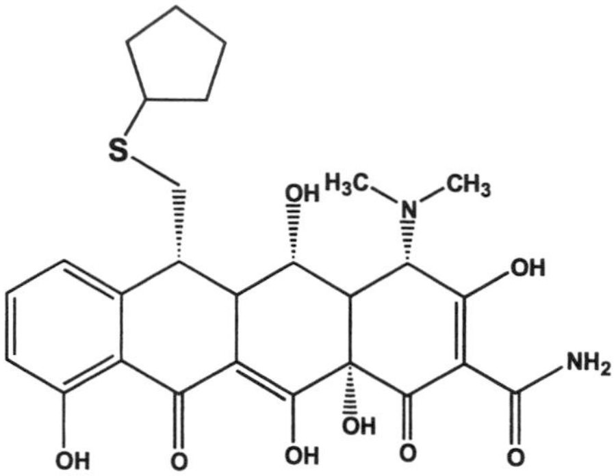

I

13-CPTC

Figure 41. Chemical structure of 13-cyclopentylthio-5-OH tetracycline (13-CPTC).

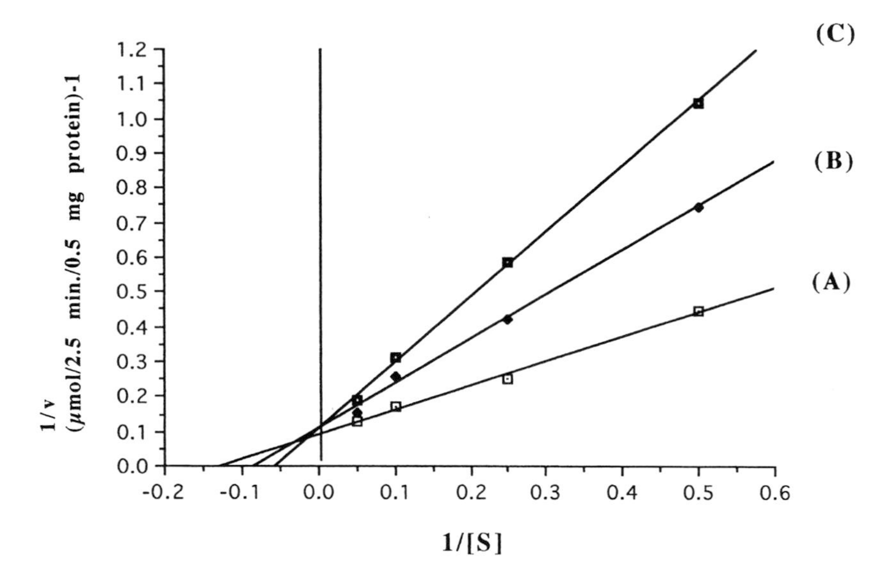

Figure 42. Lineweaver-Burke plot of inhibition of ^{3}H-Tc by the efflux protein inhibitor 13-CPTC in everted membrane vesicles. Increasing amounts of 13-CPTC were added at 0.9 μM (B) and 3.0 μM (C) while no 13-CPTC (A) indicated competitive inhibition.

from one of high affinity and translocation ability to one of decreased affinity for tetracycline and low translocation ability.

In whole cells against susceptible *E. coli*, ^{3}H-Tc accumulation was not hampered by the presence of 13-CPTC (Fig. 43, A). However, in resistant *E. coli*

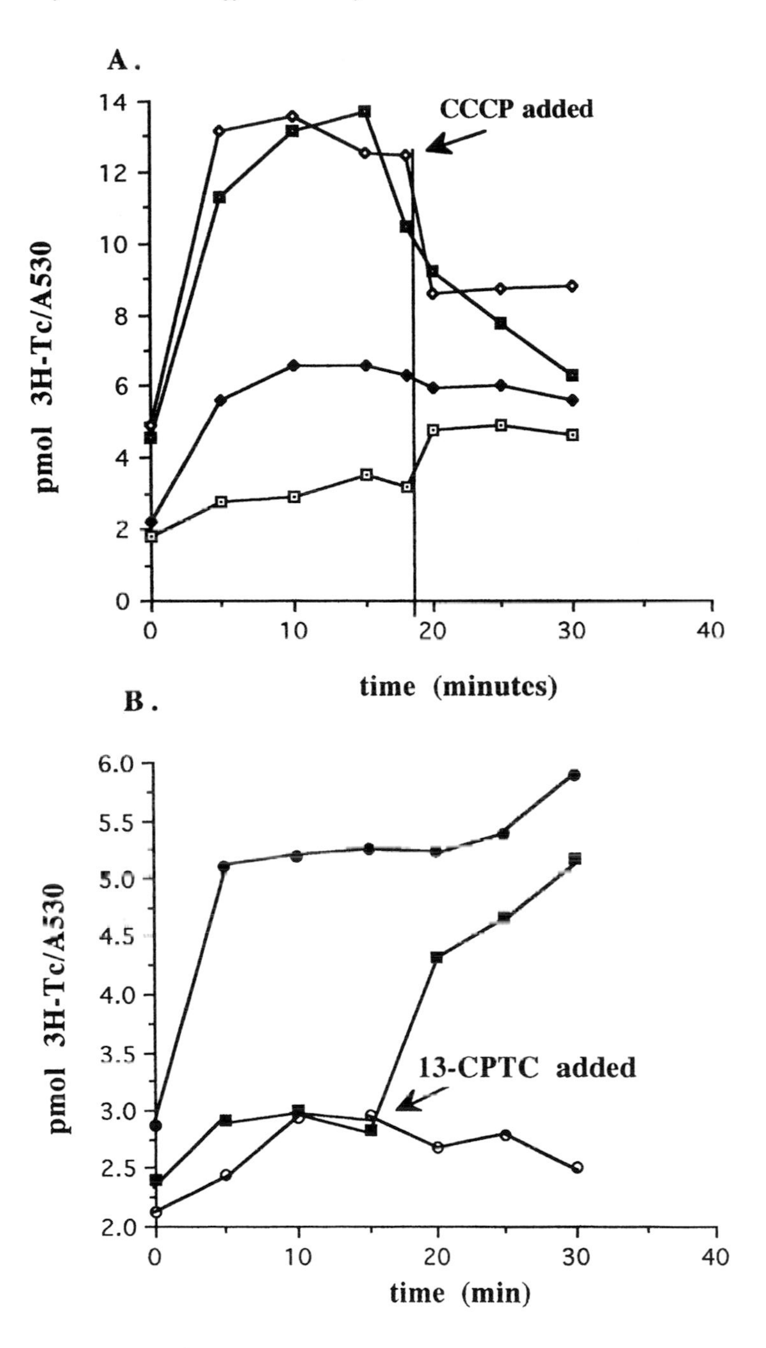

Figure 43. Accumulation of ^{3}H-Tc in *E. coli* ML308-225 without (■) and with (□) 13-CPTC (40 μM) and *E. coli* D1-209 Tet(B) without (○) and with (●) 13-CPTC 40 μM (A). Accumulation of ^{3}H-Tc in *E. coli* D1-209 Tet(B) preincubated with 13-CPTC (●), or without 13-CPTC (○). A rapid rise in accumulation occurs after addition of 40 μM of efflux protein inhibitor 15 min after glucose was added (■) (B).

with Tet(B), adding 13-CPTC caused a rapid uptake of ^{3}H-Tc, indicating that efflux was now inhibited (Fig. 43, B). Also, resistant *E. coli* Tet(B) cells readily accumulate tetracycline when they were pre-incubated with 13-CPTC.

The antibacterial effects of 13-CPTC alone and in combination with doxycycline show that alone 13-CPTC is hardly effective (Tab. 5), exhibiting MICs of between 6.25 and 25 µg/mL against Gram-negative bacteria. 13-CPTC was found, however, to be active against Gram-positive bacteria bearing Tet K and Tet L efflux determinants. Furthermore, combinations of 13-CPTC with doxycycline exhibit synergy, especially the efflux-mediated tetracycline resistance due to Tet(B), indicating a reversal of resistance, where both the compounds are 2-fold or more potent when combined than when used alone.

These findings demonstrated that 13-CPTC acts as a potent inhibitor of the Tet(B) protein, binding competitively with tetracycline at the putative binding site of efflux. We hypothesize that the ability of 13-CPTC as an efflux inhibitor to affect the growth of tetracycline-resistant bacteria is due to competitive inhibition of the efflux protein. By combining subinhibitory amounts of an efflux inhibitor and a tetracycline, it is possible to demonstrate a pronounced inhibitory effect both on molecular processes and bacterial growth. Similarly, the addition of 13-CPTC to bacteria actively effluxing tetracycline results in rapid accumulation of tetracycline, reversing efflux and ultimately antibiotic resistance.

Our research findings suggest that the tetracycline molecule may be synthetically modified toward the development of an effective tetracycline block-

Table 5. Minimum inhibitory concentrations of doxycyclines and 13-cyclopentylthio-5-OH Tc (13-CPTC) against tetracycline susceptible and resistant strains alone and in combination

	[a]MIC, µg/mL		
	Doxycycline	13-Propylthio I	13[(3-chloropropyl)]
susceptible			
E. coli ML308-225	0.5		20
E. coli D31m4	0.5		0.8
S. aureus (RN450)	0.25		0.5
E. faecalis (ATCC9790)	0.25		0.8
drug alone			
E. coli ML308-225 Tet A	50		50
E. coli ML308-225 Tet B	>50		50
E. coli D31m4 Tet B (LPS-)	12.5		6.25
S. aureus (RN4250) Tet K	12.5		1.56
E. faecalis (ATCC9790) Tet L	1.56		0.39
In combination: doxycycline/analogue			
E. coli ML308-225 Tet A			3.12/1.56 synergy
E. coli ML308-225 Tet B			1.56/1.56 synergy
E. coli D31m4 Tet B (LPS-)			1.56/0.39 synergy
S. aureus (RN4250) Tet K			3.12/0.78 additive
E. faecalis (ATCC9790) Tet L			1.56/0.19 additive

[a] Amount of compound to inhibit growth after 18 h at 37 °C. NE = no effect.

ing agent and efflux protein inhibitor. By blocking the Tet(B) efflux protein we have demonstrated the ability to modulate antibiotic transport and tetracycline resistance phenotypes, changing the resistant cell to a state of susceptibility by using therapeutically relevant concentrations of a tetracycline.

In view of the widespread antibiotic resistance among bacterial species, this approach offers methods and insights for the development of compounds that may one day be used to reverse tetracycline resistance and antibiotic resistance clinically.

The past and present uses of tetracyclines

While the tetracyclines in the past have had success as antibiotics against microorganisms, we have seen that they also have agonist, antagonist and physiological properties against a wide variety of proteins, enzymes and macromolecular targets in eukaryotic cells. Tetracyclines, as microbial secondary metabolites, function to change their environment and establish territorial boundaries against both prokaryotes and eukaryotes. Thus, it becomes clearly evident from studying the tetracyclines that they are pharmacologically and chemically "promiscuous" agents, capable of interacting with numerous molecular receptors and modifying cellular responses.

We have studied the number of articles published on tetracyclines, which totals 4, 976 over the last 10 years (1990–2000), and over 13,000 since 1948, the year tetracyclines were discovered and announced, and have found that technological fronts are emerging in bacteriology, dental and enzymatic pharmacology, genetic techniques and transgenics, and cellular physiology of eukaryotic cells. In other diverse fields of research, such as neurology, reproductive biology and virology the utility and activity of the tetracyclines are being discovered and are also emerging as new technological fronts.

While other chemical classes of antibiotics have considerably more citations over the same 10-year time period (i.e., β-lactams, 14,805; cephalosporins, 9,695), they are primarily within the discipline of bacteriology, and none have shown the wide discipline diversity or have demonstrated such cross-utility as have the tetracyclines.

Tetracyclines, as dynamic entities, possess unique chemical and biological characteristics that may explain their ability to interact and modify so many different cellular targets, receptors and cellular properties. The discovery of new uses of tetracyclines and their novel biological properties against both prokaryotes and eukaryotes is currently underway by the contributors to this manuscript and numerous scientists throughout the world.

Acknowledgements
Studies of tetracycline efflux inhibition were supported by the Upjohn Company, Kalamazoo, MI and Tufts University School of Medicine, Boston, MA. I thank Dr. S. B. Levy for continued support and guidance, Dr. L. McMurry for many years of helpful discussions, and Dr. M. Ismail for careful review

of this manuscript and background reading. I also thank D. Messersmith and J. Dumornay for the NMR spectra and Dr. B. Podlogar for help with the molecular modeling studies. I also would like to thank the librarians at the National Library of Medicine, Bethesda, MD and M. I. T. Hayden library for their help and assistance, and Ms. S. Cabal at the University of Mississippi for locating Benjamin Duggar's photograph and other archival information.

References

1 Doetsch RN (1960) *Microbiology: historical contributions from 1776 to 1908*. Rutgers University Press, N.J., xii p 23

2 Wainwright M (1990) *Miracle cure: the story of antibiotics*. Basil Blackwell Ltd, Oxford, pp 89–93

3 Fleming A (1955) Foreword to: Penicillin: A Symposium. *The Practitioner* 174: 5

4 Fleming A (1929) On the antibacterial action of cultures of a *Penicillium*, with special references to their use and isolation of *B. influenzae. Brit J Exp Pathol* 10: 226–236

5 Florey HW (1944) Penicillin: Its development for medical uses. *Nature* 153: 41–42

6 Florey HW, Chain E, Heatley NG, Jennings MA, Sanders AG, Abraham EP, Florey ME (1949) *Antibiotics*, 2 vol. Oxford University Press, London, New York, Toronto, p 673

7 Waksman SA (1965) A quarter-century of the antibiotic era. *Antimicrob Agents Chemother* 5: 9–19

8 Waksman SA, Flynn JE (1973) History of the word 'antibiotic'. *J Hist Med Allied Sci* 28: 284–286

9 Duggar BM (1948) Aureomycin. *Ann N Y Acad Sci* 51: 177–181

10 Duggar BM (1949) Aureomycin and preparation of the same. *US patent 2, 482, 055* (American Cyanamid)

11 Finlay AC, Kane JH (1950) Terramycin and its production. *US patent 2,516, 080* (Chas Pfizer Co)

12 Hochstein FA, Stephens CR, Conover LH, Regna PP, Pasternack R, Brunings KJ, Woodward RB (1953) The structure of terramycin. *J Amer Chem Soc* 75: 5455–5475

13 Stephens CR, Conover LH, Pasternack R, Hochstein FA, Moreland WT, Regna PP, Pilgrim FJ, Brunings KJ, Woodward RB (1954) The structure of aureomycin. *J Amer Chem Soc* 76: 3568–3575

14 Conover LH, Moreland WT, English AR, Stephens CR, Pilgrim FJ (1953) Terramycin XI Tetracycline. *J Amer Chem Soc* 75: 462

15 McCormick JRD (1967) Tetracyclines: *In*: D Gottlieb, PD Shaw (eds): *Antibiotics, vol 2-Biosynthesis*. Springer-Verlag, Berlin, Heidelberg, New York, pp 113–122

16 Hostalek Z, Tinterova M, Jechova V, Blumauerova M, Suchy J, Vanek Z (1969) Regulation of biosynthesis of secondary metabolites. I. Biosynthesis of chlortetracycline and tricarboxylic acid cycle activity. *Biotechnol Bioeng* 11: 539–548

17 Behal V, Prochazkova V, Vanek Z (1969) Regulation of biosynthesis of secondary metabolites. II. Fatty acids and chlortetracycline in *Streptomyces aureofaciens. Folia Microbiol* 14: 112–116

18 McCormick JR, Jensen ER, Arnold N, Corey HS, Joachim UH, Johnson S, Miller PA, Sjolander NO (1968) Biosynthesis of tetracyclines. XI. The methylanthrone analog of protetrone. *J Amer Chem Soc* 90: 7127–7129

19 McCormick JR, Jensen ER (1969) Biosynthesis of the tetracyclines. XII. Anhydrodemethylchlortetracycline from a mutant of *Streptomyces aureofaciens. J Amer Chem Soc* 91: 206

20 Hostalek Z, Tinterova M, Jechova V, Blumauerova M, Suchy J, Vanek Z (1969) Regulation of biosynthesis of secondary metabolites. I. Biosynthesis of chlortetracycline and tricarboxylic acid cycle activity. *Biotechnol Bioeng* 11: 539–548

21 Brown, JM, McNeil MM, Desmond, EP (1999) *Nocardia, Rhodococcus, Gordona, Actinomadura, Streptomyces*, and other *Actinomycetes* of medical importance. *In*: PA Murray (ed.): *Manual of clinical microbiology*, 7th edition. ASM Press, Washington D.C., pp 370–398

22 Perlman D, Heuser L (1962) Process for producing 6-demethyl-6-hydroxy tetracyclines. *US patent 3,466,093*

23 Wells JS, O'Sullivan J, Aklonis C, Ax HA, Tymiak AA, Kirsch DR, Trejo WH, Principe P (1992) Dactylocyclines, novel tetracyclines derivatives produced by a *Dactylosporangium* sp. I. Taxonomy, production, isolation and biological activity. *J Antibiot* 45: 1892–1898

24 Tymiak AA, Ax HA, Bolgar MS, Kahle AD, Porubcan MA, Andersen NH (1992) Dactylocyclines, novel tetracycline derivatives produced by a *Dactylosporangium* sp. II. Structure elucidation. *J Antibiot* 45: 1899–1906

25 Garmaise DL, Chu DT, Bernstein E, Inaba M, Stamm JM (1979) Synthesis and antibacterial activity of 2'-substituted chelocardin analogues. *J Med Chem* 22: 559–564

26 Mitscher LA, Juvarkar JV, Rosenbrook W Jr, Andres WW, Schenk J, Egan RS (1970) Structure of chelocardin, a novel tetracycline antibiotic. *J Amer Chem Soc* 92: 6070–6071

27 Pullman B (1988) Binding affinities and sequence selectivity in the interaction of antitumor anthracyclines and anthracenediones with double stranded polynucleotides and DNA. *In*: JW Lown (ed.): *Anthracycline and anthracenedione-based anticancer agents*. Elsevier Science Publishers, Amsterdam, Oxford, New York, Tokyo, pp 371–392

28 Hatsu M, Sasaki T, Watabe H, Miyadoh S, Nagasawa M, Shomura T, Sezaki M, Inouye S, Kondo S (1992) A new tetracycline antibiotic with antitumor activity. I. Taxonomy and fermentation of the producing strain, isolation and characterization of SF2575. *J Antibiot* 45: 320–324

29 Hatsu M, Sasaki T, Gomi S, Kodama Y, Sezaki M, Inouye S, Kondo S (1992) A new tetracycline antibiotic with antitumor activity. II. The structural elucidation of SF2575. *J Antibiot* 45: 325–330

30 Horiguchi T, Hayashi K, Tsubotani S, Iinuma S, Harada S, Tanida S (1994) New naphthacenecarboxamide antibiotics, TAN-1518 A and B, have inhibitory activity against mammalian DNA topoisomerase I. *J Antibiot* 47: 545–556

31 International Union of Pure and Applied Chemistry

32 Stezowski JJ (1976) Chemical-structural properties of tetracycline derivatives. 1. Molecular structure and conformation of the free base derivatives. *J Amer Chem Soc* 98: 6012–6018

33 Hughes LJ, Stezowski JJ, Hughes RE (1979) Chemical-structural properties of tetracycline derivitives. 7. Evidence for the coexistence of the zwitterionic and nonionized forms of the free base in solution. *J Amer Chem Soc* 101: 7655–7657

34 Duarte HA, Carvalho S, Paniago EB, Simas AM (1999) Importance of tautomers in the chemical behavior of tetracyclines. *J Pharm Sci* 88: 111–120

35 Valcavi U, Brandt A, Corsi GB, Minoja F, Pascucci G (1981) Chemical modifications in the tetracycline series. *J Antibiot* 34: 34–39

36 Sigler A, Schubert P, Hillen W, Niederweis M (2000) Permeation of tetracyclines through membranes of liposomes and *Escherichia coli. Eur J Biochem* 267: 527–533

37 Gottstein WJ, Minor WF, Cheney LC (1958) Carboxamido derivatives of the tetracyclines. *J Amer Chem Soc* 81: 1198–1201

38 Siedel W, Soder A, Lindner F (1958) Die Aminomethylierung der Tetracycline. Zur Chemie des Reverin. *Munch Med Wochenschr* 17: 661–663

39 Hughes D, Wilson WL, Butterfield AG, Pound NJ (1974) Stability of rolitetracycline in aqueous solution. *J Pharm Pharmacol* 26. 79–80

40 Tamorria CR, Esse RC (1965) Alkoxyalkyltetracyclines. *J Med Chem* 8: 870–872

41 Stephens C, Beereboom JJ, Rennhard HH, Gordon PN, Murai K, Blackwood RK, Schach von Wittenau M (1963) 6-Deoxytetracyclines. IV. Preparation, C-6 stereochemistry, and reactions. *J Amer Chem Soc* 85: 2643–2652

42 Beereboom J, Butler K (1962) Hydrolysis of 2-decarboxamido-2-cyano-6-deoxytetracycline derivatives. *US Patent 3,069,467*

43 Doerschuk A, Bitler B, McCormick J (1955) Reversible isomerizations in the tetracycline family. *J Amer Chem Soc* 77: 4687

44 Stephens C, Conover LH, Gordon PN, Pennington FC, Wagner RH, Brunings KJ, Pilgrim FJ (1956) Epitetracycline – the chemical relationship between tetracycline and "Quatrimycin". *J Amer Chem Soc* 78: 1515–1519

45 McCormick JRD, Fox SM, Smith LL, Bitler BA, Reichenthal J, Origoni VE, Muller WH, Winterbottom R, Doershuk AP (1956) Studies of the reversible epimerization occurring in the tetracycline family. The preparation, properties, and proof of structure of some 4-epitetracyclines. *J Amer Chem Soc* 78: 3547–3548

46 Yuen PH, Sokoloski TD (1977) Kinetics of concomittant degradation of tetracycline to epi tetracycline, anhydrotetracycline, and epianhydrotetracycline in acid phosphate solution. *J Pharm Sci* 66: 1648–1650

47 Blanchflower JW, McCracken RJ, Haggen AS, Kennedy GD (1997) Confirmation assay for the determination of tetracycline, oxytetracycline, chlortetracycline and its isomers in muscle and kidney using liquid chromatography-mass spectrometry. *J Chromatogr B* 692: 351–360

48 Boothe JH, Bonvicino GE, Waller CY, Petisi JP, Wilkinson RW, Broschard RB (1958) Chemistry of the tetracycline antibiotics. I. Quaternary derivatives. *J Amer Chem Soc* 80: 1654–1657
49 Heggie W, Santos P, Galindro Jose (1999) A process for the preparation of 4-(des-dimethylamino)-tetracyclines. *European Patent Application EP 0 962 552 A1*
50 McCormick JRD, Fox SM, Smith LL, Bitler BA, Reichenthal J, Origoni VE, Muller WH, Winterbottom R, Doershuk AP (1957) Studies of the reversible epimerization occurring in the tetracycline family. The preparation, properties and proof of structure of some 4-epi-tetracyclines. *J Amer Chem Soc* 78: 2849–2858
51 Green A, Boothe JH (1960) Chemistry of the tetracycline antibiotics. III. 12a-Deoxytetracycline. *J Amer Chem Soc* 82: 3949–3953
52 Blackwood RK, Rennhard HH, Stephens CR (1960) Some transformations at the 12a-position of the tetracyclines. *J Amer Chem Soc* 82: 745–746
53 Blackwood R, English A (1970) Structure-activity relationships in the tetracycline series. *Adv Appl Microbiol* 13: 237–266
54 Blackwood RK (1962) Tetracycline derivatives and process producing the same. *US Patent 2,976,318*
55 Bernardi L, De Castiglione R, Colonna V, Masi P, Mazzoleni R (1975) Tetracycline derivatives. I. Esters of 5-oxytetracyclines: chemistry and biological activity. *Il Farmaco – Ed Sci* 29: 902–909
56 Valcavi U, Campanella G, Pacini N (1963) Pyrazole derivatives of tetracycline and chlortetracycline. *Gazz Chim Ital* 93: 916–928
57 McCormick JR, Jensen ER, Johnson K, Sjolander NO (1967) Biosynthesis of the tetracyclines. IX. 4- Aminodedimethylaminoanhydrodemethylchlortetracycline from a mutant of *Streptomyces aureofaciens*. *J Amer Chem Soc* 90: 2201–2202
58 Walton VC, Howlett MR, Selzer GB (1970) Anhydrotetracycline and 4-epianhydrotetracycline in market tetracyclines and aged tetracycline products. *J Pharm Sci* 59: 1160–1164
59 Cullen SI, Crounse RG (1965) Cutaneous pharmacology of the tetracyclines. *J Invest Dermatol* 45: 263–268
60 Breen K, Schenker S, Heimberg M (1972) The effect of tetracycline on the hepatic secretion of triglyceride. *Biochim Biophys Acta* 270: 74–80
61 Jones CC (1973) Megaloblastic anemia associated with long-term tetracycline therapy. Report of a case. *Ann Intern Med* 78: 910–912
62 Oliva B, Gordon G, McNicholas P, Ellestad G, Chopra I (1992) Evidence that tetracycline analogs whose primary target is not the bacterial ribosome cause lysis of *Escherichia coli*. *Antimicrob Agents Chemother* 36: 913–919
63 Blackwood RK, Beereboom JJ, Rennhard HH, Schach Von Wittenau M, Stephens CR (1961) 6-Methylenetetracyclines. I. A new class of tetracycline antibiotics. *J Amer Chem Soc* 83: 2773–2775
64 Blackwood RK, Beereboom JJ, Rennhard HH, Schach Von Wittenau M, Stephens CR (1963) 6-Methyenetetracyclines. III. Preparation and properties. *J Amer Chem Soc* 85: 3943–3953
65 Stephens CR, Beereboom JJ, Rennhard HH, Gordon P, Murai K, Blackwood RK, Schach Von Wittenau M (1963) 6-Deoxytetracyclines. IV. Preparation, C-6 stereochemistry, and reactions. *J Amer Chem Soc* 85: 2643–2652
66 Villax I, Page P (1985) Process for the preparation of a-6-deoxytetracyclines. *US Patent 4,500,458*
67 Nelson ML, Park BH, Andrews JS, Georgian VA, Thomas RC, Levy SB (1993) Inhibition of the tetracycline efflux protein by 13-thiosubstituted-5-hydroxy-6-deoxytetracyclines. *J Med Chem* 36: 370–377
68 Bitha P, Hlavka J, Boothe J (1970) 6-Fluorotetracyclines. *J Med Chem* 13: 89–92
69 Hlavka J, Schneller A, Krazinski H, Boothe JH (1962) The 6-deoxytetracyclines. III. Electrophilic and nucleophilic substitution. *J Amer Chem Soc* 84: 1426–1430
70 Church R, Schaub R, Weiss M (1971) Synthesis of 7-dimethylamino-6-demethyl-6-deoxytetracycline (Minocycline) via 9-nitro-6-demethyl-6-deoxytetracycline. *J Org Chem* 36: 723–725
71 Zambrano RT, Butler K (1962) Reductive alkylation process. *US Patent 3,483,251*
72 Sum PE, Lee V, Testa RT, Hlavka JJ, Ellestad G, Bloom JD, Gluzman Y, Tally F (1994) Glycylcyclines. 1. A new generation of potent antibacterial agents through modification of 9-aminotetracyclines. *J Med Chem* 37: 184–188
73 Sum PE, Petersen P (1999) Synthesis and structure-activity relationship of novel glycylcycline derivatives leading to the discovery of GAR-936. *Bioorganic and Medicinal Chemistry Letters* 9: 1459–1462

74 Barden TC, Buckwalter BL, Testa RT, Petersen PJ, Lee VJ (1994) Glycylcyclines. 3. 9-aminodoxycyclinecarboxamides. *J Med Chem* 37: 3205–3211
75 Webster's New Collegiate Dictionary (1976) defines promiscuous as: *adj* composed of all sorts of things, not restricted to one class or sort
76 Fuwa I, Okuda J (1966) Inhibitory action of tetracyclines on polynucleotide phosphorylase. *J Biochem* 59: 95–103
77 Kromery V, Kellen J (1966) Changes in some enzymes of bacterial electron transport accompanying resistance to oxytetracycline. *J Bacteriol* 92: 1264–1266
78 Waring MJ (1965) The effects of antimicrobial agents on ribonucleic acid polymerase. *Mol Pharmacol* 1: 1–13
79 Park JT (personal communication, unpublished data)
80 Colaizzi JL, Knevel AM, Martin AN (1965) Biophysical study of the mode of action of the tetracycline antibiotics. Inhibition of metalloflavoenzyme NADH cytochrome oxidoreductase. *J Pharm Sci* 54: 1425–1436
81 Gale EF, Folkes JP (1953) The assimilation of amino-acids by bacteria. Actions of antibiotics on nucleic acid and protein synthesis in *Staphylococcus aureus*. *Biochemistry* 53: 493–498
82 Hash JH, Wishnick M, Miller PA (1964) On the mode of action of the tetracycline antibiotics in *Staphylococcus aureus*. *J Biol Chem* 239: 2070–2078
83 Day LE (1966) Tetracycline inhibition of cell-free protein synthesis. I. Binding of tetracycline to components of the system. *J Bacteriol* 91: 1917–1923
84 Day LE (1966) Tetracycline inhibition of cell-free protein synthesis. II. Effect of the binding of tetracycline to components of the system. *J Bacteriol* 92: 197–203
85 Laskin AI, May Chan W (1964) Inhibition by tetracyclines of polyuridylic acid directed phenylalanine incorporation in *Escherichia coli* cell-free systems. *Biochem Biophys Res Commun* 14: 137–142
86 Morgan T, Ribush N (1972) The effect of oxytetracycline and doxycycline on protein metabolism. *Med J Australia* 1: 55–58
87 Miller GH, Khalil SA, Martin AN (1971) Structure-activity relationships of tetracyclines. I. Inhibition of cell division and protein and nucleic acid syntheses in *Escherichia coli*. *J Pharm Sci* 60: 33–40
88 Atherly AG (1974) Specific inhibition of ribosomal RNA synthesis in *Escherichia coli* by tetracycline. *Cell* 3: 145–151
89 Beliavskaya IV, Navashin SM, Gryaznova NS, Sazykin YO (1975) Uptake of oxytetracycline and minocycline by *E. coli* cells and their effect on protein synthesis. *In*: S Mitsuhashi et al. (eds): *Drug-inactivating enzymes and antibiotic resistance*. Avicenum, Prague, pp 231–233
90 Crowfoot PD, Esfahani M, Wakil SJ (1972) Relation between protein synthesis and phospholipid synthesis and turnover in *Escherichia coli*. *J Bacteriol* 112: 1408–15
91 Nakao M, Kitanaka E, Ochiai K, Nakazawa S (1972) Cell wall synthesis by *Staphylococcus aureus* in the presence of protein synthesis inhibitory agents. II. Electron microscopic study. *J Antibiot* 25: 469–470
92 Hirashima A, Childs G, Inouye M (1973) Differential inhibitory effects of antibiotics on the biosynthesis of envelope proteins of *Escherichia coli*. *J Mol Biol* 79: 373–389
93 Last JA (1969) Studies on the binding of tetracycline to ribosomes. *Biochim Biophys Acta* 195: 506–514
94 Le Goffic F, Moreau N, Langrene S, Pasquier A (1980) Binding of antibiotics to the bacterial ribosome studied by aqueous two-phase partitioning. *Anal Biochem* 107: 417–423
95 Bodley JW, Zieve FJ (1969) On the specificity of the two ribosomal binding sites: studies with tetracycline. *Biochem Biophys Res Commun* 36: 463–468
96 Streltsov SA, Kukhanova MK, Krayevsky AA, Beljavskaja IV, Victorova LS, Gursky GV, Treboganov AD, Gottikh BP (1974) Binding of oxytetracycline to *E. coli* ribosomes. *Mol Biol Rep* 1: 391–396
97 Tritton TR (1977) Ribosome-tetracycline interactions. *Biochemistry* 16: 4133–4138
98 Kolesnikov IV, Protasova NY, Gudkov AT (1996) Tetracyclines induce changes in accessibility of ribosomal proteins to proteases. *Biochimie* 78: 868–873
99 Fey G, Reiss M, Kersten H (1973) Interaction of tetracylines with ribosomal subunits from *Escherichia coli*. A fluorometric investigation. *Biochemistry* 12: 1160–1164
100 Epe B, Woolley P (1984) The binding of 6-demethylchlortetracycline to 70S, 50S and 30S ribosomal particles: a quantitative study by fluorescence anisotropy. *EMBO J* 3: 121–126

101 Cooperman BS (1980) Photolabile antibiotics as probes of ribosomal structure and function. *Ann N Y Acad Sci* 346: 302–323
102 Goldman RA, Cooperman BS, Strychar, WA, Williams BA, Tritton TR (1980) Photoincorporation of tetracycline into *Escherichia coli* ribosomes. *FEBS Lett* 118: 113–118
103 Goldman RA, Hasan T, Hall CC, Strycharz WA, Cooperman BS (1983) Photoincorporation of tetracycline into *Escherichia coli* ribosomes. Identification of the major proteins photolabeled by native tetracycline and tetracycline photoproducts and implications for the inhibitory action of tetracycline on protein synthesis. *Biochemistry* 22: 359–368
104 Hasan T, Goldman RA, Cooperman BS (1985) Photoaffinity labeling of the tetracycline binding site of the *Escherichia coli* ribosome. The uses of a high intensity light source and of radioactive sancycline derivatives. *Biochem Pharmacol* 34: 1065–1071
105 Maxwell IH (1967) Partial removal of bound transfer RNA from polysomes engaged in protein synthesis *in vitro* after addition of tetracycline. *Biochim Biophys Acta* 138: 337–346
106 Maxwell IH (1968) Studies of the binding of tetracycline to ribosomes *in vitro*. *Mol Pharmacol* 4: 25–37
107 Sarkar S, Thach RE (1968) Inhibition of formylmethionyl-transfer RNA binding to ribosomes by tetracycline. *Proc Natl Acad Sci USA* 60: 1479–1486
108 Craven GR, Gavin R, Fanning T (1969) The transfer RNA binding site of the 30 S ribosome and the site of tetracycline inhibition. *Cold Spring Harbor Symp Quant Biol* 34: 129–137
109 Semenkov YP, Makarov EM, Makhno VI, Kirillov SV (1982) Kinetic aspects of tetracycline action on the acceptor (A) site of *Escherichia coli* ribosomes. *FEBS Lett* 144: 125–129
110 Brown CM, McCaughan KK, Tate WP (1971) Two regions of the *Escherichia coli* 16S ribosomal RNA are important for decoding stop signals in polypeptide chain termination. *Nucl Acid Res* 21: 2109–2115
111 Lodmell JS, Tapprich WE, Hill E (1983) Evidence for a conformational change in the exit site of the *Escherichia coli* ribosome upon tRNA binding. *Biochemistry* 32: 4067–4072
112 Oliva B, Chopra I (1992) Tet determinants provide poor protection against some tetracyclines: further evidence for division of tetracyclines into two classes. *Antimicrob Agents Chemother* 36: 876–878
113 Oliva B, Gordon G, McNicholas P, Ellestad G, Chopra I (1992) Evidence that tetracycline analogs whose primary target is not the bacterial ribosome cause lysis of *Escherichia coli*. *Antimicrob Agents Chemother* 36: 913–919
114 Rasmussen B, Noller HF, Daubresse G, Oliva B, Misulovin Z, Rothstein DM, Ellestad GA, Gluzman Y, Tally FP, Chopra I (1991) Molecular basis of tetracycline action: identification of analogs whose primary target is not the bacterial ribosome. *Antimicrob Agents Chemother* 35: 2306–2311
115 Franklin TJ (1964) The effect of chlortetracycline on the transfer of leucine and 'transfer' ribonucleic acid to rat-liver ribosomes *in vitro*. *Biochem J* 90: 624–628
116 Yeh SD, Shils ME (1966) Tetracycline and incorporation of amino acids into proteins of rat tissues. *Proc Soc Exp Biol Med* 121: 729–734
117 Greenberger NJ (1967) Inhibition of protein synthesis in rat intestinal slices by tetracycline. *Nature* 214: 702–703
118 Reboud AM, Dubost S, Reboud JP (1982) Photoincorporation of tetracycline into rat-liver ribosomes and subunits. *Eur J Biochem* 124: 389–396
119 Bread NS Jr, Armentrout SA, Weisberger AS (1974) Inhibition of mammalian protein synthesis by antibiotics. *Pharmacol Rev* 21: 213–245
120 Ganguly R, Pennock DG, Kluge RM (1984) Inhibition of protein synthesis and lymphokine production by tetracycline. *Allerg Immunol* 30: 104–109
121 van den Bogert C, Kroon AM (1981) Tissue distribution and effects on mitochondrial protein synthesis of tetracyclines after prolonged continuous intravenous administration to rats. *Biochem Pharmacol* 30: 1706–1709
122 Pershadsingh HA, Martin AP, Vorbeck ML, Long JW Jr, Stubbs EB Jr (1982) Ca^{2+}-dependent depolarization of energized mitochondrial membrane potential by chlortetracycline (aureomycin). *J Biol Chem* 257: 12 481–12 484
123 van den Bogert C, van Kernebeek G, de Leij L, Kroon AM (1986) Inhibition of mitochondrial protein synthesis leads to proliferation arrest in the G1-phase of the cell cycle. *Cancer Lett* 32: 41–51
124 Van den Bogert C, Melis TE, Kroon AM (1989) Mitochondrial biogenesis during the activation

of lymphocytes by mitogens: the immunosuppressive action of tetracyclines. *J Leukocyte Biol* 46: 128–133
125 Spooner PR (1990) Oxytetracycline inhibition of mitochondrial protein synthesis in bovine lymphocytes infected with *Theileria parva* or stimulated by mitogen. *Parasitology* 101: 387–393
126 Hamai N, Nakamura M, Asano A (1997) Inhibition of mitochondrial protein synthesis impaired C2C12 myoblast differentiation. *Cell Struct Funct* 22: 421–431
127 Physicians' Desk Reference, 54th edition (2000) Medical Economics Co, Montvale, NJ pages 2371, 1543, 1537, 1528, 2371, 2384, 3164
128 McMurry LM, Petrucci RE Jr, Levy SB (1980) Active efflux of tetracycline encoded by four genetically different tetracycline resistance determinants in *E. coli. Proc Natl Acad Sci USA* 77: 3974–3977
129 Burdett V (1996) Tet(M)-promoted release of tetracycline from ribosomes is GTP dependent. *J Bacteriol* 178: 3246–3251
130 Levy SB, McMurry LM, Burdett V, Courvalin P, Hillen W, Roberts MC, Taylor DE (1989) Nomenclature for tetracycline resistance determinants. *Antimicrob Agents Chemother* 33: 1373–1374
131 Levy SB, McMurry L (1974) Detection of an inducible membrane protein associated with R-factor mediated tetracycline resistance. *Biochem Biophys Res Commun* 56: 1060–1068
132 Yamaguchi A, Ono N, Akasaka T, Noumi T, Sawai T (1990) Metal-tetracycline/H^+ antiporter of *Escherichia coli* encoded by a transposon, Tn10. The role of the conserved dipeptide, Ser65-Asp66, in tetracycline transport. *J Biol Chem* 265: 15 525–15 530
133 Eckert B, Beck CF (1989) Topology of the transposon Tn10-encoded tetracycline resistance protein within the inner membrane of *Eschericia coli. J Biol Chem* 264: 11 663–11 670
134 Rosen BP, McClees JS (1974) Active transport of calcium in everted membrane vesicles in *Eschericia coli. Proc Natl Acad Sci USA* 71: 5942–5046
135 Nakamura T, Hsu CM, Rosen BP (1986) Cation/Proton antiport systems in *Eschericia coli. J Biol Chem* 261: 678–683
136 Nelson ML, Park BH, Levy SB (1994) Molecular requirements for the inhibition of the tetracycline antiport protein and the effect of potent inhibitors on the growth of tetracycline-resistant bacteria. *J Med Chem* 37: 1355–1361
137 Veerloop A, Hoogenstraaten W, Tipker J (1976) Development and application of new steric substituent parameters in drug design. *In*: EJ Ariens (ed.): *Drug design*, Vol VIII, Chapter 4. Academic Press, New York, pp 165–209
138 Nelson ML, Levy SB (1999) Reversal of tetracycline resistance mediated by different tetracycline efflux determinants by an inhibitor of the Tet(B) antiport protein. *Antimicrob Agents Chemother* 43: 1719–1724

Tetracyclines in Biology, Chemistry and Medicine
ed. by M. Nelson, W. Hillen and R.A. Greenwald

Proton and metal ion binding of tetracyclines

Siegfried Schneider

Institut für Physikalische und Theoretische Chemie, Friedrich-Alexander-Universität Erlangen-Nürnberg, Egerlandstr. 3, D-91058 Erlangen, Germany

Dedicated to Professor Dr. Friedrich Dörr on the occasion of his 80th birthday

Introduction

An aspect of major importance in understanding the pharmaco-kinetical properties of drugs and the mechanism of their action is that of acid-base equilibria. According to the definition of Lowry and Broensted, acids are proton donors and bases are proton acceptors. Upon deprotonation the acid yields the conjugate base, whereas protonation converts the base into its conjugate acid. An orally administered drug will most likely change its protonation state during its passage through the acidic stomach and the alkaline intestinal tract. This change will influence its solubility in an aqueous environment, its permeability through membranes and, last but not least, its binding constant to the receptor site.

In case of tetracyclines (cf. Tab. 1), the situation is complicated by two additional facts. First, most derivatives contain three functional groups which can be subject to protonation-deprotonation equilibria (Fig. 1). The conformation of the drug is furthermore dependent on the protonation state. Second, it is well established that tetracyclines bind to many proteins preferentially, if not exclusively, as complexes with divalent metal ions like Mg^{2+} or Ca^{2+}. In a solution that contains, besides the tetracyclines, di- or trivalent metal ions, one therefore finds several types of complexes which differ from each other in degree of protonation of the tetracycline, in metal: ligand ratio and in conformation and/or metal binding site. The relative content of the different complexes varies with pH and ionic strength of the solution, the nature and concentration of the metal ions and to a certain degree also with the history of the solution.

Because of the importance of these dependencies for the practical application of tetracyclines as antibiotics [1], e.g., the bioavailability of these drugs in blood plasma, numerous studies have been performed in the past to elucidate them. Unfortunately, the experimental conditions were seldom similar enough to allow a direct comparison of the presented results, and the transfer of conclusions was not always justified.

Table 1. Structures of widely used "normal" tetracyclines

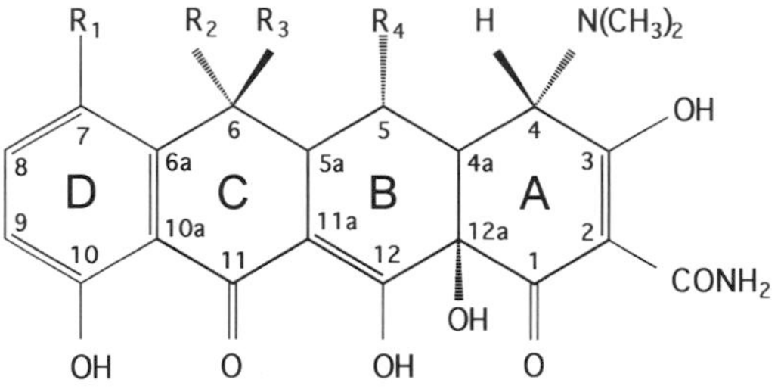

	R_1	R_2	R_3	R_4
Tetracycline	H	CH_3	OH	H
Chlortetracycline	Cl	CH_3	OH	H
Oxytetracycline	H	CH_3	OH	OH
Demeclocycline	Cl	H	OH	H
Methacycline	H	CH_2		OH
Doxycycline	H	CH_3	H	OH
Minocycline	$N(CH_3)_2$	H	H	H

Several critical reviews have been published around 1985 covering the chemistry [2], the physico-chemical properties [3, 4] and pharmacological aspects [5–7] of tetracyclines. This review will therefore concentrate on research done during the past 15 years concerning topics like the relation between conformation and state of protonation or metal ion chelation. New experimental results will be discussed in comparison with results from quantum-chemical calculations and molecular dynamics simulations.

Chemical structures and adopted geometries of normal tetracyclines

Tetracyclines comprise four annelated six-membered rings usually denoted as A, B, C and D. Their stereochemistry is very complex since carbon atoms 4, 4a, 5, 5a, 6 and 12a (see Tab. 1) can, in principle, be asymmetrically substituted. The orientation of the various substituents in natural and semi-synthetic derivatives was proven by synthesis (for a review see, e.g., reference [5]) and/or crystal structure determination. When crystallised from non-aqueous media, tetracyclines are supposed to exist in the neutral form [8, 9]. However, when crystallised from aqueous media, the zwitterion is found (Fig. 1) in which a proton is transfered from O3 to N4 [10]. According to our experience, the zwitterion is formed independently of the pH of the aqueous starting solution (Leypold, private communication).

Moreover, the molecules are not flat but adopt a conformation in which, due to a rotation about the C4a–C12a bond, the conjugated system extending from

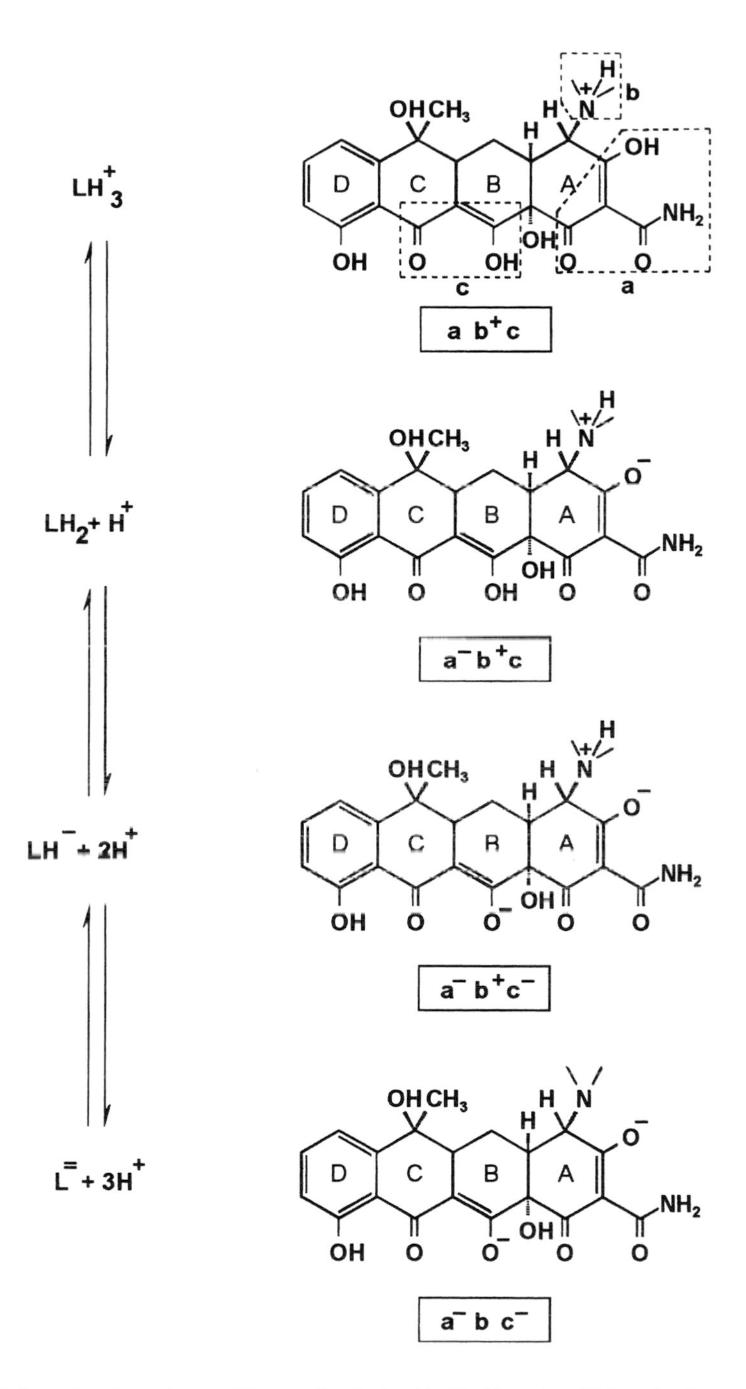

Figure 1. Deprotonation scheme of Tetracycline hydrochloride (the anion Cl^- is omitted for simplicity). The possible sites for protonation/deprotonation equilibria are denoted a, b and c in contrast to the rings A, B, C and D.

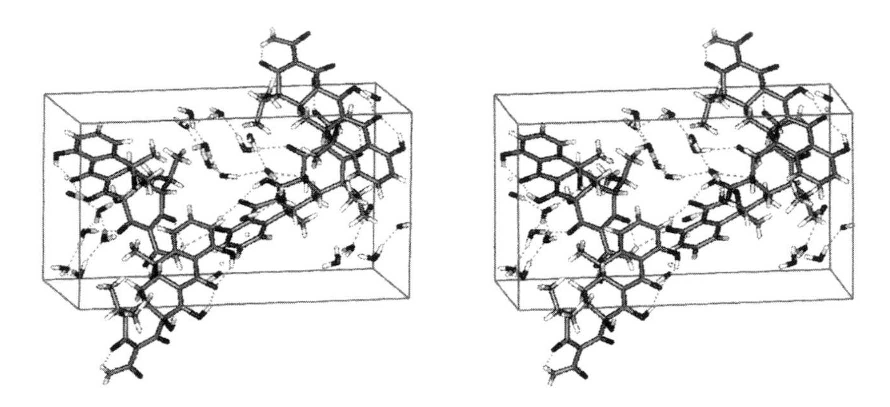

Figure 2. Stereo view of the unit cell of Tetracycline hexahydrate crystallised from acidic solution of Tetracycline hydrochloride. Note the network of hydrogen bonded water molecules. Tetracycline exists as zwitterion in the "extended" conformation (dimethyl amino group lies below the plane spanned by the BCD ring system). (By courtesy of C. Leypold)

O1 to O3 (later on termed A chromophore) lies above and the dimethyl amino group below the plane defined by the π-conjugated system in rings B, C and D (later on called BCD chromophore) (Fig. 2). This conformation is usually called the A conformation [11] or the extended conformation [12]. Tetracylines in the neutral form adopt another conformation, usually termed B conformation or twisted conformation, because the amino group swings in the direction of O6. It was proposed that the latter conformation is stabilised by a strong intramolecular hydrogen bond formed between the proton at O3 and the amide oxygen [8, 9].

Since the unit cell of tetracycline hexahydrate crystals contains 4 tetracycline and 24 water molecules which form a strong and extensive network of hydrogen bonds (Fig. 2), one can speculate whether the conformations found in the crystals are determined by crystal packing forces and whether the same conformers are the dominant species in solution. Conformational studies in solution were performed initially with Oxytetracycline, which behaves very similarly to Tetracycline itself (see below). Hughes et al. [13] recorded CD spectra in ethanol/water mixtures over a wide range of compositions. The spectra provided strong evidence for a solvent-dependent equilibrium between the zwitterionic form and the non-ionised (neutral) form. The zwitterion predominates in water-rich solutions, whereas the non-ionised form is the major species when the ethanol content exceeds ca. 85% (v/v). The second solution form was also postulated to explain the ^{1}H-NMR data of Oxytetracycline in non-aqueous solvents [14]. Natural abundance ^{15}N-NMR has been used to study several tetracyclines in the solid state and in solution [15]. Because the chemical shifts of ^{15}N are sensitive to the site of protonation in the A-ring, it could be shown by comparison that the structural integrity (site of protonation) of the investigated tetracyclines is maintained in $(CD)_3SO$ solutions.

Quantum-chemical model calculations

Gottschalk and Clark [16] recently published, in continuation of previous work [17], the results of model calculations for Tetracycline in pharmacologically important protonation states employing both semi-empirical and DFT methods. For the neutral, zwitterionic and anionic molecule in the gas phase, 2 or 3 conformers were found to lie within an energy range of 40 $kJ \cdot mol^{-1}$ (Tab. 2). Inclusion of solute-solvent interaction within the so-called COSMO-model (conductor-like solvation model [18]) caused in general only small adjustments in the overall geometry of the various conformers. It is evident from inspection of Figure 3 that the arrangement of the A- and B-rings in the most stable configuration in water (for each case shown in the lower picture) resembles more closely the experimentally determined extended conformation. The second, less stable conformer should therefore be related to the twisted conformation. In the mono-deprotonated form an additional conformer is found

Table 2. Calculated relative energies (in $kJ \cdot mol^{-1}$) of various conformers of Tetracycline (shown in Fig. 3) in the gas phase and in aqueous medium (from reference [19])

Conformer	neutral form gas/solution	zwitterionic form gas/solution	anionic form gas/solution
1	0/0	0/0	0/0
2	19.2/15.0	19.2/31.0	18.0/30.5
3			107.8/34.3

whose geometry is quite different and therefore cannot be related to any experimentally known geometry [19].

The tautomerism of Tetracycline in the four different states of protonation was studied on a semi-empirical level by Duarte et al. [20]. In the fully protonated species LH_3^+, the extended conformation is favoured by about 10 $kJ \cdot mol^{-1}$ *versus* the twisted one. The most stable tautomer is the one with the proton of site a (Fig. 1) bound to the oxygen of the amide group instead of to O3. The calculated stabilisation of the zwitterionic species in aqueous solution is not large enough to make it the most stable form of LH_2. The non-ionised form is predicted to adopt the twisted conformation, whereas Gottschalk and Clark propose the extended conformation. For the anion, LH^-, and the dianion, $L^=$, the twisted conformation is again favoured by about 3 $kJ \cdot mol^{-1}$. It is not obvious why the predicted geometries are not in accordance with the results published by Gottschalk and Clark. The explicit inclusion of the specific hydrogen-bonding interaction with the first layer solvent molecules will, without doubt, improve the theoretical results.

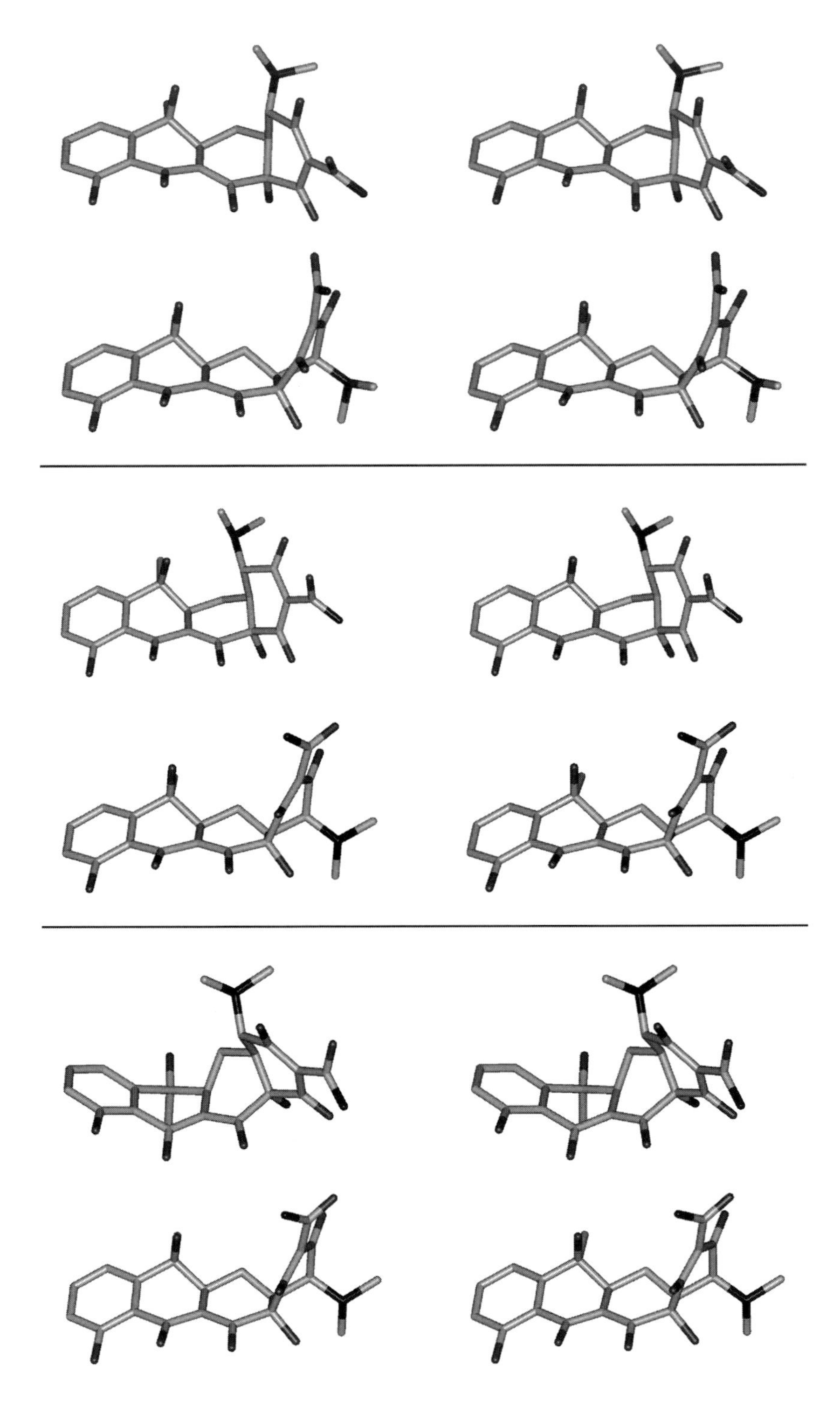

Thermal and photochemical degradation of tetracyclines

It has been known for a long time that tetracyclines show a phototoxic response (e.g., [21]). It is still not clear whether it is caused by the drug itself or by products generated either as primary products or as the result of consecutive thermal reactions (Scheme 1). The formation of Dedimethylaminotetracycline in the absence of oxygen [22] and Anhydrotetracycline (ATC) in aqueous solutions at pH < 7.5 has long been known, as well as the epimerisation in acidic solution (formation of epi-tetracycline or ETC) [23]. In the presence of oxygen and at high pH, quinone derivatives (QTC) are formed [24, 25]. Photolysis in aqueous solution (pH about 6.4, $\lambda > 330$ nm) yields Lumitetracycline (LTC) with high efficacy [26, 27]. Strongly fluorescing isotetracyclines (ITC) are formed in thermal reactions at high pH [28]. The formation of degradation products can perturb spectroscopic investigations in several ways. A minor effect will usually be caused by the signal reduction due to a decrease in tetracycline concentration. More harassing is if the signal contribution from the (photo-) product(s) surpasses that of the tetracyclines under investigation. This can easily be the case in fluorescence studies in view of the low fluorescence quantum yields of the free tetracyclines. Similarly, some of the photoproducts displayed in Scheme 1 could contribute significantly to the CD signal and deceive one into observing structural changes of the drug, which are actually not occurring. Last but not least, it should be mentioned that our attempts to record preresonance Raman spectra of tetracyclines in solution failed because of fluorescence from unidentified photoproducts. It was only possible to record good solution spectra with NIR excitation. They matched well with the preresonance solid state Raman spectra (Brehm, private communication).

Proton binding in normal tetracyclines

Tetracycline exhibits three macroscopic acidity constants, the assignment of which is now widely accepted. In aqueous solution, the first deprotonation of the fully protonated species, termed LH_3^+ (or ab^+c) in Figure 4, occurs at O3 ($pK_a = 3.1 \div 3.5$). This step leads immediately to the zwitterionic form of the neutral compound ($LH_2^\pm$ or a^-b^+c), which represents a tautomer to the non-ionised form LH_2^o or abc. In view of the large difference to $pK_{a2} = 7.2 \div 8.5$ and $pK_{a3} = 9.0 \div 10.9$, there is no doubt about the complete deprotonation of O3 before the second deprotonation step occurs. There was, however, a great deal of discussion about the proper assignment of the second and third macroscopic acidity constant to the two relevant functional groups, namely the

Figure 3. Stereo views of the two most stable conformers calculated by Gottschalk and Clark [16] for the neutral (top), zwitterionic (middle) and anionic form (bottom) of Tetracycline (AM1 Hamiltonian, COSMO-model for consideration of solute-solvent interaction). (By courtesy of H. Lahnig).

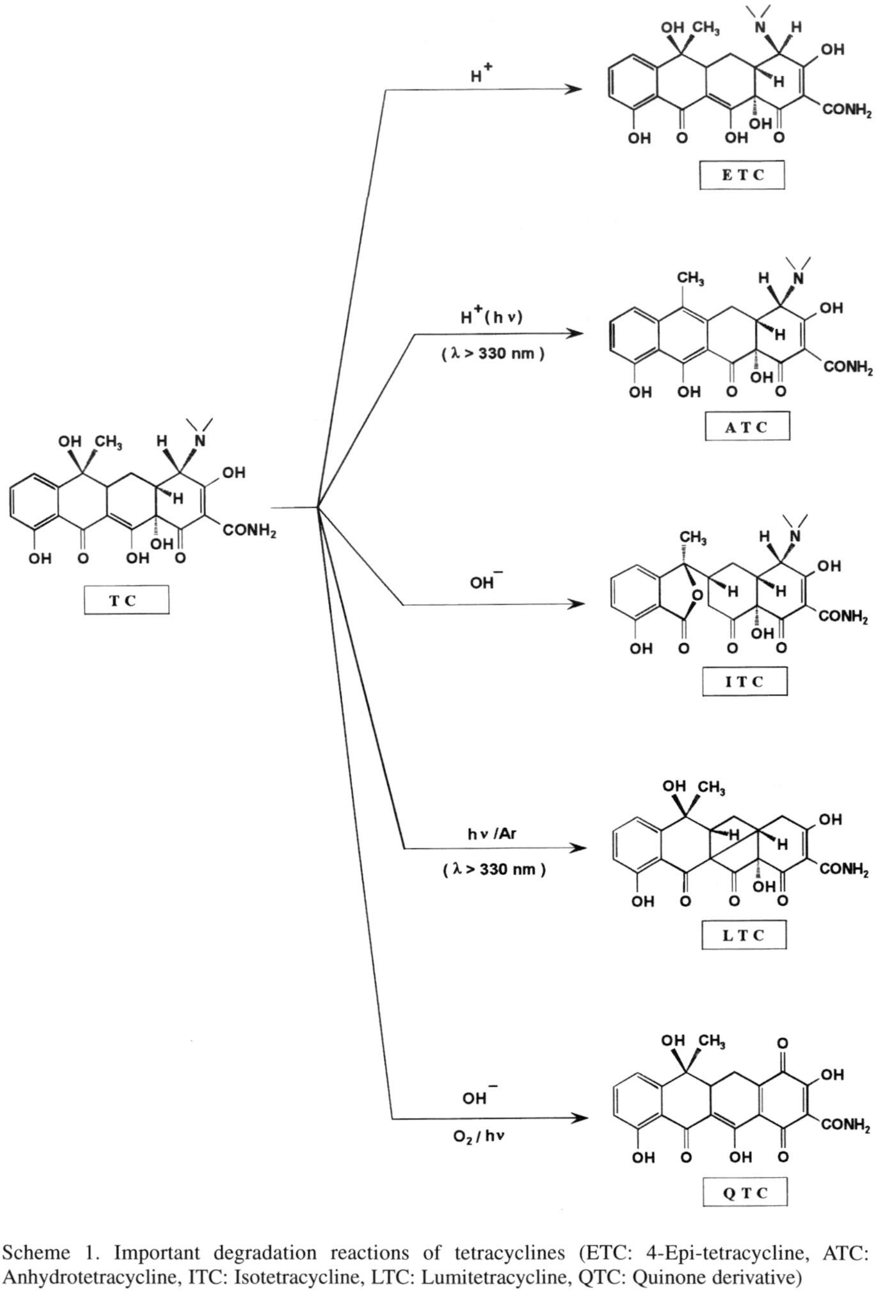

Scheme 1. Important degradation reactions of tetracyclines (ETC: 4-Epi-tetracycline, ATC: Anhydrotetracycline, ITC: Isotetracycline, LTC: Lumitetracycline, QTC: Quinone derivative)

dimethylammonium group (b) and the O11-O12 β-diketone moiety (c) of the ring BCD system. Martin [3] has presented a thorough discussion of the related experimental material and shown that pK_{a2} is dominated by O12H (c) and

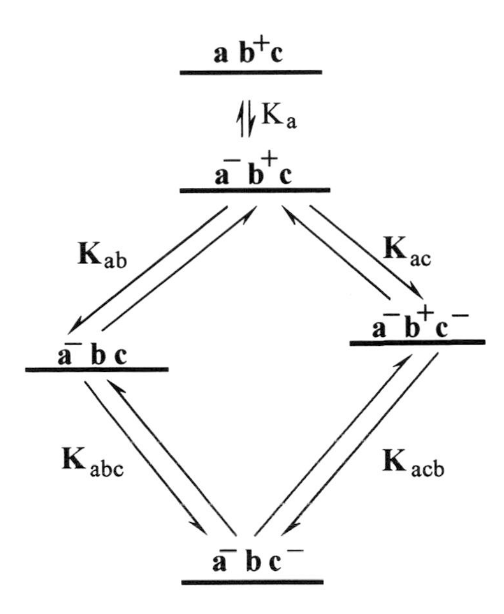

Figure 4. Definition of microconstants for the deprotonation of Tetracycline in aqueous solution (the neutral species is present only in zwitterionic form $a^- b^+ c$).

pK_{a3} by $N4^+H$ (b) deprotonation. The structure of the dianion $L^=$ is again unambiguously defined as a^-bc^-. On the basis of titration experiments with tetracycline methiodide [29], it was postulated that a fourth acidic group exists in tetracyclines, namely O10H ($pK_{a4} = 10.67$ in methanol/water mixtures (1:1)). Duarte et al. [20] postulated the existence of an isosbestic point at around 260 nm in the pH range 9 ÷ 10.5. They propose $pK_{a4} \approx 11.8$ for the deprotonation of O10H.

A more quantitative analysis of the protonation/deprotonation equilibria in tetracyclines requires the consideration of microconstants for each acidic group (Fig. 4), which depend on the charge distribution in the molecule and the solvent used. If the first deprotonation step comprises, as discussed above, that of O3H then the general scheme can be simplified for aqueous solutions to (see [3]):

$$K_1 = (H^+)\,(a^-b^+c)/(ab^+c) = K_a \quad (1)$$
$$K_2 = (H^+)\,[(a^-b^+c^-) + (a^-bc)]/(a^-b^+c^-) = K_{ab} \cdot K_{abc}/K_3 = K_{ac} \cdot K_{acb}/K_3 \quad (2)$$
$$K_3 = (H^+)\,(a^-bc^-)/[(a^-b^+c^-) + (a^-bc)] = [K_{abc}^{-1} + K_{acb}^{-1}]^{-1} \quad (3)$$

where the acidity microconstants are defined via $K_i = (H^+)\,(B_i)/(H^+B_i)$. (In reference [3] the microconstants are written with small letters. In order to avoid a confusion with rate constants, capital letters are used to endorse that here also equilibrium constants are meant.)

The validity of equations (1) ÷ (3) can easily be verified if one keeps in mind that

$$pK_a = -\log K_a = -\log(\exp(-\Delta G/RT) = 2.3\ \Delta G/RT \quad (4)$$

As illustrated in Figure 4, the change in Gibbs function is the same when going from a^-b^+c to a^-bc^- via $a^-b^+c^-$ or a^-bc. This implies that $K_{ab} \cdot K_{abc}$ equals $K_{ac} \cdot K_{acb}$. If $K_{abc} << K_{acb}$, then one gets: $pK_3 \approx pK_{abc}$ (and *vice versa*). This inequality implies also that for the route via a^-bc to be the dominant one, the change of Gibbs function between a^-b^+c and a^-bc must be smaller than between a^-b^+c and $a^-b^+c^-$. If, on the other hand, $K_{abc} \approx K_{acb}$ (a^-bc and $a^+b^+c^-$ are nearly isoenergetic) then one gets: $pK_3 \approx pK_{abc}/2 \approx pK_{acb}/2$ and analogously: $pK_2 \approx 2 \cdot pK_{ab} \approx 2 \cdot pK_{ac}$.

The above discussion shows that the microconstants cannot be deduced from the macroconstants except if additional information is available. One possibility discussed extensively by Martin [3] is to incorporate information about the pK_a values of derivatives, which lack one of the functional groups and can therefore follow only one of the considered deprotonation pathways. Table 3 summarises the values given by Martin [3] and Bhatt and Jee [30] for the microconstants appearing in our deprotonation scheme (Fig. 4). An alternative procedure would be the application of detection techniques which can distinguish between the two isomers a^-bc and $a^-b^+c^-$ and thus allow independent determination of the ratio K_{acb}/K_{abc} at a given pH from the equilibrium concentrations (a^-bc) and ($a^-b^+c^-$). The prerequisite for such a procedure is the knowledge of the properties of the individual species, which may exist in solution next to each other. Using the data of Martin (Tab. 3), the relative amount of each of the occurring species is shown in dependence on pH of the solution in Figure 5. Between about pH 4.2 and pH 7.2, the zwitterionic species $LH_2^{\pm}$ is the dominant one. Around pH 9.0 about 80% of the molecules are present as anionic species LH^-. Assuming the microscopic acidity constants given in reference [3], about 80% of the LH^- species are predicted to adopt the tautomeric form $a^-b^+c^-$ (trace d) and the remaining 20% the form a^-bc (trace e).

Table 3. Microscopic acidity constants derived by different authors for the deprotonation of Tetracycline (25 °C and 0.01 M ionic strength)

Reference	pK_a	pK_{ab}	pK_{ac}	pK_{abc}	pK_{acb}
Martin [3]	3.33	8.7	7.80	8.6	9.56
Bhatt and Jee [30]	3.33	7.96	7.82	8.88	8.98

The deprotonation scheme presented in Figure 4 assumes that each species appears in only one conformation. But it has been mentioned already in a preceding section that tetracyclines are conformationally labile. They can adopt

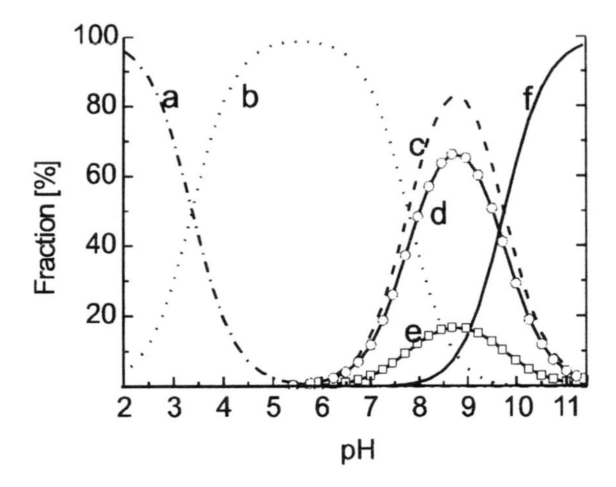

Figure 5. Relative amount F of the different species found in an aqueous solution of variable pH: LH_3^+ (a), $LH_2^\pm$ (b), LH^- (c), $LH^=$ (f); LH^- can be present as $a^-b^+c^-$ (trace d) or a^-bc (trace e). The calculation is based on the microscopic acidity constants given in reference [3].

various geometries exhibiting heats of formation which differ only by a few kJ/mol. Based on the evidence discussed above, one must assume that in the fully protonated form, tetracyclines adopt a twisted conformation (B conformation) whereas in the fully deprotonated form they adopt an extended conformation (A conformation). This implies that in a titration experiment, which starts from an acidic solution and increases the pH by addition of OH^-, the initially investigated deprotonation equilibrium is that of the twisted form B (Fig. 6). If, on the other hand, one starts with a basic solution (e.g., pH 11), then one investigates primarily the protonation behaviour of the extended conformation A. It depends clearly on the rate of addition of base and acid, respectively, whether the initial conformation is conserved during the whole titration process or whether a more or less complete change in conformation occurs at a defined pH. Experimentally, such an alteration of conformations manifests itself in spectra which change with time after addition of the titration solution. Another consequence is that the spectra must not necessarily be reproducible. A change in pH and the subsequent restoration of the starting pH can yield spectra different from that of the starting solution. According to our observation, the establishment of the thermodynamic equilibrium between the relevant conformers may occur on time-scales varying between tens of minutes and several hours.

Without further information about the differences in Gibbs function of the corresponding conformers and the activation barriers in between, it is impossible to make an estimation of the magnitude of the effects to be eventually observed in the various types of spectra. Nevertheless, it is obvious from the crude model displayed in Figure 6 that the pK_a values determined by titration experiments in the two opposite directions may differ significantly (see below). Furthermore, these values may also depend on the applied technique,

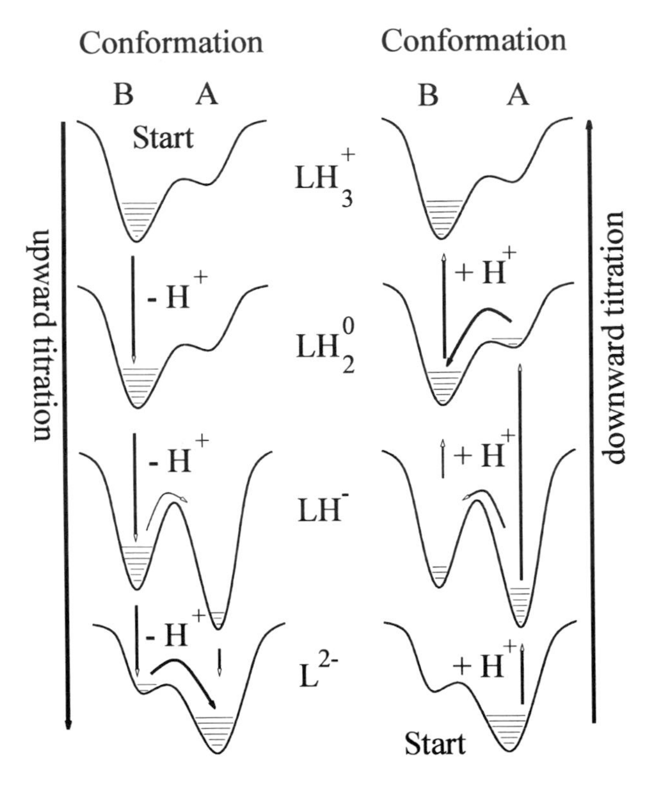

Figure 6. Modified (de-)protonation scheme for tetracyclines including the possibility of conformational changes. Conformation A corresponds to the extended, B to the twisted form.

because different conformers can contribute with different efficacy to the total signal observed (e.g., the fluorescence quantum yield most likely varies with conformation).

Evaluation of titration experiments employing UV/vis-absorption and CD spectroscopy

Because they are easy to apply, UV/vis-absorption and emission spectroscopy, as well as measurement of the circular dichroism (CD) have been used to elucidate the deprotonation scheme of tetracyclines. One advantage of these techniques is that usually the solubility is high enough to generate data sets with sufficiently high signal-to-noise ratio. Furthermore, by scanning across a sufficiently large range in wavelength, one can increase the chances that, due to larger differences in, e.g., the extinction coefficients, the evaluation procedure results in more unambiguous predictions.

According to Beer's law, the absorbance of a sample containing several species is determined by the sum of the absorbances of each species at the probing wavelength λ:

$$A\,(\lambda) = \sum \varepsilon_i(\lambda)\, c_i\, d$$

where $\varepsilon_i(\lambda)$ represents the molar extinction coefficient of species i and c_i its concentration. In principle, if there are N species in the mixture, it would be sufficient for a determination of the concentrations c_i of all N species to measure the absorbance at N different wavelengths λ_j at which the (known) molar extinction coefficients of all species differ from each other. In our case, not only is the number of contributing species unknown, but mostly also the spectra of the species to be eventually considered.

In recent years, computer programs were developed which allow analysis of the spectra of systems in which the contributing species are related by thermodynamic equilibria. In one of the programs ("Specfit" [31–34]) a factor analysis is carried out by single value decomposition on the input data (e.g., set of spectra *versus* pH) to give eigenvectors corresponding to the data and noise found in the spectra. For each proposed equilibrium model the spectra of all included species and the corresponding equilibrium constants are determined by applying a multivariate Marquardt fit to solve the mass balance equation by means of the Newton-Rhapson algorithm.

In fortunate cases, the decision which of the tested models is most likely to be the correct one can be made by relying on the criterion of the lowest χ^2. In most cases, the variation of χ^2 for different models will be so small that additional criteria must be considered to justify the preference of one or the other conceivable alternative. A good additional criterion is the shape of the spectra calculated for the involved species.

Tetracycline absorption titrations

Tetracycline is usually administered as Tetracycline hydrochloride which, dissolved in pure water, exhibits a pH between about 1.8 and 2.8, depending on concentration. Therefore many titrations start at a low pH (e.g., around 2) and increase the pH by addition of, e.g., 1.0 M sodium hydroxide. Alternatively, the titration experiments can start with pH around 10 and decrease the pH by addition of, e.g., 1.0 M hydrochloric acid. pH values below or above the given limits are usually avoided because of the occurrence of rapid degradation reactions (formation of ATC at low pH and ITC at high pH as shown in Scheme 1). In the pH range 3÷5, epimerisation occurs, the rate being dependent on the kind of buffer used [35]. In Tris buffer, which is used in our experiments, the reaction is very slow and has therefore no effect on the presented results.

As a typical example, the changes observed in the UV/vis absorption spectra upon upward titration of Tetracycline hydrochloride are shown in Figure 7.

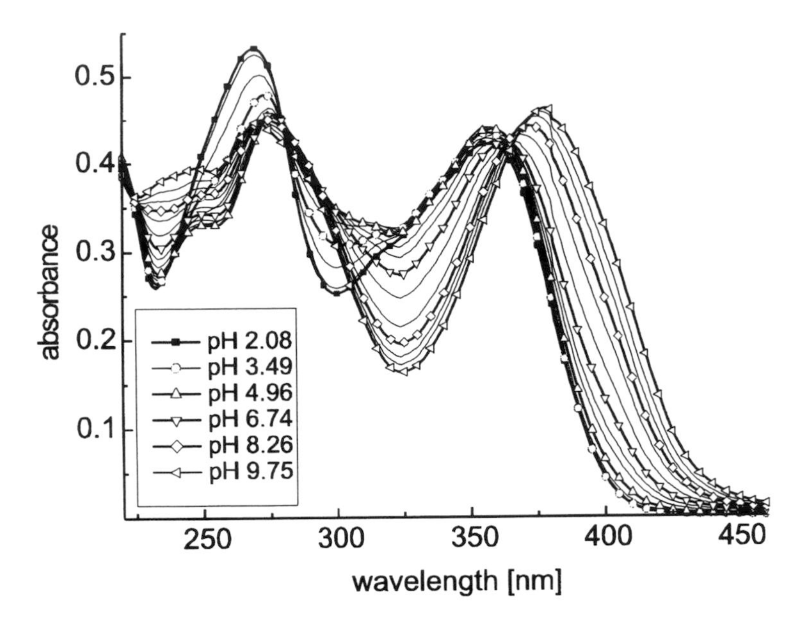

Figure 7. Absorption spectra of Tetracycline (I ≈ 0.15 M) with increasing pH. pH steps between consecutive traces are about 0.5 units. (The small kinks at about 380 nm are instrumental artefacts due to filter changes.)

One observes several isosbestic points, each of which appears only within a certain pH region. This finding provides the experimental evidence that several (consecutive) equilibria are involved. For the numerical analysis using the program "Specfit" the reaction scheme illustrated in Figure 1 is applied:

$$LH_3^+ \rightleftharpoons LH_2 + H^+ \rightleftharpoons LH^- + 2H^+ \rightleftharpoons L^= + {}^3H^+$$

It should be emphasised that in this evaluation of the experimental data only the degree of (de-) protonation is the essential characteristic quality of the species involved. This implies that, e.g., LH^- represents different conformers as well as different tautomers. If at each pH the thermodynamic equilibrium is established fast between all species representing LH^-, then the resulting absorption spectrum is unique and characteristic for this mixture of conformers and tautomers. If, on the other hand, the activation barriers for transformation are high, then the absorption spectra become dependent on time after addition of acid or base (this can explain statements in the literature that the spectra obtained in titration experiments are not reproducible).

In accordance with the above considerations, the two sets of absorption spectra recorded during upward and downward titration yield slightly different spectra for the species $L^=$, LH^-, LH_2 and LH_3^+ incorporated in the model (Fig. 8). It is obvious that the degree of protonation is the most important factor with respect to the overall appearance of the absorption spectra.

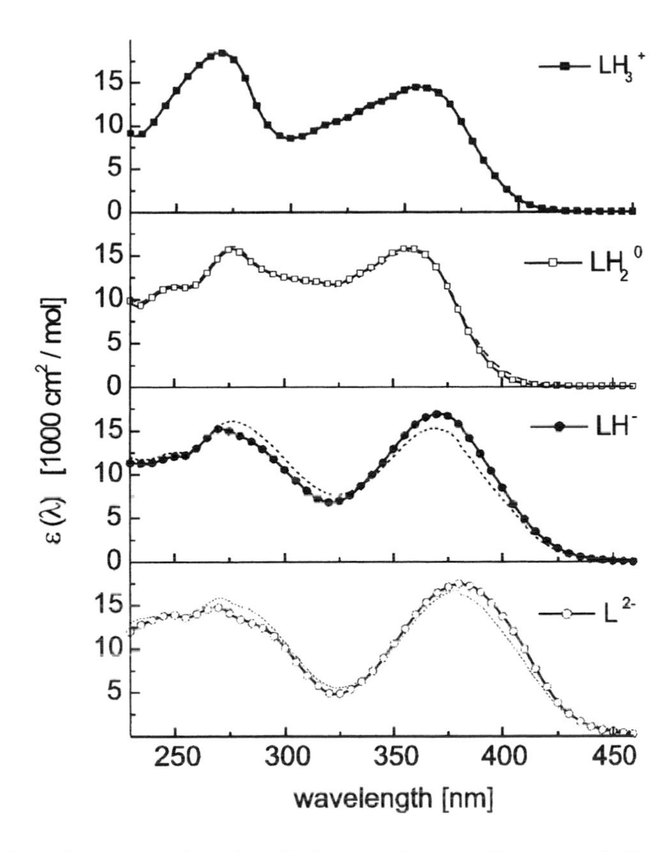

Figure 8. Absorption spectra of species $L^{=}$ (○) LH^{-} (●), LH_2 (□) and LH_3^{+} (■) derived from Tetracycline titration experiments using the program "Specfit". (Solid line: upward titration, dashed line: downward titration). Significant differences are seen only for LH^{-} and $L^{=}$.

The spectrum of the dianion $L^{=}$ exhibits a strong visible band assigned to the BCD chromophore with a maximum at about 380 nm. Upon protonation of the N4 dimethylamino group, this band shifts hypsochromically by about 10 nm, the only other significant change being a relative increase in the extinction coefficient around 250 nm where the A and the BCD chromophore absorb. The second protonation step involving the β-diketone system of the BCD chromophore shifts the long wavelength band again by about 10 nm to a shorter wavelength and makes the valley around 325 nm disappear. The final protonation of O3 has the greatest effect on the absorption in the 250–280 nm region. The change in conformation and/or site of protonation is the cause of the small differences in the species-associated spectra derived from upward and downward titration. The extinction coefficient of the long wavelength band is obviously higher in the twisted conformation (traces for LH^{-}). A more quantitative evaluation of the conformation-related spectral changes is not possible without further information about the thermodynamic parameters. We believe that the differences shown in Figure 8 are small because the titration was performed with small steps in pH (in the actual experiment the pH steps were only 0.25 units) and spectroscopic measurements were made only after thermodynamic equilibrium was nearly reached. If larger steps in pH are made, then the initial spectral differences are larger and the effect of relaxation is easily seen.

Although the absorption spectra derived from the two sets of data are fairly similar, the macroscopic acidity constants for the second and third deprotonation step differ by up to one unit (Tab. 4). This implies that the problem of conformeric and tautomeric heterogeneity plays a major role for LH_2 and LH^-, as could be deduced already from the discussion of Figure 4. Furthermore, the ratio LH_2/LH^- calculated by using either of the two sets of macroscopic acidity constants can adopt reverse values in the pH range between 7 and 8 (e.g., 1:3 and 3:1) before thermodynamic equilibrium is established.

Table 4. Macroscopic acidity constants determined for Tetracycline hydrochloride by absorption, fluorescence and circular dichroism titrations described in this work (data provided by M. Dean)

	pK_{a1}	pK_{a2}	pK_{a3}	$pk_{a2} + pK_{a3}$
Absorption (upward)	3.4	6.7	8.7	15.4
Absorption (downward)	3.4	7.5	9.7	17.2
CD (upward)	3.3	6.7	8.7	15.4
CD (downward)	3.3	7.6	9.5	17.1
Fluorescence (upward)	3.3	7.3	9.2	17.5
Fluorescence (downward)	3.4	7.2	8.8	16.0
Martin [3]	3.3	7.7	9.5	17.2

Tetracycline circular dichroism titrations

The large difference in the CD spectra of Tetracycline at pH 2 (LH_3^+) and pH 10 ($L^=$) (cf. Fig. 9a) is generally attributed to the difference in its conformation (twisted (B) at pH 2 and extended (A) at pH 10). Therefore, during the titration experiment, the change in conformation must occur. Depending on the height of the barrier in the various states of (de-) protonation, the effective transition could occur during upward and downward titration at different pH levels as indicated schematically in Figure 6. As one consequence, the CD spectra would show pronounced differences for upward and downward titration in that pH range in which conformational changes actually occur. Indeed, the CD spectra calculated for the various species from upward and downward titrations are clearly different, the pattern of the differences being similar for LH_3^+ and LH_2^o, on the one hand, and for LH^- and $L^=$, on the other hand (Fig. 9b). The evaluation yields different pK_a values from upward and down-

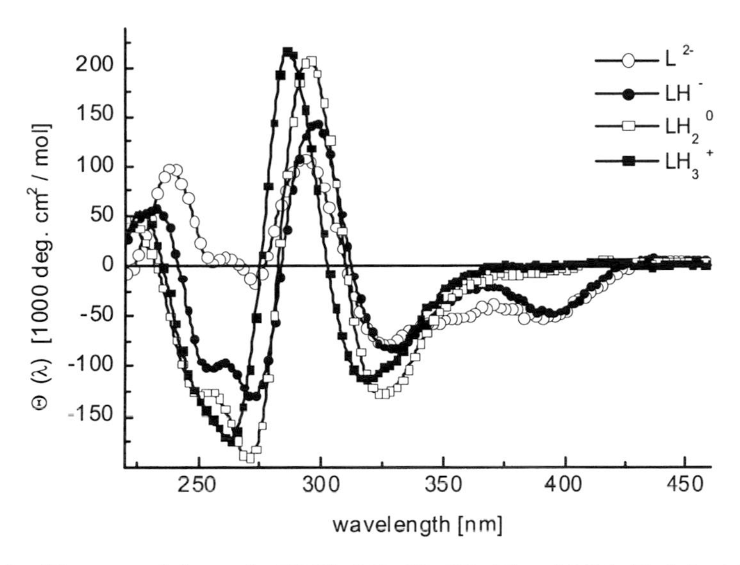

Figure 9a. CD spectra of the species $L^{=}$ (○), LH^{-} (●), LH_2 (□) and LH_3^{+} (■) derived from the Tetracycline upward titration curves.

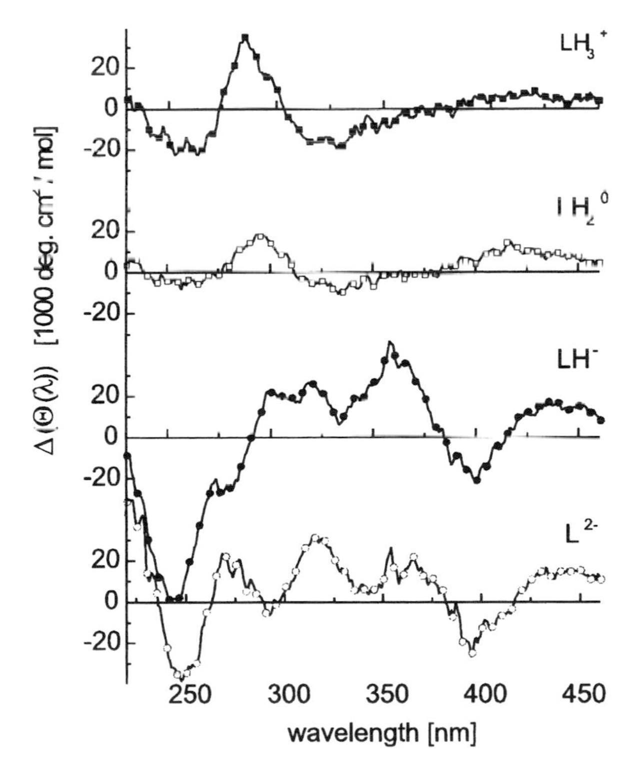

Figure 9b. Differences in the CD spectra of Tetracycline species $L^{=}$, LH^{-}, LH_2, and LH_3^{+} calculated from spectra recorded during upward and downward titration.

ward titration, the trends being similar for data derived from absorption and CD spectroscopy (Tab. 4).

Tetracycline fluorescence titrations

Upon complete deprotonation, the fluorescence yields of tetracyclines are greatly enhanced, as are those of the complexes with metal ions. It is usually argued that the intra- and intermolecular hydrogen bonds engaging the keto and enol groups of the BCD chromophore provide very effective radiationless decay channels (e.g., by excited state proton transfer [27]) thus keeping the yield for fluorescence low.

If the intramolecular hydrogen bond between O11 and O12 is eliminated, the fluorescence yield should go up significantly. Quantum-chemical model calculations show that the spatial arrangement of the mentioned groups is also different in the extended and twisted conformation (see, e.g., Fig. 3). This implies that the strength of intramolecular hydrogen bonding should vary with a change in conformation and concomitantly so should the fluorescence yield. The spectra derived from fluorescence titration experiments with Tetracycline hydrochloride (Fig. 10) nicely confirm these expectations. Starting from $L^{=}$, the first protonation step at N4 decreases the fluorescence yield by about a factor of 4 with only a small change in the emission maximum. Protonation of the dimethylamino group is thought to affect the fluorescence of the BCD chromophore by inducing a conformational change. The second protonation step at

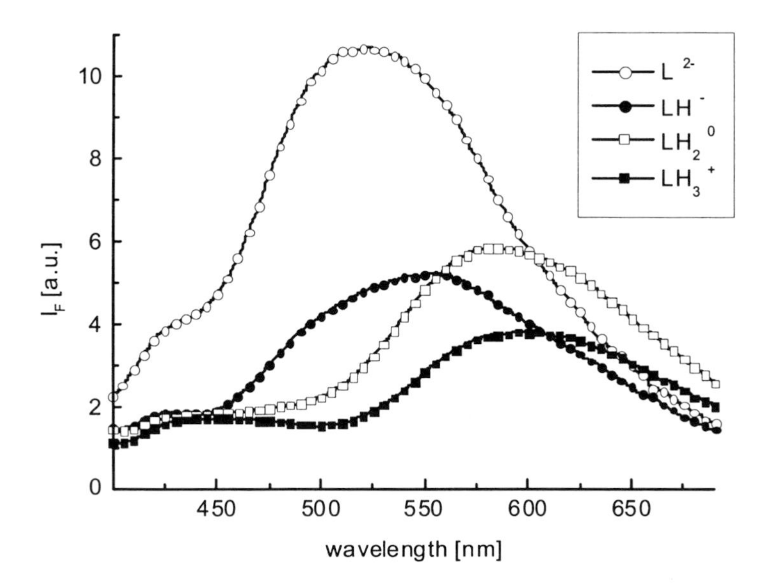

Figure 10. Fluorescence spectra of the species $L^{=}$ (○), LH^{-} (●), LH_2 (□)and LH_3^{+} (■) derived from the Tetracycline upward fluorescence titration experiments.

O11-O12 causes a significant bathochromic shift of the emission of the BCD chromophore. A smaller red shift and a further decrease in fluorescence intensity results from the third protonation step at O3. In contrast to the results obtained by absorption and CD titration, the second and third pK_a values derived from fluorescence experiments are higher in the upward than in the downward direction (Tab. 4).

Before one enters into a discussion about the origin of this difference in pK_a values, one must examine whether the experimental conditions are such that one can compare the results at all. Both in the absorption and CD titration experiments, the recorded signal is determined by the concentrations of the various species in the electronic ground state. In the fluorescence experiments, the situation is more complex because the signal depends on the number of molecules in the various fluorescing states and the corresponding fluorescence yields. The distribution of the molecules across the fluorescing states parallels their distribution in the ground state if no excited state reactions occur. Only if this prerequisite is fulfilled, are the pK_a values derived from the two different types of experiments equal. If, as mentioned previously, excited state reactions occur (e.g., deprotonation or conformational changes) then fluorescence experiments should produce different pK_a values.

Comparison of the results obtained by absorption, CD and fluorescence titrations

Inspection of Table 4 shows that all techniques yield essentially the same value for pK_{a1} in accordance with the fact that educt and product of the first deprotonation step are very well defined. The agreement between the values derived for the second and third pK_a is satisfactory if one restricts the comparison to those pairs of values which have been determined under equivalent experimental conditions employing absorption and CD titrations.

In a previous chapter we have shown that the sum of the pK_a values of the second and third deprotonation step must be a constant independent of the actual intermediates (e.g., a^-bc *versus* $a^-b^+c^-$) if the Gibbs function of the species LH_2 and $L^=$ is conserved. The difference of this sum for upward and downward titration provides additional experimental evidence that, upon titration in different directions, species with different Gibbs functions must be involved at the level of LH_2. (This conclusion is based on the assumption that the structural and geometrical heterogeneity on the level of $L^=$ should be minimal.) Furthermore, the variations in the relevant species of the type LH_2 must be such that they affect the absorption and CD spectra in a similar fashion. Since pK_{a1} is furthermore equal for titration in both directions, educt of the third protonation step and product of the first deprotonation step must be energetically very close. This implies that during downward titration, an efficient geometrical relaxation must occur on the level of LH_2 to yield as relaxation products those species which are formed upon deprotonation of LH_3^+. This relaxation can represent not only the transition from extended to twisted conformations as suggested by Lambs et al. [36] but also between different tautomeric and rotameric forms (e.g., rotation of the carboxamide group).

In the fluorescence experiments, the upward titration yields a proper sum for pK_{a2} and pK_{a3} ($\approx$17.5), but the differences of pK_{a2} and pK_{a3} determined for both titration directions are reduced. This implies that the fluorescence properties are not much altered upon the structural relaxations which were postulated at the LH_2 level based on absorption and CD measurements.

The variations of pK_{a3} with detection technique and direction of titration should be related to the tautomeric equilibrium between $a^-b^+c^-$ and a^-bc on the LH^- level. A change in conformation should not only modify the strength of intramolecular hydrogen bonding, as discussed above, but also the microscopic acidity constants for (de-)protonation at site c. Concomitantly, there must be a change in the macroscopic acidity constants if the involved microscopic reaction steps change.

Metal ion chelation by tetracyclines

It is generally assumed that the preferred sites of chelation of metal ions are the deprotonated functional groups because the dominant interaction between the metal cations and the tetracyclines should be Coulombic in nature. Consequently, spectral changes induced upon metal chelation should be similar to those caused by protonation of the same site. Metal ion binding is, however, influenced by two additional factors:

(i) It is strengthened if five- or six-membered ring chelation can occur with a basic site as anchor and

(ii) Many metal ions prefer oxygen to nitrogen donors.

Considering the three sites which are subject to protonation/deprotonation equilibria and the large number of possible donor atoms, it is immediately obvious that tetracyclines can form a great variety of complexes which differ by the number of chelated metal ions and their binding sites as well as by the number and location of protons still bound.

In principle, one can investigate the complexation behaviour of tetracyclines in several ways, which correspond to different situations in medical applications.

(i) A solution containing both tetracycline and the chosen metal ion at a fixed-concentration is titrated with either base or acid.

(ii) A solution of tetracycline is buffered for a selected pH and then titrated with increasing amounts of metal ions.

In the latter case, the intuitive model assumes the tetracycline molecules to be present as a mixture of species with different degrees of protonation, which then bind the metal ions M preferentially at the ionised functional groups. (For the sake of simplicity, the charges are omitted in the following):

$$LH_n + m \cdot M \rightleftharpoons LM_mH_n \qquad (5)$$

Since the microscopic acidity constants of the ligand will change upon metal ion binding, a fraction of the primarily formed complexes LM_mH_n will deprotonate

$$LM_mH_n \rightleftharpoons LM_mH_{n-1} + H^+ \qquad (6)$$

until the equilibrium is established in accordance with the formation constants β_{lmh} of all possible species:

$$\beta_{lmh} = (L_lM_mH_h)/(L)^l(M)^m(H^+)^h \qquad (7)$$

The latter definition includes also complexes in which one metal ion binds, e.g., two tetracycline ligands or, in case of high metal ion concentrations, one ligand binds two metal ions next to a variable number of protons. For single protonation/deprotonation equilibria, the pK_a values are given by the differences between the $\log\beta$ values of the corresponding complexes. Analogous relations hold for the metal ion binding constants in the reactions:

$$L_lM_mH_h + M \rightleftharpoons L_lM_{m+1}H_h \qquad (8a)$$
$$L_lM_mH_h + M \rightleftharpoons L_lM_{m+1}H_{h-1} + H^+ \qquad (8b)$$

Concerning the reaction mechanism, there is a difference between the concerted reaction described by equation (8b) and the consecutive reaction steps according to equations (5) and (6). It should be detectable only in time-resolved experiments at variable temperature (different activation energies) or, e.g., experiments at variable pressure (different activation volumes). For thermodynamic considerations, as made in connection with the evaluation of titration experiments, the incorporation of the equilibrium (8b) is sufficient (see, e.g., reference [37]).

The analysis presented so far characterises the formed complexes by their composition (e.g., number of metal ions and protons) which also uniquely determines the net charge of the complex as $n + 2m - 2l$. This characterisation is sufficient if one is only interested in the fraction of membrane-permeating neutral complexes present at a given pH and metal ion concentration. The differentiation of the complexes of a given composition according to the binding sites of proton(s) and metal ion(s) requires detection techniques which produce a distinct and characteristic signal for each species. Unfortunately, the results of the easily available techniques like UV/vis absorption and CD spectroscopy by themselves are not sufficient for such a differentiation. Likewise, potentiometric studies cannot provide this type of structural information [38].

Up to now, no crystal structure of a tetracycline complexed with the usually applied di- or trivalent metal ions has been published. The only exception is a complex of Oxytetracycline with mercuric chloride, whose UV spectral data are, however, atypical for metal complexes of this drug [39]. Since therefore no direct information is available on actual complexation sites, only more indirect information by various kinds of spectroscopic techniques can be exploited. Based on UV/vis absorption studies Conover was the first to assign the 11,12-β-diketone system as the primary chelation site in Oxytetracycline. Later on, Colaizzi et al. [40] suggested an involvement of the carboxamido group and the 1- or 3-enol. From CD studies Mitscher et al. [41, 42] concluded that both the BCD-chromophore, at low pH, and the A-ring, at higher pH, are engaged in complexation. Finally, Caswell and Hutchinson [43] differentiated between magnesium and calcium complexes. For the latter they suggested binding via O11, O12 and O3 (or the carboxamide function). Lambs et al. [36] confirmed this differentiation between Ca^{2+} and Mg^{2+} binding. They analysed their data allowing for metal to ligand ratios of 2: 1, 1: 1 and 1: 2 depending on the actual concentration ratio in the solution. They postulated that in the ternary complexes comprising one metal and two tetracycline molecules, the two tetracyclines are chelated at different sites (one tetracycline via N4, the other via the BCD system). The concentrations of these complexes were, however, too low for a spectroscopic characterisation. For the ternary complexes containing two Ca^{2+}, they suggested chelation via N4 and O12 and for the second ion via O12 and O1. In contrast, the first Mg^{2+} ion should bind via N4 and O3, while the second coordinates O11 and O12. ^{1}H-NMR spectroscopy provided some evidence for complexation of Ca^{2+} and Mg^{2+} at the A-ring involving the 1- or 3-enol and the carboxamido group [44].

Examination of the cited literature (and the references given therein) shows that the described experiments were mostly performed under conditions that differ so much that the provided evidence can not be combined without caution. Furthermore, most experimental techniques do not give direct structural information with the effect that the presented conclusions are seldom really straightforward. Consequently, metal chelation of tetracyclines must still be considered an unresolved problem. There is need for more systematic investigations applying really structure-specific methods.

The results of quantum-chemical calculations employing the COSMO model to simulate the solute-solvent interaction have recently been published by Gottschalk and Clark [16]. In Figures 11 and 12, stereoviews of the 5 lowest-energy complexes formed from the tetracycline zwitterion and anion, respectively, are shown for comparison. It is immediately evident

(i) that different sites of complexation are correlated with different geometries (conformations) of the tetracycline ligand and
(ii) that within an energetic range of about 40 $kJ \cdot mol^{-1}$, many of the complexation schemes discussed in the literature occur.

Figure 11. Stereo view of the five energetically most favoured complexes between Mg^{2+} and the Tetracycline zwitterion (lowest energy conformation at bottom). (By courtesy of H. Lahnig.)

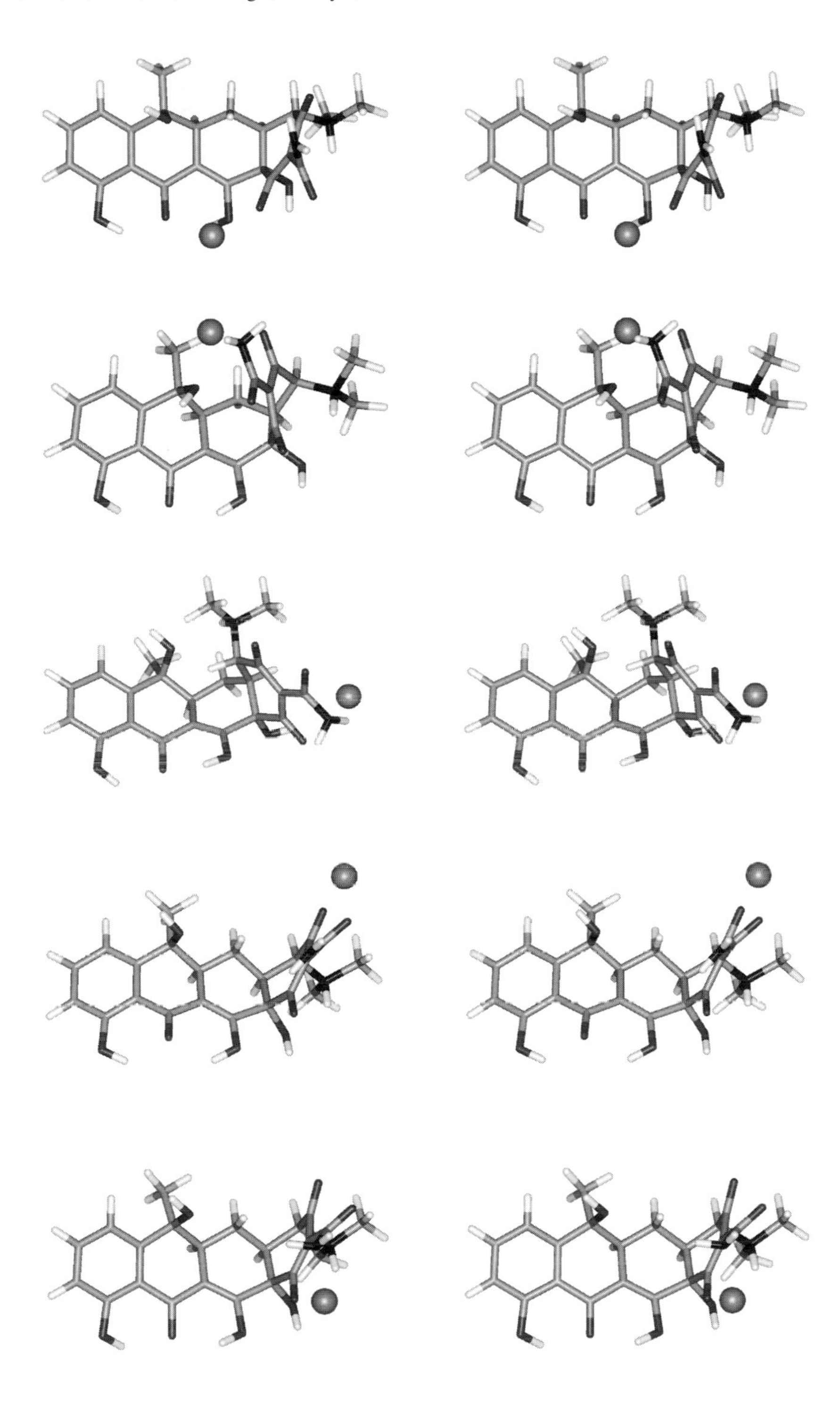

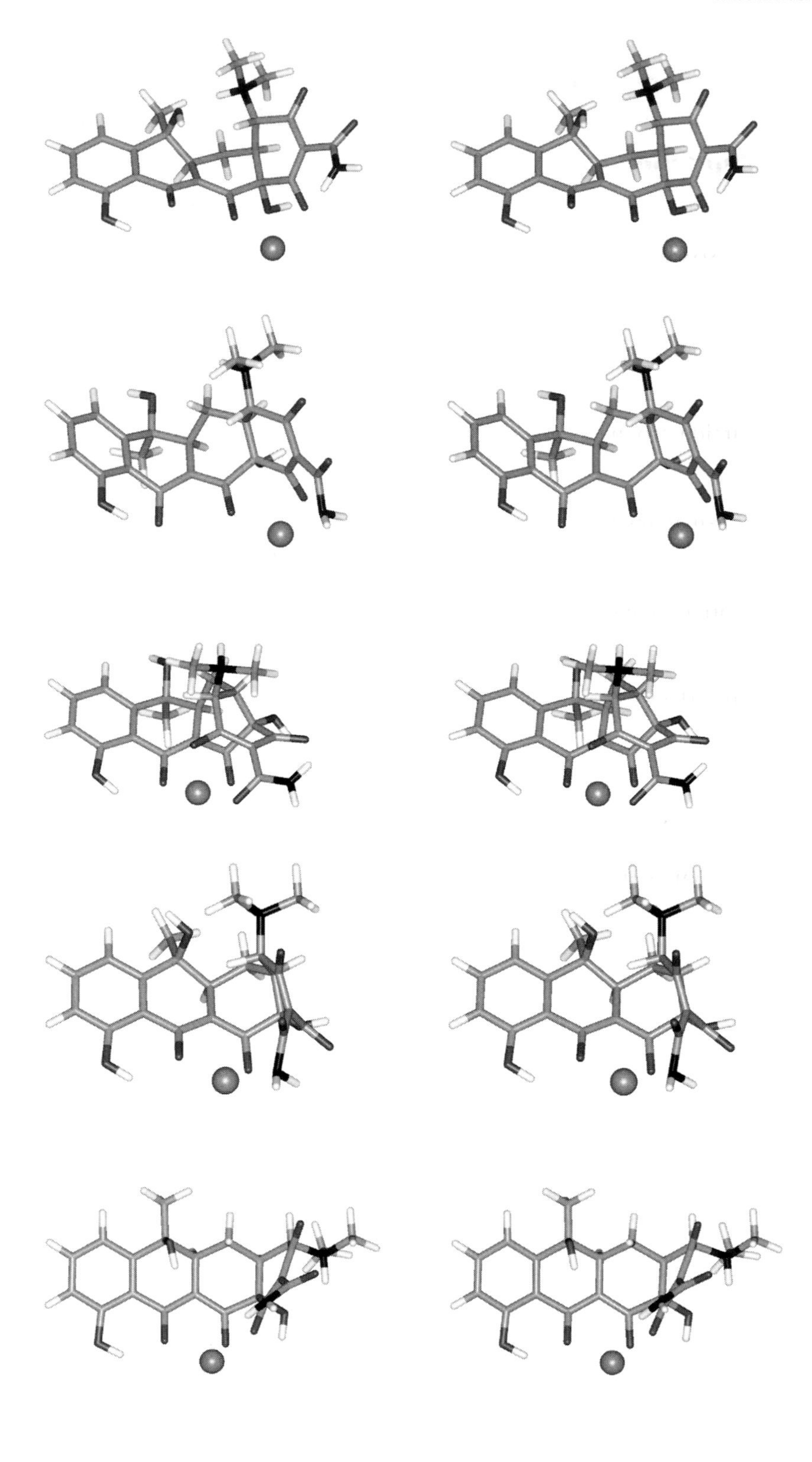

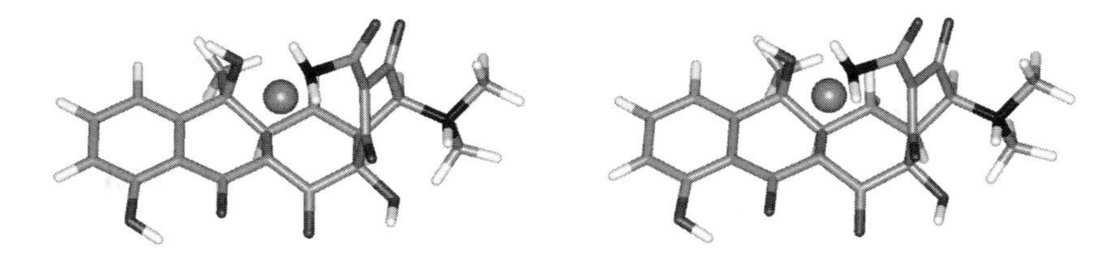

Figure 12. Stereo view of the six energetically most favoured complexes between Mg^{2+} and the Tetracycline anion (lowest energy conformation at bottom). (By courtesy of H. Lahnig)
(Continued from previous page)

The by far most stable complexes involve in both strates of protonation the carboxamido group in combination with O1 and O6, respectively. There are, however, two caveats:

(i) The employed Hamiltonian (AM1) tends to overestimate metal binding to nitrogen atoms and
(ii) Specific hydrogen bonding by water molecules in the first solvation shell is still not included.

A correct consideration of both factors could change the energetic sequence of the proposed structures. In accordance with expectation is the shift of the preferred complexation sites from the carboxamido group in combination with O1 and O3 to those with involvement of O11, O12 and O1 when going from the zwitterionic to the anionic form. Therefore it is not surprising that binding to O1 has a significant effect on the spatial arrangement of rings A and B.

Spectroscopic pH titrations of tetracycline with metal ions at a fixed ratio

Detailed experiments of this type were usually carried out with a ratio of total metal ion concentration to total ligand concentration of less than about 10. Under these circumstances only the formation of complexes containing one metal ion must be taken into account in the analysis [45]. It is furthermore advantageous if the pK_a values of the free ligands can be transferred from pH-titration experiments performed in the absence of metal ions and, provided they are known, their absorption and CD spectra, respectively, in various states of protonation. Since the distribution of conformers and tautomers belonging to each of the species $L^{=}$, LH^{-}, LH_2 and LH_3^{+} can vary with the history of the solution (cf. discussion in previous section) one gets slightly different absorption and CD spectra depending on the direction of the titration. As a consequence, the stability constants derived from both sets of spectra are also dependent on the direction of titration (cf. Tab. 5).

Table 5. Stability constants log β (L_1M_1) of complexes of various tetracyclines with Mg^{2+} and Ca^{2+} determined with the program Specfit. (The values in parenthesis are derived from data recorded during upward titration)

	M	absorption	CD	fluorescence
Tetracycline	Mg^{2+}	6.23 (4.28)	5.97 (3.66)	6.0 (4.35)
	Ca^{2+}	5.86 (5.10)	5.99 (3.94)	6.06 (5.16)
Oxytetracycline	Mg^{2+}	5.45 (4.83)		
Doxycycline	Mg^{2+}	5.94		

An important result of the experiments performed in our laboratory is that evaluation of the absorption spectra requires only the consideration of L_1M_1 type complexes. The fit of the CD spectra of the Mg^{2+} complexes needs two types of complexes, namely L_1M_1 and $L_1M_1H_1$. This implies that both Mg^{2+} and Ca^{2+} are complexed mainly by the completely deprotonated species (equation 8a) or, if ion binding occurs in the anionic form, that the last proton is easily released upon ion binding (equation 8b).

In Figure 13, the absorption and CD spectra calculated by the program "Specfit" are shown for Mg^{2+} and Ca^{2+} as complexing ions. Both types of spectra are significantly different for these two types of ions, thus providing evidence that the predominant species exhibit a different binding pattern for Mg^{2+} and Ca^{2+}. In case of Mg^{2+} binding, the complexes $[TcMg]^{\circ}$ and $[TcMgH]^{+}$ have apparently similar absorption, but different CD spectra. This can be rationalised since the deprotonation step that connects the two species occurs at N4 and alters the electron density of the chiral centre C4. As a further consequence the CD of both the A and the BCD chromophore could experience an induced effect. Noteworthy in this connection is the large difference in β values determined from absorption (and fluorescence) and CD titrations in upward direction (cf. Tab. 5).

Ca^{2+} and Mg^{2+} titrations of tetracycline at fixed pH

It was stressed above that for a given pH a distribution of conformers and/or tautomers exists in aqueous solutions of tetracyclines. Because these exhibit different spatial arrangements of the possible complexing sites, they differ in driving force for metal complexation. Consequently, the structure of the preferentially formed complexes can vary with pH of the solution. In addition, conformational changes might be suppressed by metal ion binding. That implies that non-equilibrium distributions of the various complexes could be conserved for longer times.

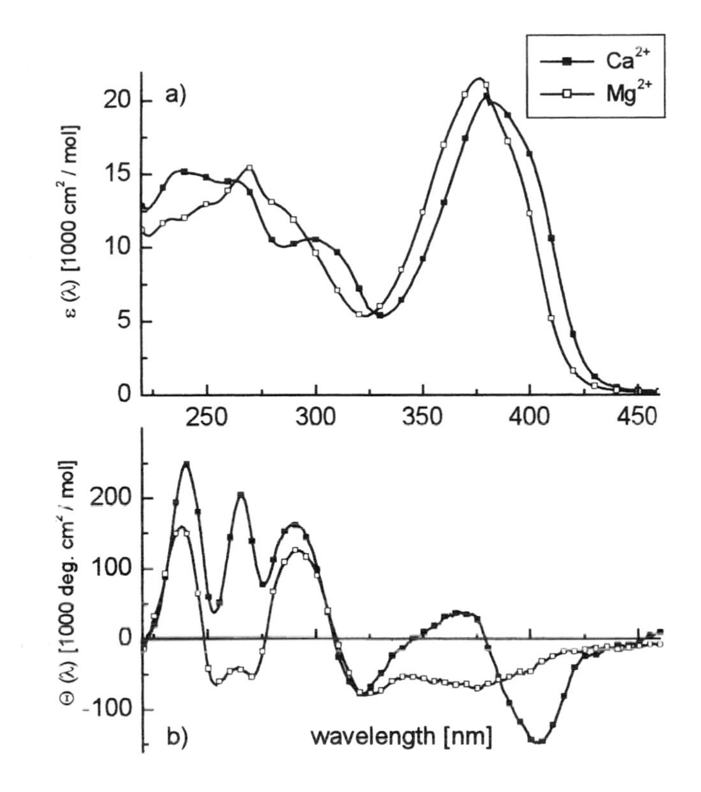

Figure 13. Absorption (a) and CD spectra (b) of the L_1M_1 type complexes with M being Mg^{2+} (□) or Ca^{2+} (■). Evaluation by means of the program "Specfit" using the data recorded during downward titration.

The distribution curves (Fig. 5) predict the anion to be the dominant species at pH 8.5 whereas the zwitterionic form dominates at pH 7.0. Since a change of conformation of the metal-free ligand is likely to occur between these two pH values, the complexation pattern could be different if the solution is buffered at one of these pH values.

Titration experiments with Mg^{2+} carried out at pH 7.0 and pH 8.5 showed essentially the same changes in the absorption spectra, thus indicating that the complexation site for the first and the second ion is conserved when switching from pH 7.0 to pH 8.5. The titration experiments with Ca^{2+} at pH 7.0 produced similar spectral changes as with Mg^{2+}, whereas at pH 8.5 a completely different development of the spectra with increasing amounts of Ca^{2+} was observed [37]. This points clearly to a different complexation pattern for Ca^{2+} at pH 8.5. Based on the characteristic features of the spectra calculated by means of "Specfit" for the $L_1M_1H_x$ and $L_1M_2H_x$ complexes (Fig. 14) the complexation sites shown in Figure 15 were suggested [37]. H_x means that the extent of deprotonation can not be determined from the set of experimental spectra. The

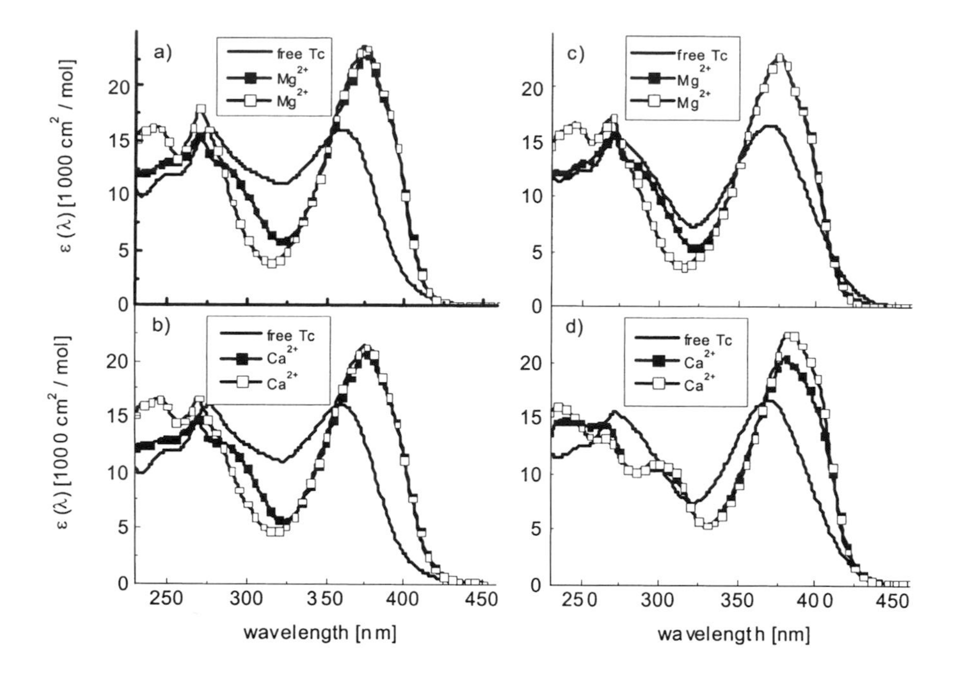

Figure 14. Absorption spectra of Tetracycline (solid line) and the $L_1M_1H_x$ (□) and $L_1M_2H_x$ complexes (■) with M = Mg^{2+} (a, c) and M = Ca^{2+} (b, d). pH was 7.0 (a, c) and 8.5 (b, d).

spectra resulting from the analysis for the complexes with one or two metal ions represent a weighted superposition of the spectra of such complexes with different numbers (x = 0 or 1) and positions of the protons. The large difference in conditional binding constants at pH 7 and pH 8.5 (Tab. 6) has an interesting consequence regarding the ratio tetracycline to metal ion required for half of the tetracycline molecules to be complexed with metal ions. The example shown in Figure 16 refers to a total tetracycline concentration $C_T = 3 \times 10^{-5}$ M. At pH 7.0, an approximately 20-fold excess of Mg^{2+} is needed, whereas at pH 8.5 a factor of 3 is sufficient for half of the tetracyclines to be bound. For Ca^{2+}, the corresponding values are a 100-fold (pH 7) and a 2-fold (pH 8.5) excess of the metal ions. Noteworthy is also that formation of bimetallic complexes starts at pH 7 at about a 100-fold excess of metal ions, whereas at pH 8.5 an approximately 10-fold excess is sufficient to induce formation of ternary complexes L_1M_2. This sensitivity of metal complexation to pH is a serious complication in various types of experiments under physiological conditions, especially if they rely on the measurement of the fluorescence intensity. Since complexation of one metal ion is usually connected with the release of one proton, increasing the metal ion concentration goes along with a decrease of pH except if a buffer is present. Because metal complexation has a large effect on the fluorescence yield, a non-linear relationship between

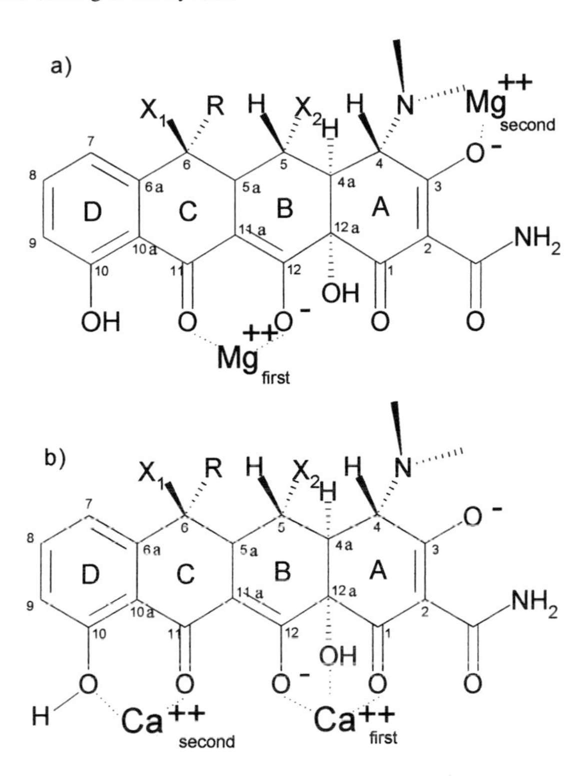

Figure 15. Proposed complexation sites for Mg^{2+} and Ca^{2+} in the Mg-type (top) and Ca-type (bottom) complexes (after reference [37]).

Table 6. Conditional binding constants for complexation of various tetracylines with Mg^{2+} and Ca^{2+}. (After Schmitt and Schneider [37])

		pH	$\log (K_1^{a)}/M^{-1})$	$\log (K_2^{b)}/M^{-1})$	type of complex[c)]
Tc	Mg^{2+}	7.0	2.88 ± 0.01	1.47 ± 0.04	Mg-type
		8.5	4.06 ± 0.01	1.79 ± 0.03	Mg-type
Tc	Ca^{2+}	7.0	2.61 ± 0.02	0.87 ± 0.12	Mg-type
		8.5	4.46 ± 0.04	2.11 ± 0.14	Ca-type
Oxy	Mg^{2+}	7.0	3.07 ± 0.01	1.56 ± 0.04	Mg-type
		8.5	4.13 ± 0.03	1.55 ± 0.10	Mg-type
Oxy	Ca^{2+}	7.0	2.61 ± 0.02	1.00 ± 0.11	Mg-type
		8.5	3.71 ± 0.06	2.15 ± 0.17	Ca-type
ATC	Mg^{2+}	7.0	3.16 ± 0.03	0.79 ± 0.09	Mg-type
		8.5	3.99 ± 0.02	0.97 ± 0.05	Mg-type
ATC	Ca^{2+}	7.0	2.92 ± 0.05	1.17 ± 0.03	nd [d)]
		8.5	3.30 ± 0.05	0.77 ± 0.12	nd [d)]

a) defined as $K_1 = (L_1M_1)/(L)(M)$
b) defined as $K_2 = (L_1M_2)/(L_1M_1)$ (M
c) structures shown in Fig. 15
d) not determined

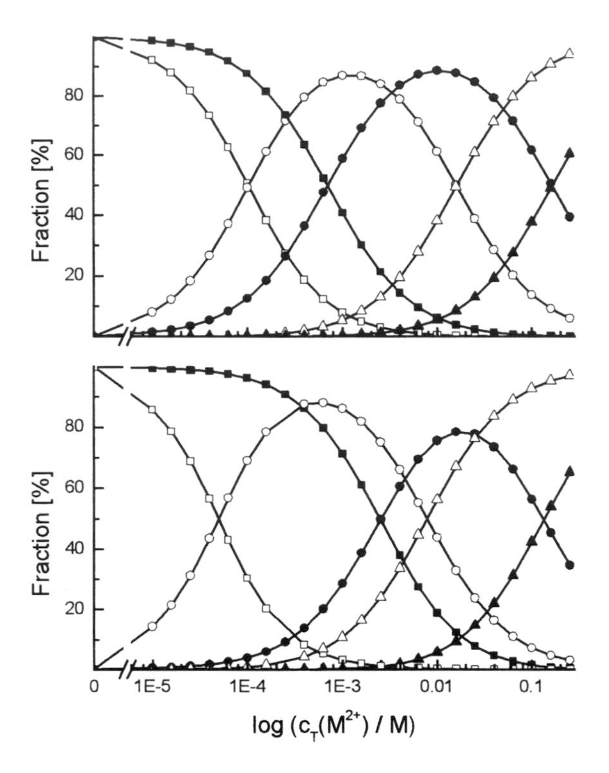

Figure 16. Comparison of the partition profiles of Tetracycline complexed with Mg^{2+} (top) and Ca^{2+} (bottom) at pH 7.0 (filled) and pH 8.5 (open symbols). Total concentration of Tetracycline is $C_T = 3 \cdot 10^{-5}$ M. Free ligand (□, ■), LM (●, ○), and LM_2 (▲, △) (after reference [37]).

metal ion concentration and fluorescence intensity is likely in non-buffered solutions.

Calorimetric studies of metal binding

It is a well-known phenomenon that two reactions can have a similar driving force (ΔG), although the corresponding changes in enthalpy (ΔH) and entropy (ΔS) are quite different (enthalpy–entropy compensation). In case of metal ion complexation, such a compensatory relationship can reflect different modes of binding and changes of solvation. Ohyama and Cowan [46] investigated the complex formation of Tetracycline with Mg^{2+}, Mn^{2+}, Ca^{2+}, and Sr^{2+} by ultra-sensitive calorimetry. Experiments were carried out at pH 9.5 to ensure complex formation between specific solution species. Independent pH titration studies showed that the protonation state of N4 had no influence on the binding constant K for magnesium ions. In Table 7, the thermodynamic parameters for Mg^{2+} and Ca^{2+} given in reference [46] are quoted. The change in sign of

Table 7. Thermodynamic parameters for binding of Mg^{2+} and Ca^{2+} by Tetracycline (pH 9.5). Data taken from reference [46]

M	T/K	$K \cdot 10^4/M^{-1}$	ΔG/kJ mol^{-1}	ΔH/kJ mol^{-1}	ΔS/JK^{-1} mol^{-1}
Mg^{2+}	278	2.77 ± 0.14	−23.7	−21.0	9.6
	308	6.70 ± 0.62	−28.4	−9.5	60.2
Ca^{2+}	278	10.8 ± 1.3	−26.7	−54.7	−100.7
	308	12.9 ± 1.8	−30.0	−36.8	−22.2

ΔS is interpreted by the authors as an effect of the distinct hydration environment for ions with higher (Mg^{2+} and Mn^{2+}) and lower (Ca^{2+} and Sr^{2+}) charge density. In the former, longer-range ordering of the solvent is achieved relative to the latter ones. After complexation with Tetracycline, the hydration sphere of Mg^{2+} and Mn^{2+} is most influenced by the decrease in local charge density and the hydrophobicity of the Tetracycline moiety. The negative ΔS value observed for Ca^{2+} and Sr^{2+} is thought to reflect the structure-forming contribution from water molecules around the hydrophobic Tetracycline ring system.

The more favourable reaction enthalpy ΔH obtained for Ca^{2+} and Sr^{2+} should reflect the easy coordination of the larger cations. Coordination of the smaller ions (Mg^{2+} and Mn^{2+}) could introduce strain on the bonds that form the six-membered ring chelation complex [47]. Finally, the authors emphasise the larger change in heat capacity upon binding of Ca^{2+} *versus* binding of Mg^{2+}. The interpretation given of the calorimetric data suggests that quantum chemical model calculations can be successful only if the first layers of solvent molecules are taken into account explicitly.

Fluorescence decay times of tetracycline complexes with Mg^{2+} and Ca^{2+}

Time-resolved fluorescence measurements for free tetracyclines are difficult because traces of impurities and/or reaction products can contribute more to the total signal intensity than the species under investigation. Elimination of the intramolecular hydrogen bond at O11–O12 via deprotonation and/or complexation of tetracycline with proper divalent metal ions leads to an increase of the fluorescence yield as discussed above. Efficient formation of the complexes requires adjustment of the pH of the solution to pH ≥ 8.0. Under these conditions, the metal-free compounds are present as a mixture of $LH_2^{\pm}$, LH^- and the stronger fluorescing dianions.

A proper fit of the fluorescence decay of Tetracycline in aqueous solution (pH 7, I = 0.1 M, λ_{det} = 550 nm) needs four exponentials (private communication by D.-Th. Marian). Since the amplitude of the fourth component increases with time, it is linked to a reaction product. The remaining three components can be assigned to the three species mentioned above. The zwitterion

$LH_2^{\pm}$ exhibits the shortest decay time ($\tau_1 = 30$ ps) due to fast radiationless decay mediated by the internal hydrogen bond O11-H-O12. The second component ($\tau_2 = 270$ ps) is assigned to LH^- and the long-lived component ($\tau_3 = 1.5$ ns) to the dianion $L^=$. The rationale behind this assignment is that upon addition of Mg^{2+}, the lifetimes of the first and third component do not change. The relative contributions β_1 and β_3 drop, however, from about 0.58 and 0.17 to nearly zero, whereas the second component increases in intensity without much change in decay time ($\tau_2' = 385$ ps). These results imply that deprotonation at O11–O12 already decreases the rate of radiationless deactivation (increase in lifetime from 30 ps to 270 ps). Complexation with Mg^{2+} apparently decreases this rate further by making the molecule less flexible (increase in lifetime from 270 ps to 385 ps).

According to Kunz [48], Tetracycline (10^{-4} M in 0.05 M Tris buffer at pH 8.0) exhibits a fluorescence decay time of about 0.4 ns when complexed with Mg^{2+}. Upon complexation with Ca^{2+}, a fluorescence lifetime of about 0.25 ns is found. It is perhaps not surprising that the complexes with epi-Tetracycline exhibit approximately the same fluorescence decay times. The great similarity in properties of Tetracycline and Oxytetracycline manifests itself also in the excited state lifetimes of the Mg^{2+} complexes (0.44 ns). Evidence for the importance of hydrogen bonding as promoter of radiationless decay, not only in Tetracycline but also in the metal chelates, can be derived from experiments in D_2O and in acetonitrile – water mixtures. Switching from H_2O (pH = 8.0) to D_2O (pD = 8.0) results in an increase of the fluorescence lifetime of the Mg^{2+} complexes from about 0.4 ns to about 1.35 ns.

Proton binding and metal ion complexation of tetracycline derivatives

Tetracycline derivatives, which possess the same three acidic groups as "normal" Tetracycline, show very similar changes in the absorption and CD spectra upon pH titration. Consequently, similar pK_a values are found (cf. Tab. 8). The same holds true for complexation with Mg^{2+} and Ca^{2+} (Tab. 6). Doxycycline, in our opinion, forms 1: 1 Mg-type complexes both with Mg^{2+}

Table 8. Macroscopic acidity constants for various derivatives of Tetracycline derived from absorption titration experiments in upward and downward direction (data from M. Dean)

	pk_{a1}	pk_{a2}	pk_{a3}	reference
Oxytetracycline	3.3	6.8	8.4	this work (up)
	3.2	7.3	9.1	this work (down)
	3.3	7.3	9.1	[3]
Doxycycline	3.0	7.7	9.1	downward
	3.4	7.7	9.7	[3]
Anhydrotetracycline	3.93	5.94	8.48	[56]

and Ca^{2+} at pH 7.0 [37]. (At higher pH, measurements could not be performed due to rapid deterioration of the sample.) According to Mitscher et al. [49] binding of the second Mg^{2+} involves C5 OH and C12a OH in 5-hydroxytetracyclines.

Proton and metal ion binding of anhydrotetracycline (ATC)

Since ATCs are important degradation products they were also studied extensively in the past to clarify how their presence can influence the antibiotic action of the parent compounds.

There is extensive spectroscopic evidence that 5,6-ATC can also adopt different conformations. It is generally assumed that one pair of conformers resembles the twisted and extended conformation described for Tetracycline. The only publication of X-ray data does unfortunately not provide co-ordinates of the atoms [50]. Accordingly, the twisted conformation should be the dominant one in neutral and acidic solution, whereas the extended conformation exists as majority species at high pH [45].

Semi-empirical quantum-chemical model calculations using the AM1 Hamiltonian and the continuum model COSMO confirmed the empirical assignment [51]. The fully protonated (LH_3^+) and the zwitterionic species ($LH_2^\pm$) adopt nearly exclusively (97.5%) the A conformation. The anionic species LH^- and the fully deprotonated species $L^=$ should be present as a mixture of both conformers. In water, the predicted ratio is 68% (A)/32% (B) for LH^- and 34% (A)/66% (B) for $L^=$.

At first glance, the complexation behaviour of ATC investigated by absorption and CD spectroscopy appears to be pronouncedly different from that of the "normal" tetracyclines. Because of the lower pK_{a2} and pK_{a3} values (Tab. 8), the extent of protonation of the N4 dimethylammonium group relative to that of the O11–O12 β-diketone system is significantly changed. The absolute values of the first conditional binding constants for the binding of Mg and Ca ions shown in Table 6 support previous suggestions of O11–O12 complexation like in Tetracycline. The much smaller value of the second binding constant K_2 favours the assumption of A-ring complexation by the second Mg^{2+}.

Complexation of tetracyclines with lanthanide ions

As mentioned before, tetracyclines chelate with a large variety of metal ions, like alkaline and alkaline earth metal ions, with transition metal ions and also with trivalent lanthanides. The lanthanide ions show luminescence (e.g., Eu^{3+} around 600 nm) with decay times on the order of 20–120 μs [52]. Sensitisation of the Eu^{3+} phosphorescence by energy transfer within the complexes can be used to monitor tetracyclines by phosphorimetric techniques with high sensitivity. The advantage of the time-gated detection mode is that fluorescence from impurities and/or degradation products can easily be suppressed, thus allowing the detection of tetracyclines with a sub-nanogram limit. The optimum tetracycline to Eu^{3+} ratio was found to be 1:1, indicating a high association constant. Since Eu^{3+} exhibits an ion radius similar to Mg^{2+}, it is assumed that the pattern of complexation is the same for both ions. Larger lanthanide ions like La^{3+} and Yb^{3+} experience a modified Coulombic interaction, thus leading to different complexation sites [53].

Tetracycline – metal ion complexes bound in the Tet repressor protein

Tetracyclines bind to repressor or activator proteins preferentially as complexes with Mg^{2+} or Ca^{2+}, which are found in blood plasma at concentrations of about 1 mM. The binding constants of the metal ion complexes in the protein are usually determined by fluorescence titration experiments. In these, a buffered solution of the protein (e.g., TetR) and the tetracycline under investigation is titrated with the chosen kind of metal ion. In view of the above, the mechanism of effector binding comprises as a first step the formation of the tetracycline – metal ion complexes and, as a second step, the incorporation of the complexes in the binding pocket. The stoichiometry of the free tetracycline –metal ion complexes formed is in principle determined by the concentration of the metal ions, whereas within the protein the tetracyclines are assumed to be complexed with one metal ion only [54].

The X-ray structure of the ternary Tetracycline – Mg^{2+} – TetR complex shows that Mg^{2+} co-ordinates at the β-diketone moiety O11–O12 [55]. Furthermore, the conformation is essentially the same as the one found in the Tetracycline hexahydrate crystals (Fig. 17). If one ignores the interactions between the effector and the protein, one can conclude that Mg^{2+} chelation at the β-diketone moiety does not much change the geometry of the (zwitterion-

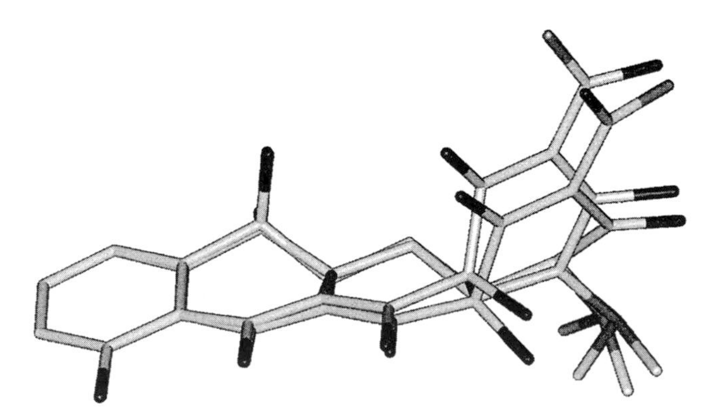

Figure 17. Overlay of the crystal structures found for Tetracycline hexahydrate (zwitterionic form) and the Tetracycline – Mg^{2+} complex embedded in the TetR protein (co-ordinates from reference [55]). (By courtesy of C. Leypold)

ic) ligand. The comparison of the CD spectra of the free and protein-bound tetracycline – Mg^{2+} complexes (Fig. 18) demonstrates that upon protein binding there are indeed only minor changes in the long wavelength range (C. Leypold, private communication).

The CD spectrum of the protein-bound Ca^{2+} complexes resembles that of the corresponding ternary complexes with Mg^{2+}, thus suggesting that the complexation site again comprises O11–O12. The CD spectrum of the free complexes in solution deviates, however, pronouncedly from that of the Mg^{2+} complexes. This is in accordance with the different chelation pattern of Ca^{2+} at pH 8.5 as discussed in the previous section.

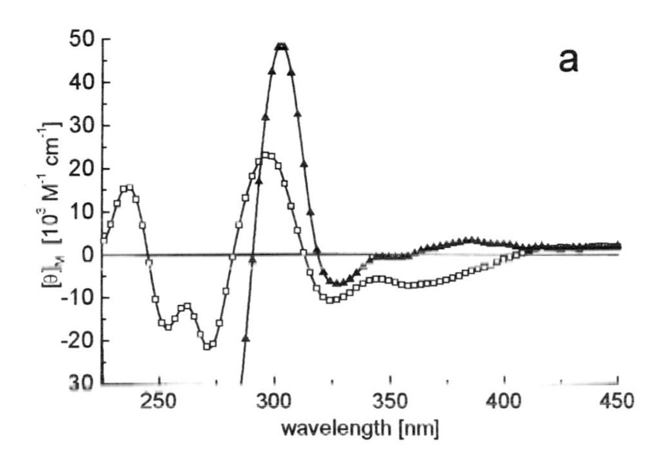

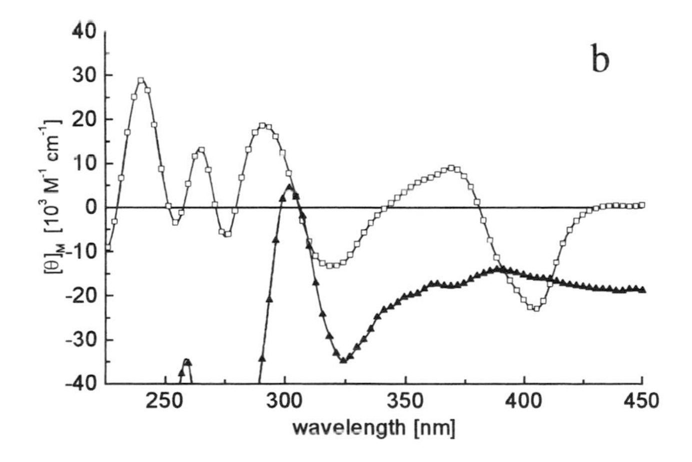

Figure 18. Comparison of the CD spectra of free (open symbols) and protein bound complexes (closed symbols) of Tetracycline with Mg^{2+} (a) and Ca^{2+} (b). The ionic strength of the solution at pH = 8.5 was fixed by addition of 0.15 M NaCl. Metal to ligand ratio was 40 for Mg^{2+} and 27 for Ca^{2+}. (By courtesy of C. Leypold)

Fluorescence kinetics of ternary complexes including the Tet repressor protein

Kunz [48] has measured the fluorescence kinetics of ternary complexes between various mutants of the Tet repressor protein TetR (tryptophan scanning analysis described by Kintrup et al. [48]), various tetracyclines and either Mg^{2+} or Ca^{2+} as complexing metal ions. He found that, except for the combination ATC and Mg^{2+}, the effector fluorescence could be fitted by a biexponential decay law.

The decay times deduced for Tetracycline and Oxytetracycline (cf. Tab. 9) must be considered as equal within experimental error. The differences between the lifetimes of the Mg^{2+} and Ca^{2+} complexes are small, thus indicating that the geometry of the complexes and their embedding in the binding pocket must be similar. Anhydrotetracycline exhibits significantly longer decay times, the shorter-lived component being less intense or even missing (Mg^{2+} complexes).

Table 9. Fit parameters derived for the effector fluorescence when imbedded in the Tet Repressor protein Tet R $(B)_{WT}$ (0.05 M Tris buffer, pH 8.0, 0.15 M NaCl, 5 mM β-mercaptoethanol: c(Tet R(B)) = 5×10^{-4} M; c(Tc) = 10^{-4} M; c(M^{2+}) = 10^{-2} M)

Tc	M	τ_1/ns	τ_2/ns	β_1 [a]	β_2 [a]
Tetracycline	Mg^{2+}	1.9	2.9	0.65	0.35
	Ca^{2+}	1.4	2.7	0.38	0.62
Oxytetracycline	Mg^{2+}	1.5	2.8	0.55	0.45
	Ca^{2+}	1.3	2.7	0.66	0.34
Anhydrotetracycline	Mg^{2+}		3.6		1.0
	Ca^{2+}	1.8	3.9	0.12	0.88

[a] relative contribution to total fluorescence $\beta_i = A_i\tau_i/\Sigma A_i\tau_i$

The origin of the observed biexponential decay is not yet clear, because there are several alternatives for an explanation:

(i) It could be the result of a structural heterogeneity of the protein and concomitantly different chromophore – protein interactions.

(ii) It could be caused by a geometrical heterogeneity of the effector molecules like the conformational heterogeneity described for the free tetracyclines.

(iii) It could result from different patterns of metal ion complexation (including the complexation of more than one ion).

It should be mentioned that within the series of single tryptophan mutants which were investigated, the fluorescence decay kinetics was, in general, fairly independent of the site of mutation. Exceptions were found only if the introduced tryptophan modified the binding pocket and/or caused steric interactions with the effector [48].

Concluding remarks

Tetracyclines appear to be like chameleons. They can readily adjust themselves to variations in their surroundings, e.g., the pH value. They can adopt a large number of different structures with relatively small differences in total energy. Therefore, it is practically impossible to find conditions under which one species is present nearly exclusively. As a consequence, it is also extremely difficult to determine experimentally the thermodynamic and spectroscopic properties of each species as defined by state of protonation or metal ion chelation, conformation of the four annelated rings, or rotation and hydrogen bonding of functional groups.

At first glance, it is annoying to see the wide spread of the values found in the literature for apparently simple characteristic parameters, e.g., the pK_a values. The differences in the numbers presented are mostly several times larger than the quoted possible errors. An explanation for this fact could be found in the above-mentioned structural heterogeneity, which tetracyclines exhibit in solution under essentially all reasonable experimental conditions. As could be demonstrated in this contribution, different detection techniques see different species with different sensitivity. Alternatively, some techniques may not be able to distinguish between two species whereas others do. Since the results of the data analyses will be governed by those species which contribute most to the recorded signals, it must not be surprising if, e.g., absorption and fluorescence spectra produce somewhat different pK_a values or stability constants.

Another important aspect, which was partly ignored in past experiments, is the kinetic one. Structural relaxation to establish the thermodynamic equilibrium will occur on different time-scales depending on whether, e.g., conformational changes with high activation barriers or changes in state of protonation or hydrogen bonding are required. Metal complexation seems to lock tetracyclines in certain conformations. As a consequence, thermal and photochemical degradation reactions can be enhanced, but also decreased, as a consequence of metal ion chelation.

From the early days of tetracycline research onward, it was suggested that it is the flexibility of these molecules that makes them such a successful and efficient drug. On the other hand, this flexibility makes it also very difficult to establish a structure – function relationship as long as its precise structure has not been determined at the site of action.

Acknowledgement
The author wants to thank the past and present coworkers for their help in preparing the manuscript, especially M. Dean, Dr. M. Kunz, C. Leypold, D.-Th. Marian and M. Schmitt. Thanks are also due to Dr. F. Heinemann (Anorgan. Institut, FAU) for providing the X-ray structures of several tetracycline species and Dr. M. Gottschalk and Dr. H. Lahnig for providing several figures.
Financial support by Deutsche Forschungsgemeinschaft, Volkswagenstiftung and Fonds der Chemischen Industrie is also gratefully acknowledged.

References

1 Mitscher LA (1978) *The chemistry of the tetracycline antibiotics*. Marcel Dekker, New York
2 Dürckheimer W (1975) Tetracycline: Chemie, Biochemie und Struktur-Wirkungs-Beziehungen. *Angew Chem* 87: 751–764
3 Martin BR (1985) Tetracyclines and Daunorubicin. *In*: *Antibiotics and their complexes*. Marcel Dekker, New York, Vol. 19: 19–52
4 Ali SL (1984) Tetracycline Hydrochloride. *In*: *Analytical Profiles of Drug Substances*. American Pharmaceutical Association, New York, Vol. 13: 597–653
5 JJ Hlavka, JH Boothe (eds) (1985) *The Tetracyclines*. Springer Verlag, Heidelberg
6 Martin AR (1998) Antibacterial antibiotics. *In*: JN Delgado, WA Remers (eds): *Wilson and Gisvold's textbook of organic medicinal and pharmaceutical chemistry*. Lippincott-Raven, Philadelphia, New York, 253–325
7 Chobra I, Hawkey PM, Hinton M (1992) Tetracyclines, molecular and clinical aspects. *J Antimicrob Chemother* 29: 245–277
8 Cioni P, Strambini GB (1996) Pressure effects on the structure of oligomeric proteins prior to subunit dissociation. *J Mol Biol* 263: 789–799
9 Ricci RW, Nesta JM (1976) Inter- and intramolecular quenching of indole fluorescence by carbonyl compounds. *J Phys Chem* 80: 974–980
10 Prewo R, Stezowski JJ (1977) Chemical-structure properties of tetracycline derivatives. 3. The integrity of the conformation of the non-ionised free base. *J Amer Chem Soc* 99: 1117–1121
11 Mitscher LA, Bonacci AC, Sokoloski TD (1968) Circular dichroism and solution conformation of the tetracycline antibiotics. *In*: GL Hobby (ed.): *Antimicrobial Agents and Chemotherapy*. American Society for Microbiology, Bethesda, 78–86
12 Gulbis J, Everett GW Jr (1976) Metal binding characteristics of the tetracycline derivatives in DMSO solution. *Tetrahedron Lett* 32: 913–917
13 Hughes LJ, Stezowski JJ, Hughes RE (1979) Chemical-structural properties of tetracycline derivatives. 7. Evidence for the coexistence of the zwitterionic and non-ionised forms of the free base in solution. *J Amer Chem Soc* 101: 7655–7657
14 Schach von Wittenau M, Blackwood RK (1966) Proton magnetic resonance spectra of tetracyclines. *J Chem Soc* 31: 613–615
15 Curtis R, Wasylishen RE (1991) A nitrogen-15 magnetic resonance study of the tetracycline antibiotics. *Can J Chem* 69: 834–838
16 Gottschalk M, Clark T (2000) Conformational analysis of tetracycline and its Mg^{2+}-chelates in the gas phase and aqueous solution. *J Mol Model*; *in press*
17 Lanig H, Gottschalk M, Schneider S, Clark T (1999) Conformational analysis of tetracycline using molecular mechanics and semi empirical MO-calculations. *J Mol Model* 5: 46–62
18 Klamt A, Schüürmann G (1993) COSMO: a new approach to dielectric screening in solvents with explicit expressions for the screening energy and its gradient. *J Chem Soc Perkin Trans 2* 799–805
19 Gottschalk M (2000) Konformationsanalyse von Tetrazyklin und seinen Mg^{2+}-Komplexen. PhD thesis, Friedrich-Alexander-Universität, Erlangen
20 Duarte HA, Carvalho S, Paniago EB, Simas AM (1999) Importance of tautomers in the chemical behavior of tetracyclines. *J Pharm Sci* 88: 111–120
21 Shea CR, Hefetz Y, Gillies R, Wimberly J, Dalickas G (1990) Mechanistic investigation of doxycycline photosensitization by picosecond-pulsed and continuous wave laser irradiation of cells in culture. *J Biol Chem* 265: 5977–5982
22 Hlavka JJ, Krazinski HM (1963) The 6-deoxytetracyclines. VI. A photochemical transformation. *J Org Chem* 28: 1422–1423
23 Hussar DA, Niebergall PJ, Sugita ET, Doluisio JT (1968) Aspects of the epimerisation of certain tetracycline derivatives. *J Pharm Pharmacol* 20: 539–546
24 Davies AK, Kellar JFM, Phillips GO, Reid AG (1979) Photochemical oxidation of tetracycline in aqueous solution. *J Chem Soc Perkin Trans 2* 369–375
25 Moore DE, Fallon MP, Burt CD (1983) Photo-oxidation of tetracycline – a differential pulse polarographic study. *Int J Pharm* 14: 133–142
26 Drexel RE, Olack G, Jones C, Chmurny GN, Santini R, Morrison H (1990) Lumitetracycline: A novel new tetracycline photoproduct. *J Org Chem* 55: 2471–2478
27 Morrison H, Olack G (1991) Photochemical and photophysical studies of tetracycline. *J Amer*

Chem Soc 113: 8110–8118

28 Schwartzman G, Wayland L, Alexander T, Furnkranz K, Selzer G (1979) Chlortetracycline hydrochloride. *In*: K Florey (ed.): *Analytical profiles of drug substances*, Bd. 8, Academic Press, 101–137

29 Rigler NE, Bag SP, Leyden DE, Sudmeier JL, Reilley CN (1965) Determination of a protonation scheme of tetracycline using nuclear magnetic resonance. *Anal Chem* 37: 872–875

30 Bhatt VK, Jee RD (1985) Micro-ionization acidity constants for tetracyclines from fluorescence measurements. *Anal Chim Acta* 167: 233–240

31 Grampp H, Maeder M, Meyer CJ, Zuberbühler AD (1985) Calculation of equilibrium constants from multiwavelength spectroscopic data – I Mathematical considerations. *Talanta* 32: 95–101

32 Grampp H, Maeder M, Meyer CJ, Zuberbühler AD (1985) Calculation of equilibrium constants from multiwavelength spectroscopic data – II SPECFIT: two user-friendly programs in BASIC and standard FORTRAN 77. *Talanta* 32: 257–264

33 Grampp H, Maeder M, Meyer CJ, Zuberbühler AD (1985) Calculation of equilibrium constants from multiwavelength spectroscopic data – III Model-free analysis of spectrophotometric and ESR titrations. *Talanta* 32: 1133–1139

34 Grampp H, Maeder M, Meyer CJ, Zuberbühler AD (1986) Calculation of equilibrium constants from multiwavelength spectroscopic data – IV Model-free least-squares refinement by use of evolving factor analysis. *Talanta* 33: 943–951

35 McCormick JRD, Fox SM, Smith LL, Bitler BA, Reichenthal J, Origoni VE, Muller WH, Winterbottom R (1957) Studies of the reversible epimerisation occurring in the tetracycline family. The preparation, properties, and proof of structure of some 4-epi tetracyclines. *J Amer Chem Soc* 79: 2849–2858

36 Lambs L, Decock – Le Reverend B, Kozlowski H, Berthon G (1988) Metal ion-tetracycline interactions in biological fluids. 9. Circular dichroism spectra of calcium and magnesium complexes with tetracycline, oxytetracycline, doxycycline and chlortetracycline and discussion of their binding modes. *Inorg Chem* 27: 3001–3012

37 Schmitt MO, Schneider S (2000) Spectroscopic investigation of complexation between various tetracyclines and Mg^{2+} or Ca^{2+}. *Phys Chem Comm* 9

38 Lambs L, Venturini M, Decock – Le Reverend B, Kozlowski H, Berthon G (1988) Metal ion-tetracycline interactions in biological fluids. Part 8. Potentiometric and spectroscopic studies on the formation of Ca(II) and Mg(II) complexes with 4-dedimehylamino-tetracycline and 6-desoxy-6-demethyl-tetracycline. *J Inorg Biochem* 33: 193–210

39 Jogun KH, Stezowski JJ (1976) Chemical-structural properties of tetracycline derivatives. 2. Coordination and conformational aspects of oxytetracycline metal ion complexation. *J Amer Chem Soc* 98: 6018–6026

40 Colaizzi JL, Knevel AM, Martin AN (1965) Biophysical study of the mode of action of the tetracycline antibiotics. *J Pharm Sci* 54: 1425–1436

41 Mitscher LA, Bonacci AC, Sokoloski TD (1968) Circular dichroism and the solution conformation of the tetracycline antibiotics. *Tetrahedron Lett* 51: 5361–5364

42 Mitscher LA, Bonacci AC, Slater-Eng B, Hacker AK, Sokoloski TD (1969) Interaction of various tetracyclines with metallic cations in aqueous solutions as measured by circular dichroism. *Antimicrob Agents Chemother* 111–115

43 Caswell AH, Hutchison JD (1971) Selectivity of cation chelation to tetracyclines: evidence for special conformation of calcium chelate. *Biochem Biophys Res Commun* 43: 625–630

44 Asleson GL, Stoel LJ, Newman EC, Frank CW (1974) NMR spectra of tetracyclines: assignment of additional protons. *J Pharm Sci* 63: 1144–1146

45 Wessels JM, Ford WE, Szymczak W, Schneider S (1998) The complexation of tetracycline and anhydrotetracycline with Mg^{2+} and Ca^{2+}: a spectroscopic study. *J Phys Chem* 102: 9323–9331

46 Ohyama T, Cowan JA (1995) Calorimetric studies of metal binding to tetracycline. Role of solvent structure in defining the selectivity of metal ion-drug interactions. *Inorg Chem* 34: 1083–3086

47 Hancock RD (1990) Molecular mechanics calculations and metal ion recognition. *Accounts of Chemical Research* 23: 253–257

48 Kunz M (2000) Untersuchungen zu den Bindungszuständen des TetR(B) und deren Struktur im Bereich der Aminosäuren. PhD thesis, Friedrich-Alexander-Universität, Erlangen

49 Mitscher LA, Slater-Eng B, Sokoloski TD (1972) Circular dichroism measurements of the tetracyclines. *Antimicrob Agents Chemother* 2: 66–72

50 Palenik GJ, Mathew M, Restivo R (1978) Structural studies of tetracyclines. Crystal and molecular structures of anhydrotetracycline hydrobromide monohydrate and 6-demethyl-7-chlorotetracycline hydrochloride trihydrate. *J Amer Chem Soc* 100: 4458–4464
51 Santos DH, Almeida de W, Zerner M (1998) Conformational analysis of the anhydrotetracycline molecule: a toxic decomposition product of tetracycline. *J Pharm Sci* 87: 190–195
52 Duggan JX (1991) Phosphorimetric detection in HPLC via trivalent lanthanides: high sensitivity time-resolved luminescence detection of tetracyclines using Eu^{3+} in a micellar post column reagent. *J Liq Chromatogr* 14: 2499–2525
53 Babushkina T, Grosheva V, Zolin V, Koreneva L (1997) Optical and NMR spectroscopy studies of complexation between tetracycline and lanthanide ions. *Russ J Coord Chem* 23: 709–711
54 Takahashi M, Degenkolb J, Hillen W (1991) Determination of the equilibrium association constant between tet repressor and tetracycline at limiting Mg^{2+} concentrations: a generally applicable method for effector-dependent high-affinity complexes. *Anal Biochem* 199: 197–202
55 Hinrichs W, Kisker C, Duvel M, Müller A, Tovar K, Hillen W, Saenger W (1994) Structure of the tet repressor-tetracycline complex and regulation of antibiotic resistance. *Science* 264: 418–420
56 Mellos Matos de SV, Beraldo H (1995) Complexes of anhydrotetracycline. 3: An absorption and circular dichroism study of the Ni(II), Cu(II) and Zn(II) complexes in aqueous solution. *J Braz Chem Soc* 4: 405–411
57 Kintrup M, Schubert P, Kunz M, Chabbert M, Alberti P, Bombarda E, Schneider S, Hillen W (2000) Trp scanning analysis of Tet repressor reveals conformational changes associated with operator and anhydrotetracycline binding. *Eur J Biochem* 267: 821–829

Section II
Tetracycline-dependent gene regulation: A versatile tool for all organisms

(Editor: Wolfgang Hillen)

Tetracyclines in Biology, Chemistry and Medicine
ed. by M. Nelson, W. Hillen and R.A. Greenwald

Gene regulation by the tetracycline-inducible Tet repressor-operator system – molecular mechanisms at atomic resolution

Winfried Hinrichs* and Christiane Fenske

* *Institut für Chemie und Biochemie, Ernst-Moritz-Arndt Universität Greifswald, Soldmannstr. 16, D-17487 Greifswald, Germany*

Introduction

At present, three-dimensional structures characterizing specific tetracycline/protein interactions are available only for the Tet repressor, TetR. This is the regulatory switch for the most common resistance mechanism against tetracyclines, Tc, in Gram-negative bacteria. Crystallographic investigations with at least 2.5 Å resolution of TetR/Tc complexes [1, 2] and the TetR/DNA complex [3] provide a clear view of endpoints for the functional allosteric pathway of this distinct regulatory system and reveal mechanisms that underlie TetR/Tc recognition and induced conformational changes forcing dissociation of the repressor/operator-DNA complex, TetR/*tetO*.

Understanding the molecular mechanisms of bacterial resistance against antibiotics is of significant clinical importance [4]. Nowadays, general interest is stimulated by the fact that TetR/*tetO* is the most efficiently inducible system of transcriptional regulation known to date. It is commonly used as a tool for selective targeted gene regulation in eukaryotic systems [5–10].

The resistance mechanism

The most frequently observed resistance against tetracycline (Tc) in Gram-negative bacteria depends on the export of a tetracycline-magnesium complex, $[MgTc]^+$ (Fig. 1), by the TetA protein. The TetA is embedded in the cytoplasmic membrane and mediates active efflux of $[MgTc]^+$ against equimolar uptake of a proton (proton motive force) [11, 12]. Regulation of expression of TetA is under tight transcriptional control of the tetracycline repressor, TetR. This unique type of resistance determinant has a central regulatory part with overlapping operators and promoters between the two genes *tetA* and *tetR*, arranged with divergent polarity [13]. In the absence of tetracycline, the two operator sites, *tetO1* and *tetO2*, are protected by TetR homodimers preventing

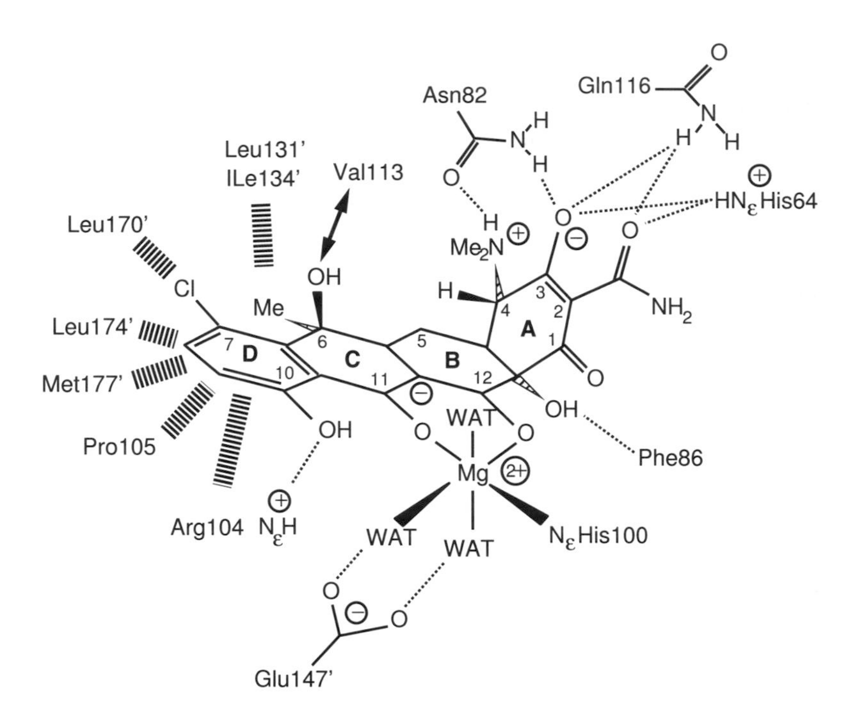

Figure 1. Chemical structure of the tetracycline-Mg^{2+} complex, $[MgTc]^+$, and schematic recognition pattern of amino acid side chains of TetR in the binding tunnel. Amino acids of the two TetR monomers are distinguished by ('), hydrogen bonds are indicated by thin and van der Waals contacts by thick broken lines. Note the unusual O-H...π interaction between the hydroxyl group at position 12a of Tc and the aromatic side chain of Phe86; the interaction between Val 113 and the hydroxyl substituent at position 6 of Tc is less favorable, indicated by a double arrow. Mg^{2+} is octahedrally coordinated by the chelating 1,3-ketoenolate (O11, O12) of Tc, by His100N_ε, and by three water molecules (WAT). Variations of substituents in the hydrophobic half of Tc retain the antibiotic activity, these are functional groups at positions of C5, C6, C7, C8 and C9, also substituents at N4 [26]. The figure shows 7-chlortetracycline [2].

transcription of the repressor itself and the membrane-residing efflux protein, TetA. The resistance mechanism is activated by $[MgTc]^+$ binding to TetR. Specific drug/repressor interactions enforce conformational changes in TetR and abolish the high affinity to the cognate operator sequences, *tetO1* and *tetO2*, allowing transcription of the controlled genes.

The mechanism is encoded by seven highly homologous resistance determinants (classes A to E, G and H), which are located on plasmids and transposons [14, 15]. The genetic organization is described in detail elsewhere; for reviews see [16, 17].

In contrast to inducer- or effector-controlled repressor-systems for regulation of metabolic pathways, where transcription has to be regulated more gently, the silent (TetR bound to *tetO*) and the active states (TetR bound to $[MgTc]^+$) of the resistance mechanism have to be exactly distinguished for two reasons. First, TetR is required to be a particularly $[MgTc]^+$-sensitive switch, because the bacteriostatic action of the drug depends on inhibition of ribosomal protein synthesis. The affinity of TetR for *tetO* is reduced by 9 orders of magnitude after binding of $[MgTc]^+$ to TetR ($K_a \sim 10^9\ M^{-1}$). This enables expression of the resistance genes before the drug binds to the small ribosomal subunit ($K_a \sim 10^6\ M^{-1}$) [18]. On the other hand, high-level expression of TetA is lethal for the bacterial cell, because unspecific cation transport interferes with the maintenance of the electrostatic potential across the cytoplasmic membrane [19]. Therefore, a well-defined repressed status of the tetA gene by strong binding of TetR to *tetO* is essential (association constant $K_a \sim 10^{11}\ M^{-1}$).

This guarantees that both situations of the resistance mechanism are efficiently distinguished. Moreover, sub-inhibitory Tc concentrations are sufficient to allow transcription of the TetR-controlled genes to proceed.

TetR/inducer complexes

Tetracycline binds to TetR as a chelate complex with a divalent metal ion, e.g., $[MgTc]^+$. The complex carries one net positive charge and is also the target of the TetA protein [12, 20]. The chemical structure of the tetracycline-Mg^{2+} complex is shown in Figure 1. The 1,3-ketoenolate moiety of the Tc rings B and C is deprotonated under physiological pH conditions and coordinates divalent metal ions with high affinity [21]. The binding constant to Mg^{2+} is about $10^3\ M^{-1}$ and even higher for other divalent metal ions, e.g., Fe(II) [22]. Crystal structure data are available for TetR/$[MgTc]^+$ complexes with different Tc analogs: tetracycline, 7-chlorotetracycline and 9–(N,N–dimethylglycylamido)–6–demethyl–6–deoxy–tetracycline [1, 2, 23]. The TetR/Tc-complexes in the absence of divalent metal ions are biologically not relevant but, nevertheless, of crucial importance in analyzing the structural properties of TetR [24]. The different states of Tc- and/or $[MgTc]^+$-complexation are useful in rationalizing the steps sequentially leading to complete induction.

Tetracycline/TetR contacts

The tetracycline analogs differ from each other only by substituent variations. The molecules are clearly separated into a hydrophobic and a hydrophilic half (see Fig. 1). Modifications of hydrophilic substituents abolish bacteriostatic interaction with the prokaryotic ribosome [25].

Variations in the hydrophobic half are permissible and provide a high probability of antibacterial activity [26]. Obviously, TetR is exactly designed to mimic the ribosomal target, because non-antibiotic Tc analogs are not able to induce TetR, even if they have a reasonable binding constant to TetR [27]. The same decrease of inducibility is observed for TetR-variants with single amino acid exchange of residues, which are involved in $[MgTc]^+$-recognition and -induced allosteric transitions [28].

A first group with hydrophilic amino acid residues anchors and positions Tc ring A by hydrogen bonding at the far end of the binding tunnel (His64, Asn82, Phe86 and Gln116) (see Fig. 1). Remarkable is the aromatic hydrogen bond between a hydroxyl group (O12a) of Tc and the aromatic system of Phe86, which is strictly conserved in all TetR classes. Another group includes side chains of conserved His100, Thr103 and Glu147', all involved in direct or water-mediated Mg^{2+}-coordination.

The hydrophobic region in the binding tunnel is coated by type-conserved Val113, Leu131, Ile134, Leu170', Leu174', Met177', exclusively involved in non-polar van-der-Waals contacts to the hydrophobic part of $[MgTc]^+$ with substituents at positions 6 to 9 and the corresponding framework of Tc-rings C and D.

The Tet repressor protein

The TetR of the resistance determinant class B, TetR(B), has been characterized extensively with molecular biological and biochemical methods, but unfortunately crystallographic results are based on TetR of class D, TetR(D). However, these two repressor proteins share about two-thirds of sequence identity; crystal structure investigation at medium resolution of TetR(B) revealed identical polypeptide folding with TetR(D) [1]. Therefore, we can assume that the discussed sequence-specific interactions of TetR with $[MgTc]^+$ or *tetO* will be identical in both classes. Even the oligonucleotide sequences of the corresponding *tetO* only differ in one base pair. Taking together all information on TetR, this is the best-characterized example for specific tetracycline/protein interaction and one of the best-understood inducible repressor systems [29].

The length of the polypeptide chains of TetR within the seven classes varies between 207 and 219 amino acid residues. Sequence identity is high, ranging between 45 and 75%. The first X-ray crystal structure analysis of the TetR(D) in complex with $[MgTc]^+$ (the induced state) [1] revealed a polypeptide chain consisting of only 207 amino acid residues out of the 218 residues corresponding to the gene nucleotide sequence. Posttranslational deletion of the N-terminal Met1 and loss of 10 C-terminal residues during preparation were clearly identified in the electron density maps and verified by matrix-assisted laser desorption/ionization mass spectrometry of dissolved crystals [30]. The advantage of the C-terminal truncation of TetR(D) is that it has only one C-terminal residue more than the highly homologous TetR(B), supporting the similarity of both proteins.

A genetically truncated TetR(D) was used with the C-terminal amino acids 209 to 218 deliberately deleted. This variant allows an improved purification protocol [31] with enhanced crystallization properties compared to previous preparations of wild-type TetR(D) [2, 30]. These deletions have no effect on the biological activity of TetR(D). Quantification of *in vivo* repression and induction efficiencies of wild-type and the genetically truncated TetR(D) show that functionality is not impaired by the truncation [32].

Topology of the TetR homodimer

The quaternary structure of TetR is known to be a very stable homodimer in solution [32, 33]. In the crystal structures the local dyad of the homodimer coincides with the symmetry of twofold rotation axes and both monomers are identical. The all-helical TetR monomer is folded into 10 α-helices (α1 to α10 for one monomer and α1' to α10' for the other one); see Figure 2. The TetR molecule is clearly divided into two N-terminal DNA-binding domains and a regulatory core domain with a globular shape, which provide two $[MgTc]^+$ binding pockets.

The DNA-binding domains consist of the N-terminal three-helix-bundles (α1 to α3 and α1' to α3'), with α2 as the supporting and α3 as the recognition

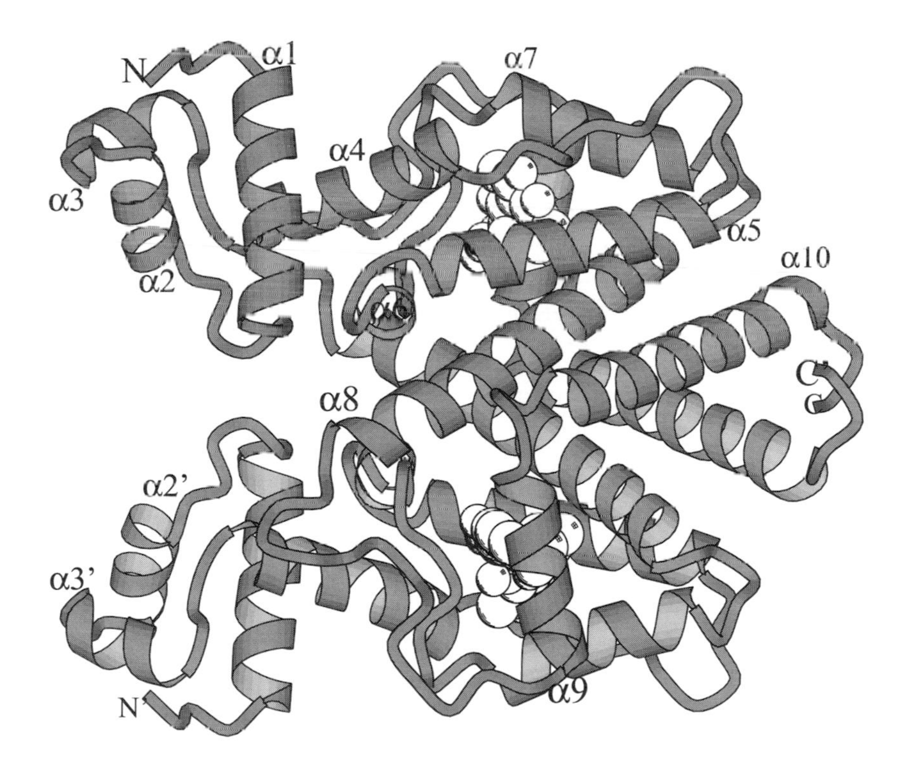

Figure 2. Overall structure of the TetR homodimer. The monomers are related by a horizontal two-fold rotation axis coinciding with the long dimension of the TetR dimer. The helices α are depicted as ribbons and Tc as space-filling models. In one monomer the helices are labeled α1 to α10 from the labeled N- to the C-terminus. The helix-turn-helix motifs of the DNA-binding domains are α2, α3 and α2', α3'. Both monomers are clearly separated, without intertwining of the two polypeptide chains.

helices of the classical helix-turn-helix motif, HTH [34, 35]. The center-to-center separation of the recognition helices α3, α3', which determines the ability of TetR to bind to *tetO*, is controlled by helices α4, α4' (amino acid residues 48 to 63); they link the DNA-binding domains to the $[MgTc]^+$ binding sites in the regulatory core domain. The N-terminal part of helix α4 closes the hydrophobic core of the DNA-binding domain. This rigid arrangement is essential for the molecular mechanisms forcing dissociation of the TetR/*tetO* complex after inducer-binding to TetR.

The scaffold of the regulatory core domain is a central four-helix-bundle consisting of the antiparallel helices α8, α10 crossing the dyad-related helices α8', α10' at an angle of ~50° [2]. If the three-dimensional structures of free TetR [36], of different TetR/$[MgTc]^+$-complexes and of the complex TetR/*tetO* are superimposed, these four α-helices and α5, α5' are identified as the most rigid and structurally best-conserved part. These six helices constitute a rigid scaffold that determines the architecture and function of the core domain of TetR. The core domain harbors two tunnel-like $[MgTc]^+$ binding cavities formed by α5 to α8, α8' and α9' (and their symmetry-related counterparts α5' to α8', α8 and α9 for the other tunnel) [1, 2]. In the induced repressor, the TetR/$[MgTc]^+$-complex, the two $[MgTc]^+$ are completely buried in the binding tunnels of TetR. Both polypeptide chains simultaneously provide amino acid residues for binding of each $[MgTc]^+$, which is a common situation observed for other effector-controlled regulatory DNA binding proteins [37, 38]. This is in agreement with increased thermal stability of TetR after inducer-binding [39].

The regulatory core domain is clearly separated into two subunits. At the periphery of the TetR core, helix α9 merely wraps around the "other" half in a hand-shake fashion, closing the tunnel after inducer-insertion by a lateral shift. This exchange of secondary structure elements for oligomer assembly was denoted as domain-swapping [40].

The entrance to the $[MgTc]^+$ binding tunnel is located between helix α9' and the interhelical loops between helices α6 and α7 and helices α8' and α9', respectively [36]. The entrance is sufficiently large for the inducer $[MgTc]^+$ to enter with Tc-ring A head-on to ensure site-specific contacts to charged, hydrophilic and hydrophobic amino acid side chains in the binding tunnels of the TetR-homodimer.

The loop connection (about 12 amino acid residues) between helices α8 and α9 is highly flexible, indicated by disorder in all TetR crystal structures. The function of this oligopeptide supports the mobility of α9. The loop is the least conserved region and the only part where deletions or insertions are found. The amino acid composition of this loop gives rise to a net negative charge of the loop in all TetR sequences, indicating a supporting role in $[MgTc]^+$-capture. TetR-mutants with deletions in this loop show reduced $[MgTc]^+$ affinity, correlating with lack of inducibility. In contrast, sequence variations by substitutions to alanine show minor effects [41]. A sufficient length of this loop is important for the transition of the open TetR structure to the closed $[MgTc]^+$-induced conformation.

In the TetR-homodimer, the $[MgTc]^+$ binding sites are separated by about 25 Å, and each is located far away (at roughly 33 Å distance) from the DNA-recognition helices α3, α3', like the effector-binding sites in the purine repressor, PurR, or the lactose repressor, LacI [42, 43]. A direct interaction of the effector molecules with operator DNA is impossible, in contrast to, e.g., the tryptophan repressor, TrpR [44].

Molecular contacts in the crystal packing

The relative orientation of the homodimers in the crystal lattice of TetR(D) and TetR(B) are different with space groups $I4_122$ and $P4_12_12$, respectively. Any speculations about artificial effects caused by contacts of TetR homodimers in the crystal packing can be ignored for the discussion of allosteric mechanisms required for induction. In the crystal lattice, neighboring TetR(D) homodimers are arranged in a head-to-head fashion facing the DNA-binding domains [2]. In contrast, TetR(B) homodimers have head-to-tail contacts. A TetR-variant with TetR(B)-sequence of the DNA-binding domain, but with TetR(D) core, crystallizes isomorphously to TetR(D) [36]. Thus, different packing is in principle determined by contacts of the rigid core domains and not by the DNA-binding domains.

Molecular structure of the TetR/operator DNA complex

We have to review the structure of the TetR/*tetO*-complex to discuss the allosteric pathway of conformational changes induced by Tc-binding. The crystallographic results are in agreement with sequence-specific TetR/*tetO* interactions proposed by genetic and biochemical studies [45–49], but reveal additional features. The high association constant requires high structural complementarities of *tetO* and the DNA-binding domains of TetR.

The twofold symmetry of TetR is maintained in the DNA-bound complex (Fig. 3). The HTH-motifs bind to the corresponding major groove of the palindromic operator, but TetR does not recognize the minor groove. All base pairs (bp) of the used 15mer operator fragment, except the central three pairs, are engaged in TetR binding. The central base pair (bp 0) is required as a spacer for the half-operators and does not contribute to sequence specificity.

No water molecules are incorporated into the protein/DNA interface. This is in remarkable contrast to the well-accepted role of water molecules for both the specificity and affinity of protein-DNA interactions [50]. A high-entropy term can be assumed for the association constant of TetR to *tetO* caused by release of water at the TetR and *tetO* interfaces. There is no 'empty' space in the interface region that could be filled by disordered water molecules.

The N-termini of the recognition helices α3, α3' point towards the palindromic center of *tetO* while the helix axes are aligned parallel to the major

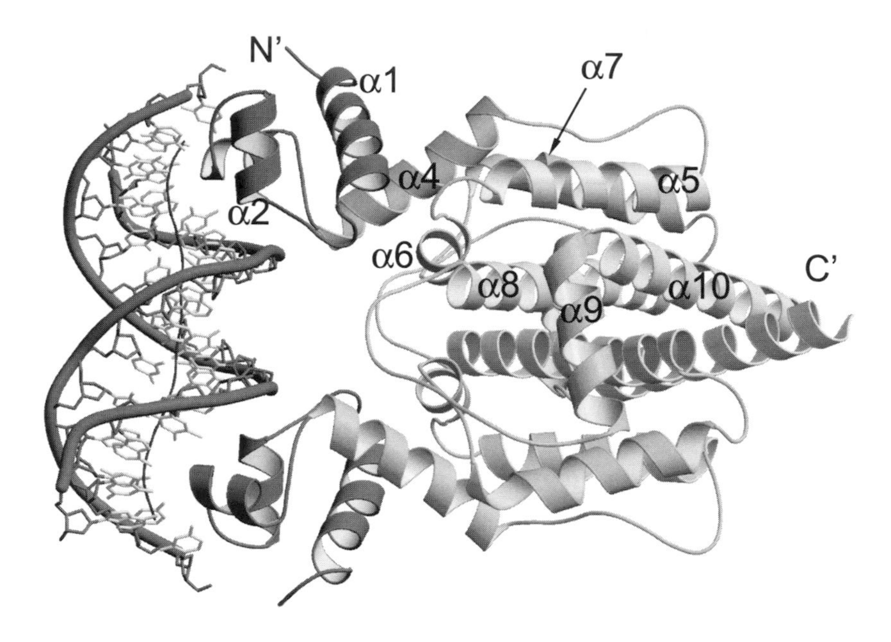

Figure 3. Structure of the TetR/*tetO*-complex. The α-helices of the homodimeric TetR are represented and labeled as in Fig. 2. The complex is rotated around the horizontal dyad axis with respect to Fig. 2 to show the positioning of the helix-turn-helix motifs in the major groove of *tetO*. The curvature of the 15 bp operator fragment is indicated by a central thin line, the phosphate-ribose backbone is represented by two tubes for the sugar-phosphate backbones with the attached nucleosides. The central third of the *tetO*-fragment is straight, the curvature of each half operator is caused by TetR binding to the major groove. Figure adapted from [3].

groove, forming an angle of ~33° with respect to the axis of the DNA double helix (Fig. 3). At the recognition site, the major groove is widened to 14–14.5 Å (canonical B-DNA 11.7 Å), whereas at the central base pair, the major groove on the opposite side of the TetR interface is narrowed to 9.5 Å. These distortions are associated with corresponding changes of the minor groove resulting in partial unwinding of the DNA duplex, which increases the helical repeat to 38 Å (B-DNA 34 Å). Each half-operator is kinked away from TetR at bp 2, but within the 15mer-*tetO* fragment the base pairs at the ends are almost parallel to each other, because the kink is compensated along bp 3 to 6 by bending towards the protein.

Several hydrogen bonds cause this kink at the G2 position. The phosphate groups attached to the 3'- and 5'-positions of the ribose are hydrogen-bonded to amino acid side chains (Thr26, Thr27, Tyr42 and Lys48) and NH groups of the peptide main chain (Thr27, Lys48). In a similar way, the 3'- and 5'-phosphate groups at the bp-7 of the anti-parallel operator strand are hydrogen-bonded by side-chain (Thr40, His44) and main-chain contacts (Glu37N of the HTH turn). In this manner hydrogen bonds to phosphate groups determine the region of each distorted half-operator recognized by TetR.

The kink is further stabilized by sequence-specific hydrogen bonds between both purine bases G2 and A3 and the side chains of Arg28 and Gln38, respectively. In the bent half-operator, all residues of helix α3 contribute to sequence-specific oligonucleotide recognition, except Leu41, which is part of the hydrophobic core stabilizing the three-helix-bundle. Unexpectedly, the recognition helix α3 (residues 38 to 44) is about one turn shorter compared to other prokaryotic regulatory proteins [2]. Intensive operator interaction of α3 is indicated by N-terminal distortion to form a 3_{10}-helical turn. TetR/*tetO* interactions of this short α3 are supported by residues which are not in the HTH sequence, but sterically close to the binding motif. These hydrogen bonds are formed between *tetO* phosphate groups and main-chain peptide N-H groups and side chains of amino acid residues (Thr26, Lys48).

Allosteric transitions induced by tetracycline binding

The distance and relative orientation of the HTH-motifs differ between the induced and non-induced status of TetR, because the center-to-center separation of the recognition helices α3, α3' increases from 36.6 Å in the operator complex to 39.6 Å after inducer-binding. Obviously, a sequence of structural changes is necessary to transfer the induction signal to the DNA-binding domain, because the $[MgTc]^+$-binding sites are at least 33 Å apart from the *tetO*-binding interface. Specific events of the allosteric pathway can be deduced by comparison of the DNA-bound TetR and the induced form, the TetR/$[MgTc]^+$-complex. After passing the entrance of the binding tunnel of each TetR monomer, specific $[MgTc]^+$-recognition triggers conformational changes, which result in the reorientation of α4, required for a changed affinity of TetR for the operator-DNA.

At the far end of the binding tunnel Tc-ring A is anchored by hydrogen bonds between its functional groups and the side chains of His64 (C-terminus of α4), Asn82, Phe86 (both on α5) and Gln116 (α7) (Figs 4, 5). This region can be assumed as being the first specific recognition target of TetR, because these amino acids remain in identical position after inducer-binding [36]. In contrast, the Tc-chelated Mg^{2+}-ion displaces His100 and Thr103 of the short helix α6 (residues 96 to 102) by 1.9 and 3.9 Å, respectively. This is facilitated by a shift of helix α6 in its C-terminal direction by 1.5 Å and a 'peeling off' of its C-terminal turn to form a type II β-turn (residues 100 to 103).

This induced fit identifies the binding site of Tc-chelated Mg^{2+} as the initiation point for conformational rearrangements associated with induction. Helix α6 (with residues Val99, Thr103) is in van-der-Waals contact with helix α4 (with residues Leu52, Ala56); these helices intersect at an angle of about –130° (globin fold). The translation of α6 forces the central part of α4 to shift in the same direction, because the arising void at the hydrophobic contact surface of these two helices has to be filled. Because helix α4 is C-terminal fixed to the regulatory domain with His 64 anchored to Tc, the N-terminus swings

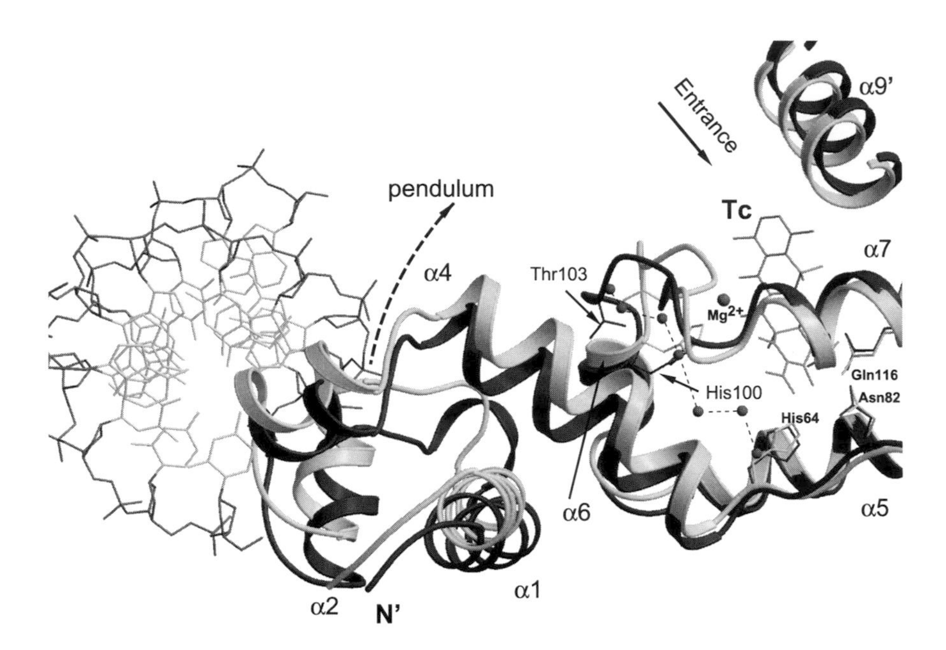

Figure 4. Comparison of TetR in induced (light gray) and DNA-bound form (dark gray). View down the operator axis, after 90° rotation around the horizontal axis with respect to Fig. 3. For clarity, only helices α1 to α8 of one monomer and α9' of the dyad-related monomer are shown, Tc is the 7-chlortetracycline [2]. Gray spheres connected by stippled lines represent the water zipper formed between α4 and the loop α6⌒α7. Induction is triggered by $[MgTc]^+$-binding, causing a shift of helix α6 and leading to a 5° rotation of helix α4. This pendulum motion with the swivel approximately at His 64 translates the DNA-binding domain with helix α3 along the major groove of *tetO* and increases the distance between helices α3 and α3' abolishing the affinity of TetR to *tetO*. Figure adapted from [3].

in a pendulum-like motion by about 5° (Fig. 4). All these events obey the constraint of the two-fold rotational symmetry inherent in the TetR homodimer. As a consequence of the hinge-like rotations of α4 and α4', the respective N-terminal DNA-binding domains are shifted apart, increasing the separation of the recognition helices α3, α3' by 3 Å. The shift of the recognition helices along the major groove disrupts the contacts between the DNA-binding domain and the respective half-operator, causing dissociation of the TetR/*tetO*-complex.

Stabilization of theTetR/$[MgTc]^+$ complex

The induced conformation of TetR is stabilized by an extended network of hydrogen bonds to prevent operator binding. The octahedral coordination sphere of the Tc-chelated Mg^{2+}-ion is completed by three meridional water ligands and the imidazole of His100. The carboxylate group of Glu147' (on helix α8' of the scaffold) forms hydrogen bonds to two Mg^{2+}-coordinating water molecules and to Gly102N of the induced β-turn at the C-terminus of helix α6.

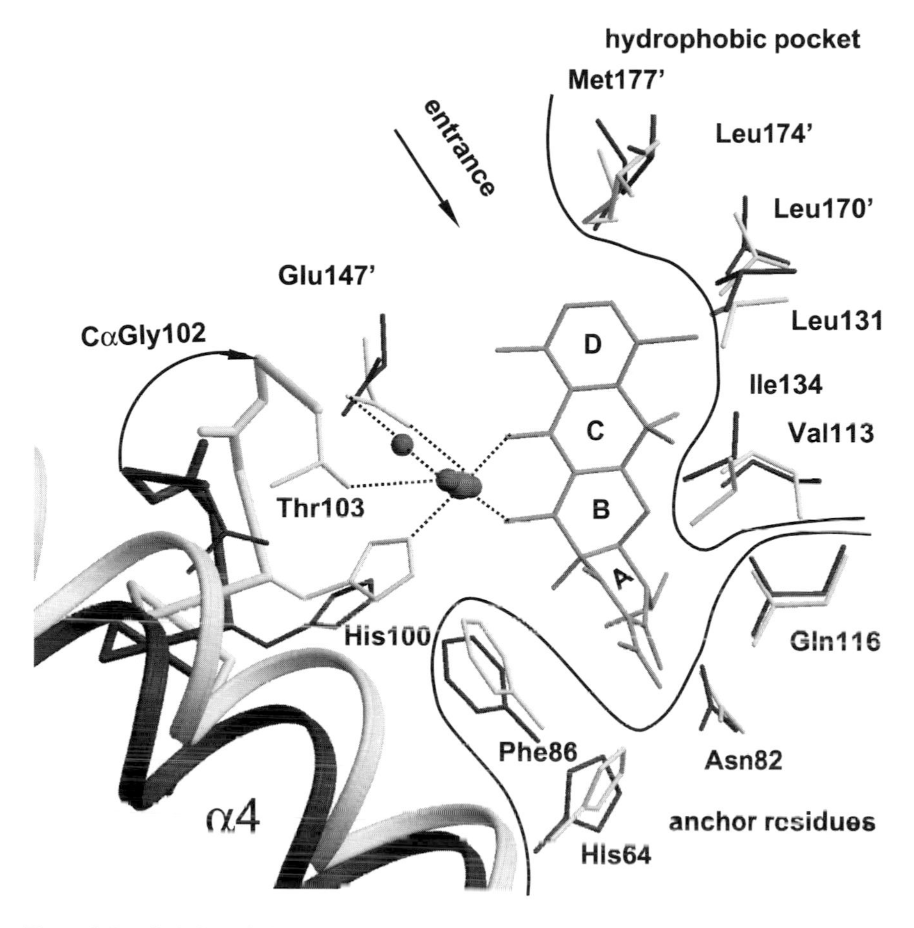

Figure 5. Detailed view of Fig. 4 at the inducer binding site. In the induced TetR, the Tc framework is anchored via ring A by hydrogen bonding to residues His 64, Asn 82, Phe 86 and Gln 116. The hydrophobic part of Tc is recognized by side chains of helices α7, α8 and α9'. Coordination of Mg^{2+} triggers translation of helix α6 and unwinding of its C-terminal turn, shifting Gly102 and Thr103 by 4.8 and 3.9 Å, respectively. His100Nε coordinates directly to Mg^{2+}, whereas Thr103Oγ is bound to Mg^{2+} via a water ligand. Figure adapted from [3].

Thr103 is part of the new β-turn and stabilizes with its hydroxyl group the β-turn with an additional hydrogen bond to the carbonyl oxygen of His100 and to the third aqua-ligand of Mg^{2+}. The coordination of Mg^{2+} is essential for the mechanism of induction, because in metal-free Tc-complexes of TetR, helix α6 remains unchanged [24].

The formation of the new β-turn at the C-terminus of helix α6 requires displacement of all residues of the loop connection between helices α6 and α7 (up to 4.8 Å for residues 102 to 106). Adjacent to this β-turn, the loop with residues 104–106 is reoriented to form a part of the hydrophobic pocket by methylene side chains of Arg104 and Pro105 for the Tc-ring D. The hydrophobic pocket for Tc is completed by the helical repeat of helix α9' with residues Leu170', Leu174' and Met177'. Helix α9' is laterally shifted, closing the entrance of the binding tunnel like a "sliding door" and finally placing its C-terminal Asp178' in an appropriate position to form a salt bridge to Arg104.

This closing mechanism identified the opening at helix α9 as the entrance of the binding tunnel. The other opening is not impaired by the inducer-binding process and remains rigid, because no structural differences in all TetR states are observed. The inserting $[MgTc]^+$ displaces approximately 20 water molecules in the binding tunnel, indicating a remarkable contribution to the entropy of the system. The narrow shape of the binding tunnel suggests that this opening at the far end is required for water release and was termed "open backdoor" [36].

A chain of water molecules along α4 links the inducer-binding site to the DNA-binding domain – the water zipper

A cooperative water-mediated hydrogen-bonding network links the $[MgTc]^+$-binding site to helix α4 and tightens the N-terminal DNA-binding domain in position, preventing binding to the major groove of *tetO*; see Figure 6.

The carbonyl oxygen Oε of the amide side chain of Gln109 (N-terminal turn of helix α7) forms hydrogen bonds both to the peptide backbone NH106 and, via a water molecule, to one aqua-ligand of the Mg^{2+}-coordination sphere. Gln109Nε is part of a zigzag-chain with 8 water molecules hydrogen-bonded to peptide carbonyl oxygens 53, 56, 57, 63 and 64 of helix α4. This hydrogen-bonding network fixes the $[MgTc]^+$-binding site to α4. In the inducer-free TetR structures, this water network is not observed because several anchor points associated with $[MgTc]^+$ are missing. The inducer-free structure is more open/loose with higher flexibility around the empty binding pocket. Therefore, the chain of water molecules along helix α4 was termed "water zipper" [36].

How to explain the reverse TetR?

Using random mutagenesis, a mutant of the class B TetR was identified which shows increased, instead of decreased, affinity to *tetO* upon inducer/effector binding. This reverse TetR is commonly used to regulate gene expression in higher eukaryotes [5]. The reverse phenotype depends on mutations which restrict the repressor to a non-inducible conformation (G102D, L101S) and on mutations which lock the DNA-binding domains in the position necessary for operator-binding (D95N, G102D).

Mutations at position Gly102 are not inducible [28]. Any variation of this residue causes sterical hindrance and interferes directly with the formation of the β-turn next to α6 [36]. This is supported by the L101S exchange, which probably stabilizes the C-terminus of α6, preventing β-turn formation.

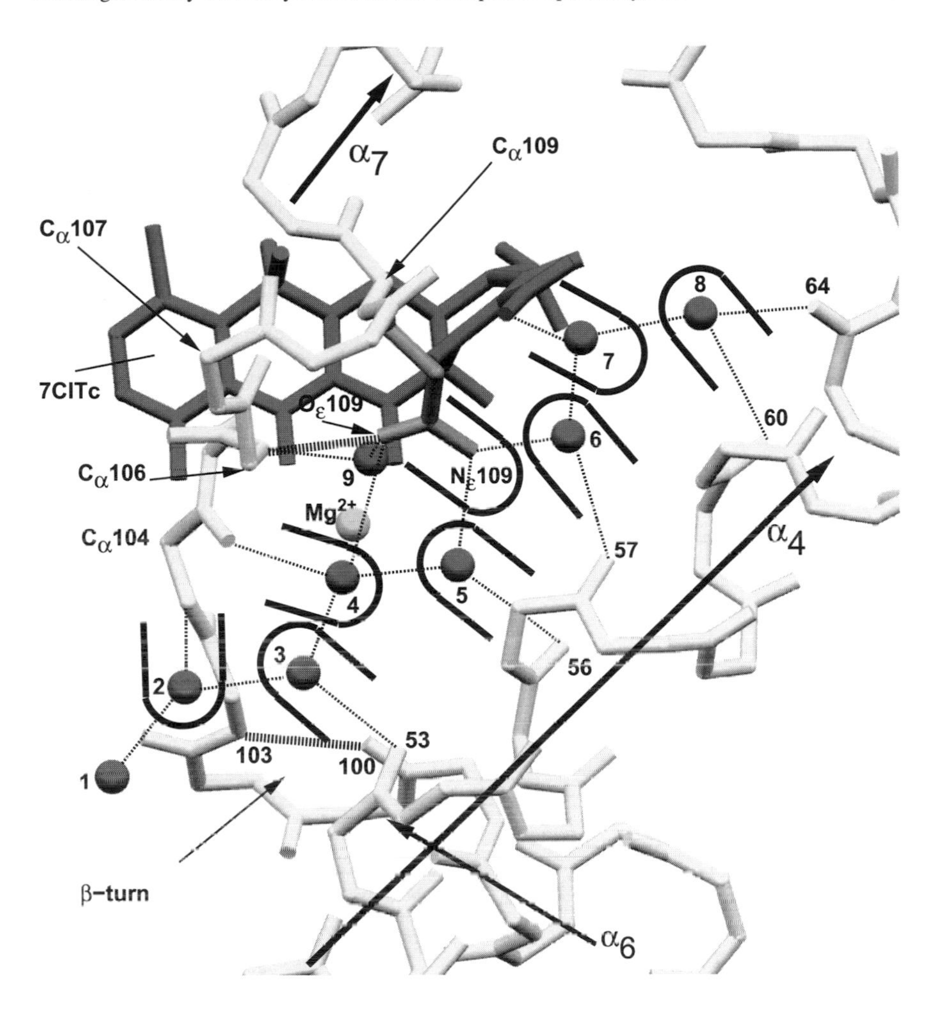

Figure 6. The water zipper. In the induced structure, the DNA-binding domains are tightly locked in a conformation unable to bind to the operator DNA. This rigidity is supported by an extended network of hydrogen bonds (dotted lines) which stabilize the positions of α4 to α6 after induction. The network is becoming possible only after the loop segment connecting α6 and α7 moves into position following $[MgTc]^+$-recognition and induction, which starts closing the indicated "water zipper" with β-turn formation of residues His100–Thr103. A zigzag-chain of water molecules (dark spheres, marked by U-shapes) forms, on one side, hydrogen bonds to carbonyl oxygens (53, 56, 57, 60 and 64) of α4. On the opposite side, this water chain forms hydrogen bonds to the amide of Tc at ring A via water 7, peptide NH104 (water 2) and O104 (water 4). The side chain of Gln109 has a central position in this network through Gln109Nε, bridging the water molecules 5 and 6. Gln109Oε forms hydrogen bonds to NH106 and water 9, which is in turn hydrogen-bonded to a water ligand of Mg^{2+} (not shown). Figure adapted from [36].

Tc-binding in the absence of β-turn formation does not induce TetR. This is shown by Mg^{2+}-free TetR/Tc complexes, which are in the non-induced conformation despite the bound Tc [24].

The substitutions D95N and G102D are located at both ends of α6 and are in conserved positions within all TetR classes. Both substitutions stabilize the orientation of the DNA-binding domain required for *tetO*-binding; see Figure 7. The side chain of D102 is in a position to form a salt bridge to Arg49 at the N-terminus of α4, fixing the relative orientation of helices α4 and α6. The D95N exchange allows a hydrogen bond to the C-terminus of α1 (Glu23).

The most potent effectors for the reverse TetR are Tc analogs, which lack the hydroxyl group at position 6 (Fig. 1) [5]. This hydroxyl group is in unfavorable contact with the hydrophobic side chain of Val113 and modifications at position 6 enhance Tc binding to the hydrophobic part of the binding tunnel and presumably stabilize the protein core [23, 27] supported by entropy effects, because each Tc replaces about 20 water molecules in the binding tunnels [36].

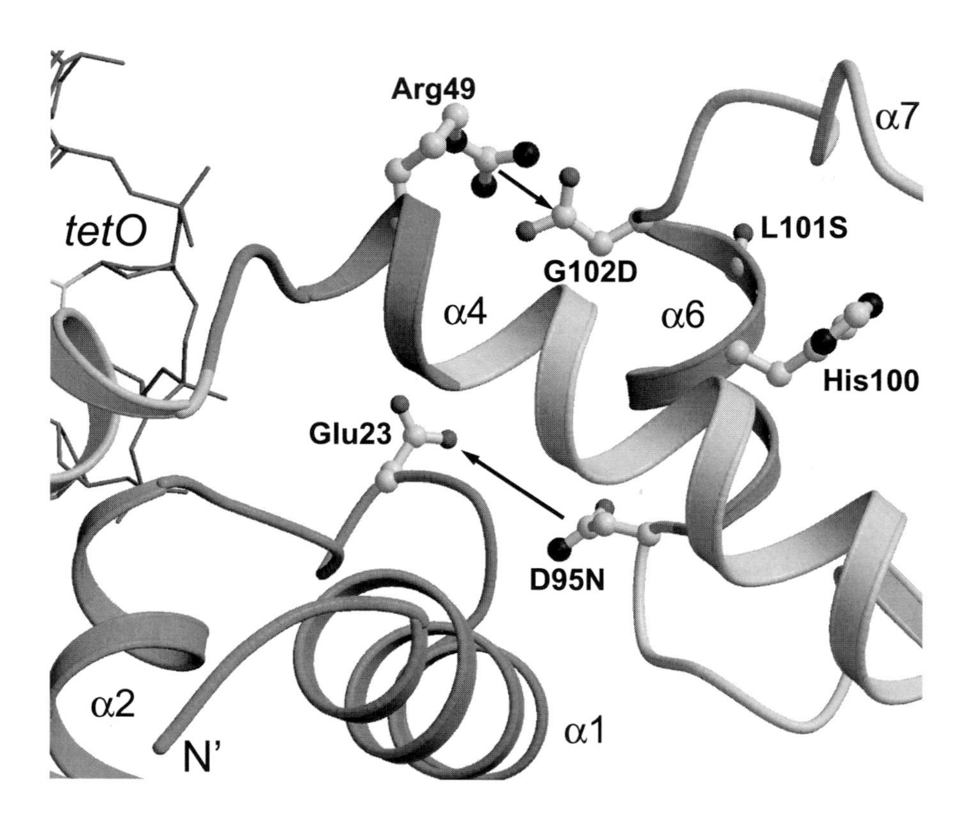

Figure 7. Same view as Figures 4, 5. Modeling study of the reverse TetR using the TetR/*tetO* structure [3]. The amino acid residues of mutated positions at both ends of α6 define the relative orientation of the DNA-binding domain and α4 in the *tetO*-binding orientation. The exchange G102D allows formation of a salt bridge to Arg 49 at the N-terminus of α4. The N-terminal part (D95N) of α6 is in position to form a hydrogen bond to Glu23 (adjacent to α1), stabilizing the orientation of the DNA-binding domain. Figure adapted from [3].

Acknowledgments
Synchrotron beam time allocations and support at the EMBL-outstation at DESY, Hamburg, LURE, *Paris*-Orsay, and the SRS, Daresbury, are gratefully acknowledged. The Deutsche Forschungsgemeinschaft and the Fonds der Chemischen Industrie gave long-term support to the project. Particular thanks are to Wolfram Saenger (Free University of Berlin) for providing the possibility to develop the TetR-project in his laboratory. This is based on a fruitful collaboration with the group of Wolfgang Hillen (University of Erlangen-Nürnberg). Claudia Alings, Heike Roscher, the Ph.D. students Martina Düvel, Caroline Kisker, Alexander Müller and Peter Orth working on the TetR project, contributed major efforts.

References

1 Hinrichs W, Kisker C, Düvel M, Müller A, Tovar K, Hillen W, Saenger W (1994) Antibiotic resistance: Structure of the Tet repressor-tetracycline complex and mechanism of induction. *Science* 264: 418–420
2 Kisker C, Hinrichs W, Tovar K, Hillen W, Saenger W (1995) The complex formed between Tet repressor and tetracycline-Mg^{2+} reveals mechanism of antibiotic resistance. *J Mol Biol* 247: 260–280
3 Orth P, Schnappinger D, Hillen W, Saenger W, Hinrichs W (2000) Structural basis of gene regulation by the tetracycline inducible Tet repressor-operator system. *Nat Struct Biol* 7: 215–219
4 Speer BS, Shoemaker NB, Salyers AA (1992) Bacterial resistance to tetracycline: mechanisms, transfer, and clinical significance. *Clin Microbiol Rev* 5: 387–399
5 Gossen M, Freundlieb S, Bender G, Müller G, Hillen W, Bujard H (1995) Transcriptional activation by tetracyclines in mammalian cells. *Science* 268: 1766–1769
6 Shockett PE, Schatz DG (1996) Diverse strategies for tetracycline-regulated inducible gene expression. *Proc Natl Acad Sci USA* 93: 5173–5176
7 Freundlieb S, Baron U, Bonin AL, Gossen M, Bujard H (1997) Use of tetracycline-controlled gene expression systems to study mammalian cell cycle. *Methods Enzymol* 283: 159–173
8 Förster K, Helbl V, Lederer T, Urlinger S, Wittenberg N, Hillen W (1999) Tetracycline-inducible expression systems with reduced basal activity in mammalian cells. *Nucl Acid Res* 27: 708–710
9 Baron U, Schnappinger D, Gossen M, Hillen W, Bujard H (1999) Generation of conditional mutants in higher eukaryotes by switching between the expression of two genes. *Proc Natl Acad Sci USA* 96: 1013–1018
10 Blau HM, Rossi FM (1999) Tet B or not tet B: advances in tetracycline-inducible gene expression. *Proc Natl Acad Sci USA* 96: 797–799
11 McMurry L, Petrucci RE, Levy SB (1980) Active efflux of tetracycline encoded by four genetically different tetracycline resistance determinants in *Escherichia coli*. *Proc Natl Acad Sci USA* 77: 3974–3977
12 Yamaguchi A, Udagawa T, Sawai T (1990) Transport of divalent cations with tetracycline as mediated by the transposon Tn*10*-encoded tetracycline resistance protein. *J Biol Chem* 265: 4089–4813
13 Bertrand KP, Postle K, Wray LV Jr, Reznikoff WS (1983) Overlapping divergent promoters control expression of Tn*10* tetracycline resistance. *Gene* 23: 149–56
14 Levy SB (1984) Resistance to Tetracyclines. *In*: LE Bryan (ed.): *Antimicrobial Drug Resistance.* Academic Press, New York, pp 191–204
15 Levy SB (1988) Tetracycline resistance determinants are widespread. *Amer Soc Microbiol News* 54: 418–421
16 Hillen W, Berens C (1994) Mechanisms underlying expression of Tn*10* encoded tetracycline resistance. *Annu Rev Microbiol* 48: 345–369
17 Schnappinger D, Hillen W (1996) Tetracyclines: antibiotic action, uptake, and resistance mechanisms. *Arch Microbiol* 165: 359–369
18 Epe B, Woolley P (1984) The binding of 6-demethylchlortetracycline to 70S, 50S and 30S ribosomal particles: a quantitative study by fluorescence anisotropy. *EMBO J* 3: 121–6
19 Eckert B, Beck CF (1989) Overproduction of transposon Tn*10*-encoded tetracycline resistance protein results in cell death and loss of membrane potential. *J Bacteriol* 171: 3557–3559
20 Takahashi M, Altschmied L, Hillen W (1986) Kinetic and equilibrium characterization of the Tet-repressor-tetracycline complex by fluorescence measurements; evidence for divalent ion requirements and energy transfer. *J Mol Biol* 187: 341–348

21 Jogun KH, Stezowski JJ (1976) Coordination and conformational aspects of oxytetracycline metal ion complexation. *J Amer Chem Soc* 98: 6018–6026
22 Ettner N, Metzger JW, Lederer T, Hulmes JD, Kisker C, Hinrichs W, Ellestad GA, Hillen W (1995) Proximity mapping of the Tet repressor-tetracycline-Fe^{2+} complex by hydrogen peroxide mediated protein cleavage. *Biochemistry* 22: 22–31
23 Orth P, Schnappinger D, Sum PE, Ellestad GA, Hillen W, Saenger W, Hinrichs W (1999) Crystal structure of Tet repressor in complex with a novel tetracycline, 9-(N,N-dimethylglycylamido)-6-demethyl-6-deoxy-tetracycline. *J Mol Biol* 285: 455–461
24 Orth P, Saenger W, Hinrichs W (1999) Tetracycline chelated Mg^{2+}-ion initiates helix unwinding for Tet repressor induction. *Biochemistry* 38: 191–198
25 Hlavka JJ, Boothe JH (eds) (1985) *Handbook of Experimental Pharmacology, The Tetracyclines.* Springer-Verlag, Berlin-Heidelberg, 332–334
26 Sum PE, Lee VJ, Testa RT, Hlavka JJ, Ellestad GA, Bloom JD, Gluzman Y, Tally FP (1994) Glycylcyclines. 1. A new generation of potent antibacterial agents through modification of 9-aminotetracyclines. *J Med Chem* 37: 184–188
27 Degenkolb J, Takahashi M, Ellestad GA, Hillen W (1991) Structural requirements of tetracycline-tet repressor interaction: determination of equilibrium binding constants for tetracycline analogs with tet repressor. *Antimicrob Agents Chemother* 35: 1591–1595
28 Müller G, Hecht B, Helbl V, Hinrichs W, Saenger W, Hillen W (1995) Characterisation of non-inducible Tet repressor mutants suggests conformational changes necessary for induction. *Nat Struct Biol* 2: 693–703
29 Matthews KS, Falcon CM, Swint-Kruse L (2000) Relieving repression. *Nat Struct Biol* 7: 184–187
30 Kisker C (1994) Antibiotika Resistenz: Röntgenstrukturanalyse des Tetracycline-Repressors und molekularer Mechanismus der Resistenz Regulation. Ph.D. Thesis, Freie Universität Berlin
31 Ettner N, Müller G, Berens C, Backes H, Schnappinger D, Schreppel T, Pfleiderer K, Hillen W (1996) Fast large scale purification of tetracycline repressor variants from overproducing *Escherichia coli* strains. *J Chromatogr A* 742: 95–105
32 Schnappinger D, Schubert P, Pfleiderer K, Hillen W (1998) Determinants of protein-protein recognition by four helix bundles: changing the dimerization specificity of Tet repressor. *EMBO J* 17: 535–543
33 Schnappinger D, Schubert P, Berens C, Pfleiderer K, Hillen W (1999) Solvent-exposed residues in the Tet repressor (TetR) four-helix bundle contribute to subunit recognition and dimer stability. *J Biol Chem* 274: 6405–6410
34 Brennan RG, Matthews BW (1989) The helix-turn-helix DNA binding motif. *J Biol Chem* 264: 1903–1906
35 Harrison SC (1992) A structural taxonomy of DNA-binding domains. *Nature* 353: 715–719
36 Orth P, Cordes F, Schnappinger D, Hillen W, Saenger W, Hinrichs W (1998) Conformational changes of the Tet repressor induced by tetracycline trapping. *J Mol Biol* 279: 439–447
37 Schevitz RW, Otwinowski Z, Joachimiak A, Lawson CL, Sigler PB (1985) The three-dimensional structure of trp repressor. *Nature* 317: 782–786
38 Weber IT, Steitz TA (1987) Structure of a complex of catabolite gene activator protein and cyclic AMP refined at 2.5 Å resolution. *J Mol Biol* 198: 311–326
39 Wagenhöfer M, Hansen D, Hillen W (1988) Thermal denaturation of engineered tet repressor proteins and their complexes with tet operator and tetracycline studied by temperature gradient gel electrophoresis. *Anal Biochem* 175: 422–432
40 Bennett MJ, Schlunegger MP, Eisenberg D (1995) 3D Domain swapping: A mechanism for oligomer assembly. *Protein Sci* 4: 2455–2468
41 Berens C, Schnappinger D, Hillen W (1997) The role of the variable region in Tet repressor for inducibility by tetracycline. *J Biol Chem* 272: 6936–6942
42 Lewis M, Chang G, Horton NC, Kercher MA, Pace HC, Schumacher MA, Brennan RG, Lu P (1996) Crystal structure of the lactose operon repressor and its complexes with DNA and inducer. *Science* 271: 1247–1254
43 Schumacher MA, Choi KY, Lu F, Zalkin H, Brennan RG (1995) Mechanism of corepressor-mediated specific DNA binding by the purine repressor. *Cell* 83: 147–155
44 Otwinowski Z, Schevitz RW, Zhang RG, Lawson CL, Joachimiak A, Marmorstein RQ, Luisi BF, Sigler PB (1988) Crystal structure of trp repressor/operator complex at atomic resolution. *Nature* 335: 321–329

45 Heuer C, Hillen W (1988) Tet repressor-tet operator contacts probed by operator DNA-modification interference studies. *J Mol Biol* 202: 407–415

46 Sizemore C, Wissmann A, Gülland U, Hillen W (1990) Quantitative analysis of Tn*10* Tet repressor binding to a complete set of tet operator mutants. *Nucl Acid Res* 18: 2875–2880

47 Wissmann A, Baumeister R, Muller G, Hecht B, Helbl V, Pfleiderer K, Hillen W (1991) Amino acids determining operator binding specificity in the helix-turn-helix motif of Tn*10* Tet repressor. *EMBO J* 10: 4145–4152

48 Baumeister R, Helbl V, Hillen W (1992) Contacts between Tet repressor and tet operator revealed by new recognition specificities of single amino acid replacement mutants. *J Mol Biol* 226: 1257–1270

49 Helbl V, Berens C, Hillen W (1995) Proximity probing of Tet repressor to tet operator by dimethylsulfate reveals protected and accessible functions for each recognized base-pair in the major groove. *J Mol Biol* 245: 538–548

50 Schwabe JW (1997) The role of water in protein-DNA interactions. *Curr Opin Struct Biol* 7: 126–134

Tetracyclines in Biology, Chemistry and Medicine
ed. by M. Nelson, W. Hillen and R.A. Greenwald

Regulation of gene expression in yeast and plants by tetracycline-dependent regulatory systems

Elisabeth Pook

Lehrstuhl für Mikrobiologie, Institut für Mikrobiologie, Biochemie und Genetik, Friedrich-Alexander-Universität Erlangen-Nürnberg, Staudtstr. 5, D-91058 Erlangen, Germany

Introduction

Promoters that allow conditional regulation of gene expression have become essential tools for the study of gene function in eukaryotes. Studies relying on comparison of genetically altered and wild-type organisms can lead to interpretation of differences not directly connected with the investigated gene. Regulatable promoters allow comparison of genetically identical organisms, in which expression of one gene may be turned on or off. Varying the transcription levels of a gene is also important for expression of toxic gene products or essential genes leading to a lethal phenotype when knocked out.

Several regulatable gene expression systems developed for use in eukaryotic cells are based on endogenous regulators and promoters that respond to inducing agents. A general disadvantage of these systems are their pleiotropic effects since inducers such as heavy metal ions or substances involved in cell metabolism or regulation processes are used. These effects impair the straightforward interpretation of results.

The features of an ideal inducible gene expression system are an inert inducer that does not interfere with other endogenous cellular processes and a specific regulator that binds its target sequence and regulates exclusively at this promoter. Dose-dependence for quantitative control and reversibility are further claims of a regulatable expression system.

Employment of heterologous regulators generally provides a greater specificity than is achieved with endogenous regulatory proteins, because they target only those genes that have been engineered to contain their binding site. Bacterial regulators are therefore suitable tools for regulation in eukaryotes. Because of the evolutionary distance between eubacteria and eukaryotes, the probability of interference of a bacterial regulator with other processes in eukaryotic cells is minute.

The tet regulatory system

The Tn*10* encoded Tet repressor is an inducible bacterial regulator of transcription. It regulates the expression of tc resistance in Gram-negative bacteria by binding to two *tet*-operators (*tetO*) in dependence on the presence or absence of the effector tc. Tight repression in the absence of the antibiotic and high sensitivity for the inducer tc mark this system. The inducer tc is a small effector molecule that shows high membrane permeability. Due to the high affinity of TetR for tc ($K_A = 3*10^9\ M^{-1}$) low, non-antibiotic doses of tc or its derivative doxycycline (dox) can be used, which do not show any discernible effect on host cells. TetR was shown to function as an effective regulator in different organisms.

The regulation principle of activation is used in the tc-controlled system described by Bujard and Gossen [1]. The system consists of a tet-regulatable promoter and the Tet-transactivator (tTA), the activity of which is controlled by tc. By fusion to the viral activation domain VP16, the function of the repressor TetR was changed to an activator that controls transcription at the corresponding tc-controllable promoters containing *tetO* sequences (see chapter by Gossen & Bujard).

The tc-regulatable system has become widely used in mammalian cells. Variants of the primary system based on the tTA improved its applicability: While the tTA-based system is tc-repressible, the development of the reverse tTA (rtTA) resulted in a tc-inducible system [2]. rtTA binds in the presence of tc to the $tetO_7$-containing promoter, but not in the absence of inducer. Therefore rtTA allows faster induction of gene expression than tTA and is more suitable for approaches that require cell growth under repressed conditions.

In order to reduce background activity of the tTA-based regulation systems, tc-regulatable trans silencers (tTS) were developed. The tTS carry silencer domains instead of activation domains fused to TetR. By combination of tTS and rtTA tc-regulatable systems were created, which are actively repressed by tTS in the absence of tc and activated by rtTA upon addition of tc [3, 4].

The suitable features of the Tet regulatory system and its continuous optimization converted it into one of the most powerful tools for modifiable gene expression. In addition to mammalian cells, the tet-system works effectively in many different organisms, like plants, yeast and protozoa. This article will give an overview of the progress of tc-regulatable systems in plants and yeast and thereby demonstrate its broad applicability.

Tc-dependent regulation in yeast

Yeast cells are frequently used as experimental systems to investigate eukaryotic cell biology, because they resemble higher eukaryotes in many aspects and are easily genetically manipulated. Yeast also plays an important role in technical processes for expression of higher eukaryotic genes. Finally, the complete sequencing of the *S. cerevisiae* genome reveals a large number of new genes with unknown function that are studied by conditional expression of their gene products. For these reasons the use of regulatable promoters for gene expression in yeast is the subject of many investigations.

Tc-regulatable gene expression systems have been employed for *Schizosaccharomyces pombe*, *Saccharomyces cerevisiae* and *Candida glabrata*. In these systems different TetR-based regulators are used for repression as well as for activation of transcription. The corresponding regulatable promoters are adapted to the requirements of the respective organisms.

Repression of transcription by TetR

The bacterial Tet repressor emerged as an effective regulator in *S. pombe* and *S. cerevisiae.* In *S. pombe* the *tet* regulatory elements were used in combination with the strong plant cauliflower mosaic virus (CaMV) 35S promoter to generate an inducible heterologous gene expression system. The system had originally been developed for gene regulation in tobacco plants [5]. It carries three *tetO* regions, one upstream and two downstream of the TATA box. Stringent repression was obtained by binding of TetR to the promoter region. Induction levels of up to 400-fold were obtained after addition of tc.

In *S. cerevisiae* TetR regulates pol III-driven transcription of the $tRNA^{Glu}$ suppressor gene up to 50-fold [6]. Insertion of only one *tetO* at –7 upstream of the suppressor gene generated a TetR-regulatable promoter. It was suggested that repression in this system is due to an interaction of TetR with RNA pol III in a way that it is arrested or inactivated. Another suggestion was the masking of binding sites for important transcription factors. Precise localization of *tetO* seems to be essential for successful regulation, as another construct in which *tetO* was inserted in –46 upstream of the suppressor gene failed to regulate transcription. A similar result had been obtained with the CaMV 35S promoter in plants, where repression was only seen when *tetO* was inserted less than 5 bp upstream of the TATA-box [7].

These examples show that the bacterial TetR works as a repressor of transcription in yeast at class II as well as at class III promoters.

Regulation of transcription by tc-dependent transregulators

The principle of transcriptional activation is more suitable for regulation in eukaryotes, than repression by a bacterial repressor, because only partial occupancy of binding sites by an activator can lead to efficient transcriptional activation, but repression is only achieved by complete occupation of binding sites by a repressor. In addition, higher cellular concentrations of repressors are needed in comparison with activators, because they have to compete with endogenous transcription factors for binding to the promoter.

Based on the observation that TetR is active in *S. cerevisiae* [6], a tTA-controlled regulation system was developed for this organism by Gari et al. [8]. The original (*tetO*$_7$)hCMV TATA promoter, established for mammalian cells (see chapter by Gossen & Bujard), seemed to be non-functional in *S. cerevisiae.* Substitution of the TATA region by the endogenous *CYC1* TATA region resulted in a promoter which allowed gene expression to be upregulated 200-fold by tTA in dependence on dox. Expression of the Tet regulatory system on a multicopy vector led to 1000-fold induction, which is similar to the regulation efficiency of the commonly used *GAL1-10* promoters.

For a better understanding of the function of *tetO*-containing promoters, two additional constructs containing only one (*tetO*) or two (*tetO*$_2$) *tet*-operator regions were compared. Both promoter constructs yielded only half of the induction levels compared to the *tetO*$_7$ promoter. These data show that there is no linear correlation between the number of *tet*-operators and the induction factor.

Insertion of one or two lambda cI linkers between the TetR and VP16 moieties (tTA* and tTA**) resulted in a significant increase of activation in dependence on the number of inserted linkers. The price of increased induction factors obtained by overexpression of tTA, or by the use of tTA**, was a slight reduction in growth rates of the yeast cultures. This negative effect was reversed by the addition of dox to the growth medium. Since it had been shown previously that high expression of TetR itself does not influence growth of *S. cerevisiae* [6], this phenomenon must be due to a negative effect of the VP16 domain expressed in high amounts or in combination with a long linker.

Regardless of the negative effects observed by overexpression of tTA, the episomal *tetO*-directed expression system was successfully used to establish a yeast screen for isolation of tTA mutants with improved properties [9] and for overexpression and characterization of unknown yeast genes [10].

An alternative tc-regulatable system was developed by Nagahashi et al. [11]. tTAs were created by fusing TetR to the transactivation domain of GAL4 or HAP4, respectively. TetR-HPA4 was the more efficient transactivator. The tc-responsive promoter contained the original Tn*10*-borne *tetO* region fused to the *HOP1* promoter of *S. cerevisiae*. Different promoter constructs with or without the original UAS and URS sequences between the TATA region and *tetO* were compared for transcriptional activity in the presence of TetR-GAL4. UAS and URS had a negative effect on transcriptional activation, which could be due to the greater distance between *tetO* and the TATA region in the corresponding constructs. The system combining *tetO-HOP1* promoter and TetR-HAP4 allowed regulation of gene expression up to 600-fold. TetR-GAL4 also worked efficiently in *C. glabrata* [12]. Functionality was shown by expression of the chromosomal *HIS3* gene under control of the *tetO-HOP1* promoter in the presence of TetR-GAL4. Addition of tc impaired cell growth on histidine-depleted medium.

Ways to create tc-activatable regulation systems

Addition of tc or dox quickly turns off transcription of the tTA-based expression system, whereas time-consuming dilution of the inducer is necessary to turn on transcription. tTA is improper for long-term repression or rapid transcriptional activation. rtTA (see chapter by Gossen & Bujard), which activates transcription in the presence of inducer, is the more suitable regulator for such approaches. Belli et al. [13] used the original rtTA for gene regulation in *S. cerevisiae*. Unfortunately, a high level of transcriptional activation in the absence of dox caused the low induction factor of seven. Concomitant expression of tTS was used to reduce this background activity. In correspondence to the development of tTS in mammalian cells [14], new yeast-specific tTS were created for regulation in *S. cerevisiae*. For that purpose, the Ssn6 protein or a truncated version of Tup1 were fused C-terminally to TetR. Ssn6 and Tup1 act as general co-repressors of a wide number of genes in yeast by affecting nucleosome positioning [15]. Both constructs reduced the basal transcription of the (*$tetO_7$*)*CYC1* promoter to nearly non-detectable levels. Utilization of TetR instead of tTS resulted in higher background activity by at least two orders of magnitude compared to TetR-Ssn6. This result clearly demonstrates the superiority of the tTS over the bacterial repressor.

Based on tTA and tTS, two dual tc-regulatable systems were created, taking advantage of the opposite effect of tc on the normal and reverse TetR moieties (Fig. 1). One system, consisting of tTA and rtTS, activates transcription in the absence of inducer and represses upon addition. The other system regulates in the opposite way by rtTA and tTS. Background levels were decreased in both systems by the respective tTS construct to undetectable levels without affecting the maximal gene activation or growth rate of the cells. The dual activator/repressor system (tTA/rtTS) was successfully applied to the study of essential orphan genes in *S. cerevisiae* [16]. One-step substitution of the gene promoters by a cassette carrying a *$tetO_2$*-regulated promoter allowed conditional expression in dependence on the dox concentration.

Another way to bypass the limitations of rtTA – the background activity in the absence of dox – was the search for new rtTA alleles with optimized phenotypes [9]. In order to find such rtTAs by random mutagenesis, a screen was established in *S. cerevisiae* based on a plasmid harboring tTA and a gene encoding GFP^+ [17] under control of the (*$tetO_7$*)*CYC1* promoter [8]. Use of GFP allows identification of active tTAs or rtTAs by illumination with long-range UV-light. A mutant pool was created by PCR mutagenesis of the *tetR* moiety and cloned into the tTA gene of the yeast vector. rtTA variants with improved properties were isolated and optimized by successive mutagenesis. The new variants (S2, M1 and M2) show extremely low background activity in the absence of inducer and an increased sensitivity for dox. These results demonstrate the broad spectrum of rtTAs that can be created as well as the potential of the yeast screen, which provides the advantage of good transformation rates necessary for efficient screening of tTA mutant pools in a eukaryotic background.

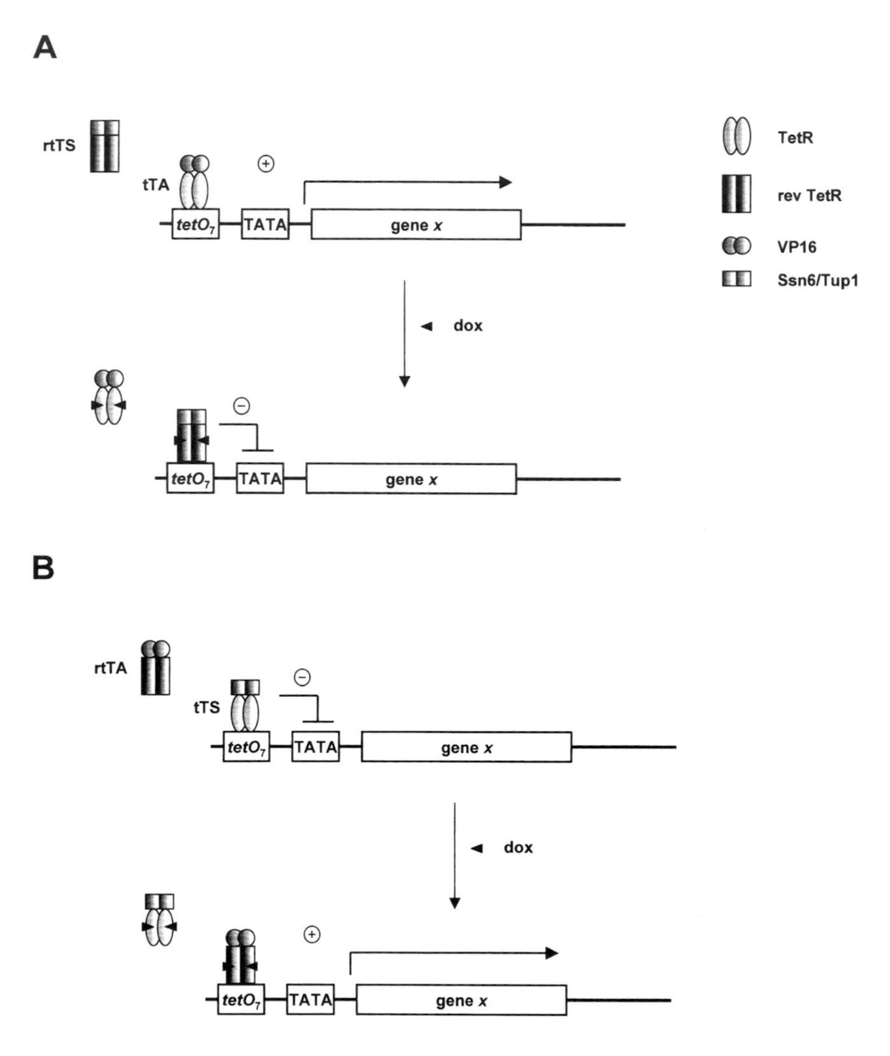

Figure 1. tc-regulatable dual systems for *S. cerevisiae*. (A) shows the tc-repressible system consisting of tTA and rtTS. In the absence of inducer tTA binds to $tetO_7$ and activates transcription. Binding of dox to both regulators leads to dissociation of tTA from the DNA and to binding of rtTS to $tetO_7$, thereby repressing transcription. (B) The tc-activatable system consists of rtTA and tTS. In the absence of inducer tTS represses transcription by binding to $tetO_7$. Addition of dox leads to dissociation of tTS from the DNA. The inducer-bound rtTA binds to $tetO_7$ and activates transcription. The domains of the regulators are indicated. The DNA is depicted as a black line, *tetO* and TATA sequences are indicated as open boxes.

The work described above clearly demonstrates the potential of TetR-based gene expression systems in yeast. The original TetR as well as the Tet-transactivators or -transsilencers were successfully used in different yeast species for gene regulation. Further optimization made the Tet regulatory system superior over other frequently used systems, because of its rapid response after induction, growth independence, and the fact that the inducer does not influence cell physiology or metabolism. In addition, the maximal induction factors of the Tet regulatory systems are as high as those obtained with GAL-based regulation, the most efficient regulation system known in yeast.

The general interest of yeasts as model organisms for the investigation of eukaryotic gene functions and as human pathogens make regulatable systems for these organisms essential for future research.

Tc-controlled gene regulation systems in plants

The first application of a tc-controlled system to regulate gene expression in eukaryotes has been carried out in plant cells. This work paved the way for subsequent employment of the tet system in other organisms. First the repressing regulation system has been established and characterized in plants and later the mammalian activating approach has been used as well.

TetR is an effective repressor in plants

The first tc-regulatable expression system was established in tobacco protoplasts [18]. The original Tet repressor-operator system was overlaid with the CaMV 35S promoter for regulation of transcription. Expression of TetR led to repression of the CaMV 35S promoter activity, which was completely abolished after addition of tc. Systematic analysis of the effect of *tetO*-bound TetR in different positions within the CaMV 35S promoter revealed the optimal arrangement of binding sites for repression [19]. Placing the *tetO* maximally 3 bp upstream or 31 bp downstream of the TATA box resulted in effective repression upon TetR binding in these positions [7, 20]. TetR seems to sterically interfere with binding of transcription factors to the TATA box, thereby inhibiting initiation of transcription. A CaMV 35S promoter containing two *tetO* downstream of the TATA box led to 80-fold repression in transgenic tobacco plants. Addition of a third *tetO* upstream of the TATA box resulted in the ‘Triple-Op’-promoter (see Fig. 2), yielding 500-fold repression in transgenic tobacco plants [5].

10^6 molecules of TetR were expressed per cell in tobacco plants. Repression is tight enough to generate plants that carry a gene with a toxic gene product under control of the TetR-regulated ‘Triple-Op’-promoter [21]. The same arrangement also works in potato and tomato plants, but could not be established in *Arabidopsis thaliana* [22].

Ulmasov et al. showed that pol III-driven transcription of tRNA suppressor genes can be regulated by TetR in transiently transfected carrot protoplasts [23]. Constructs carrying one or three *tetO* at different positions up- and downstream of the suppressor genes were assayed for repression by TetR. As in the case of class II promoters, repression factors strongly depend on number and

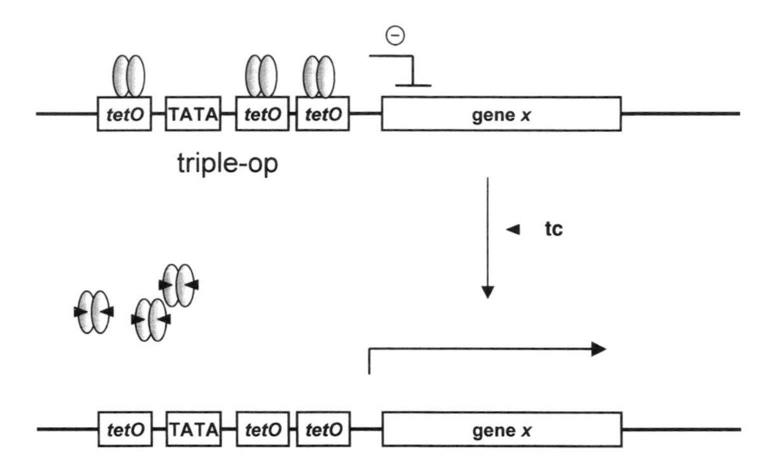

Figure 2. Schematic representation of the tc-inducible repression system. TetR is expressed constitutively and binds to three *tetO* up- and downstream of the TATA-box of the CaMV 35S-derived 'triple-op' promoter. In the absence of tc, *tetO*-bound TetR represses transcription of gene X by interference with transcription initiation. In the presence of tc, the inducer binds to TetR allowing expression of gene X. The DNA is depicted as a black line, *tetO* and TATA sequences are indicated as open boxes.

position of *tetO*. The maximum level of repression by TetR, reaching 90%, was observed with a triple *tetO* element positioned immediately upstream of the coding region of the tRNA gene. Repression was observed only at high concentrations of TetR, when the ratio of *tetR* and *tetO*-tRNA was 200:1. By incubation of the contransfected protoplasts with tc, the expression of tRNA genes was completely restored.

The tTA-based activating system

Weinmann et al. made use of the tc-dependent promoter-activating system, based on the tTA (see the chapter by Gossen & Bujard) for gene expression regulation in tobacco [24]. The target promoter for tTA was constructed by replacing the enhancer sequences of the CaMV 35S promoter by seven *tetO* and called Top10. Regulation factors of at least 700 were obtained in transgenic plants and nearly no background activity was determined after addition of tc. The repression system yielded only 500-fold induction.

tTA-Top10 mediated regulation was not stable in transgenic tobacco plants because after one year silencing of the Top10 promoter was observed. It is not clear whether the silencing is due to methylation or other suppressing effects. By removing potential methylation sites from the sequence of Top10, another promoter, designated Tax, was created [25]. Tax contains only four *tetO* sequences. This promoter was stably expressed over several generations without silencing. It showed slightly elevated background activities, but yielded 8-fold higher expression compared to Top10.

The regulation system consisting of tTA and Top10 was also established in the moss *Physcomitrella patens*, a simple model organism for higher plants [26]. In the presence of Tc negligible activity of the reporter was observed, whereas strong expression was detectable within 24 h after transfer to tc-free media.

Recently, the activating system was also established in *Arabidopsis thaliana* [27]. Plants that express tTA over at least five generations without silencing were isolated. Crosses with plants containing the Top10:GFP reporter yielded progeny in which GFP was expressed under control of tTA. The expression of GFP was tightly repressed by ten-fold lower amounts of tc compared to tobacco plants. In contrast to tobacco, silencing of Top10 does not seem to occur in moss and *Arabidopsis*.

In order to achieve inducibility by tc, rtTA [2] was introduced into tobacco and *Arabidopsis*. No regulation was obtained, although the mRNA encoding rtTA was transcribed at levels similar to that of tTA. Since no rtTA protein was detected in Western blots, it might be unstable in these plants [22]. Newly characterized rtTA alleles with improved properties in HeLa [9] also were unable to regulate plant promoters [28].

Another approach to create chemically inducible regulation in tobacco plants is described by Böhner et al. [25]. Fusion of the glucocorticoid receptor hormone-binding domain (GR HBD) to tTA rendered the regulation dependent on the presence of steroid hormones (see Fig. 3). It is suggested that GR HBD forms a complex with cellular proteins, from which it is released only in the presence of steroid hormones. Therefore, the fusion protein of tTA and GR HBD, called TGV, is functional in the presence of the inducer dexamethasone and activates transcription 150- to 520-fold from the Top10 promoter. Activation was completely downregulated by tc after 24 h. This dual regulation system shows lower regulation factors than obtained with tTA, but allows on/off switching of gene expression at defined points of time.

Repression or activation in plants?

Both regulation systems, TetR-based repression and tTA-based activation, are active in tobacco plants. Regulation by TetR allows a quick induction of transcription and is thus suitable for studies that require gene expression at a defined point in time. For studies requiring a quick turn-off of gene expression, like the determination of mRNA or protein stability, the tTA-based system is preferable.

Higher regulation efficiency is obtained with tTA and background expression is also lower in the presence of tc, compared to TetR in the absence of inducer. The low background of the activating system is due to the properties of the minimal promoter and its surrounding sequence.

A disadvantage of the repressing system is the high intracellular concentration of TetR required for efficient repression, which does not seem to be toler-

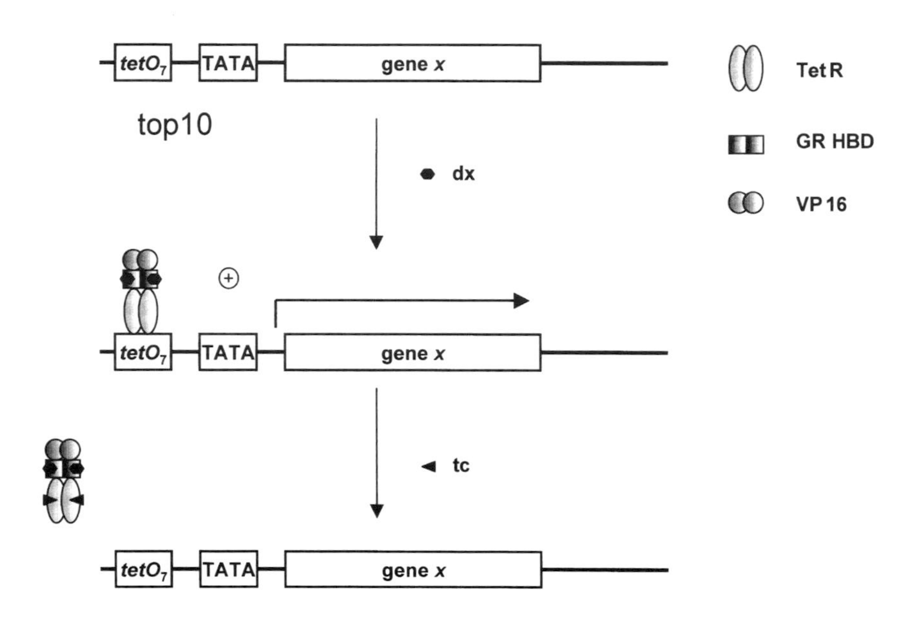

Figure 3. Schematic representation of the dx/tc dual regulation system. TGV contains TetR, GR HBD and the VP16 activation domain and is expressed constitutively. In the absence of inducer TGV is inactive and associated with cellular proteins. Binding of dexamethasone (dx) to the GR HBD moiety of TGV leads to release of the complex and allows binding to the seven *tetO* of the top10 promoter upstream of the TATA-box, thereby activating transcription by the VP16 domain. Binding of tc to the TetR moiety of TGV causes dissociation of TGV from the DNA and transcription is no longer activated. The DNA is depicted as a black line, *tetO* and TATA sequences are indicated by open boxes.

ated by several species. TetR expression in tomato caused reduced shoot dry weight and reduced leaf size [29]. To date it has not been possible to establish the TetR-based regulation system in the model organism *Arabidopsis thaliana*. In contrast, much lower concentrations of tTA are sufficient for regulation of the Top10 promoter. Tobacco plants tolerate about 50-fold less tTA than TetR [24]. Higher amounts of tTA that are obtained in transiently transformed plant cells lead to decreased transcriptional activation, probably by squelching [24].

The obvious differences in the amounts of regulatory protein needed to achieve the same effect can be rationalized by the different modes of action of TetR and tTA. One hundred percent occupancy is required for effective repression by TetR, while only 50% occupancy would be enough for activation of transcription by tTA [22]. While tTA is believed to have free access to its target sites, the repressor competes with endogenous transcription factors for binding. As a result, more TetR than tTA is needed to occupy the same number of target sites.

The limited applicability of the TetR-based regulation system, as well as the lower induction factors, clearly favor tTA for regulation in plants. Comparison of the approaches confirms the general notion that activating systems work more efficiently in higher eukaryotes than systems based on repression [22].

Application of tc and kinetics of induction

In contrast to cell cultures or animals, the inducer can be applied to plants in several ways that determine the kinetics and levels of induction. The way of tc application to whole plants is a critical parameter for efficiency and quickness of regulation. When tc is taken up in roots, it is transported throughout the whole plant without obvious effects on growth [5]. Application of tc through the roots led to the onset of protein expression after 1 day, but 10–14 days are needed for maximal induction. Alternatively tc can be applied to the leaves, from where it might be transported through the phloem into all parts of the plant. Since plants suffer from spraying with tc, this mode of application is not suitable for induction [5]. When tc is applied by vacuum infiltration into single leaves, gene expression of the inducible promoter was detected after 30 min using TetR or after 1h using tTA in combination with the corresponding promoter constructs.

Tc seems to become inactivated rather quickly in plants because the mRNA levels started to decrease 2 days after induction and reached background levels after 4 days. This may be due to the light sensitivity of tc because this is not seen when tc-treated leaves were incubated in the dark [5]. For continuous de-repression tc has to be applied at least every day.

Although the first successful use of the bacterial TetR protein to control a eukaryotic promoter was described for plants, development of new, optimized tc-controllable regulation systems proceeded much faster for application in mammalian cells. The use of tc or its more efficient derivatives has a broad potential for medical applications, such as gene therapy, but it is hard to imagine that antibiotics would be broadly used in the field. For the time being, TetR-based regulation systems are useful tools for plant gene expression in the laboratory. Nevertheless, the repressing as well as the activating Tet regulatory systems have been established in different plant species and new developments, including attempts to establish novel inducers which can be brought out in the fields, will probably make the Tet regulation system the system of choice.

References

1 Gossen M, Bujard H (1992) Tight control of gene expression in mammalian cells by tetracycline-responsive promoters. *Proc Natl Acad Sci USA* 89: 5547–5551

2 Gossen M, Freundlieb S, Bender G, Muller G, Hillen W, Bujard H (1995) Transcriptional activation by tetracyclines in mammalian cells. *Science* 268: 1766–1769
3 Forster K, Helbl V, Lederer T, Urlinger S, Wittenburg N, Hillen W (1999) Tetracycline-inducible expression systems with reduced basal activity in mammalian cells. *Nucl Acid Res* 27: 708–710
4 Freundlieb S, Schirra-Muller C, Bujard H (1999) A tetracycline controlled activation/repression system with increased potential for gene transfer into mammalian cells. *J Gene Med* 1: 4–12
5 Gatz C, Frohberg C, Wendenburg R (1992) Stringent repression and homogeneous de-repression by tetracycline of a modified CaMV 35S promoter in intact transgenic tobacco plants. *Plant J* 2: 397–404
6 Dingermann T, Frank-Stoll U, Werner H, Wissmann A, Hillen W, Jacquet M, Marschalek R (1992) RNA polymerase III catalysed transcription can be regulated in *Saccharomyces cerevisiae* by the bacterial tetracycline repressor-operator system. *EMBO J* 11: 1487–1492
7 Frohberg C, Heins L, Gatz C (1991) Characterization of the interaction of plant transcription factors using a bacterial repressor protein. *Proc Natl Acad Sci USA* 88: 10 470–10 474
8 Gari E, Piedrafita L, Aldea M, Herrero E (1997) A set of vectors with a tetracycline-regulatable promoter system for modulated gene expression in *Saccharomyces cerevisiae. Yeast* 13: 837–848
9 Urlinger S, Baron U, Thellmann M, Hasan MT, Bujard H, Hillen W (2000) Exploring the sequence space for tetracycline-dependent transcriptional activators: Novel mutations yield expanded range and sensitivity. *Proc Natl Acad Sci USA* 97: 7963–7968
10 Rodriguez-Pena JM, Cid VJ, Sanchez M, Molina M, Arroyo J, Nombela C (1998) The deletion of six ORFs of unknown function from *Saccharomyces cerevisiae* chromosome VII reveals two essential genes: YGR195w and YGR198w. *Yeast* 14: 853–860
11 Nagahashi S, Nakayama H, Hamada K, Yang H, Arisawa M, Kitada K (1997) Regulation by tetracycline of gene expression in *Saccharomyces cerevisiae. Mol Gen Genet* 255: 372–375
12 Nakayama H, Izuta M, Nagahashi S, Sihta EY, Sato Y, Yamazaki T, Arisawa M, Kitada K (1998) A controllable gene-expression system for the pathogenic fungus *Candida glabrata. Microbiology* 144: 2407–2415
13 Belli G, Gari E, Piedrafita L, Aldea M, Herrero E (1998) An activator/repressor dual system allows tight tetracycline-regulated gene expression in budding yeast [published *erratum* appears in. *Nucl Acid Res* 1998 Apr 1;26(7): following 1855]. *Nucl Acid Res* 26: 942–947
14 Deuschle U, Meyer WK, Thiesen HJ (1995) Tetracycline-reversible silencing of eukaryotic promoters. *Mol Cell Biol* 15: 1907–1914
15 Edmondson DG, Smith MM, Roth SY (1996) Repression domain of the yeast global repressor Tup1 interacts directly with histones H3 and H4. *Gene Develop* 10: 1247–1259
16 Belli G, Gari E, Aldea M, Herrero E (1998) Functional analysis of yeast essential genes using a promoter-substitution cassette and the tetracycline-regulatable dual expression system. *Yeast* 14: 1127–1138
17 Scholz O, Thiel A, Hillen W, Niederweis M (2000) Quantitative analysis of gene expression with an improved green fluorescent protein. p6. *Eur J Biochem* 267: 1565–1570
18 Gatz C, Quail P H (1988) Tn10-encoded tet repressor can regulate an operator-containing plant promoter. *Proc Natl Acad Sci USA* 85: 1394–1397
19 Gatz C, Kaiser A, Wendenburg R (1991) Regulation of a modified CaMV 35S promoter by the Tn10-encoded Tet repressor in transgenic tobacco. *Mol Gen Genet* 227: 229–237
20 Heins L, Frohberg C, Gatz C (1992) The Tn10-encoded Tet repressor blocks early but not late steps of assembly of the RNA polymerase II initiation complex *in vivo. Mol Gen Genet* 232: 328–331
21 Röder FT, Schmülling T, Gatz C (1994) Efficiency of the tetracycline-dependent gene expression system: complete suppression and efficient induction of the rolB phenotype in transgenic plants. *Mol Gen Genet* 243: 32–38
22 Gatz C (1997) Chemical control of gene expression. *Annu Rev Plant Physiol Plant Mol Biol* 48: 89–108
23 Ulmasov B, Capone J, Folk W (1997) Regulated expression of plant tRNA genes by the prokaryotic tet and lac repressors. *Plant Mol Biol* 35: 417–424
24 Weinmann P, Gossen M, Hillen W, Bujard H, Gatz C (1994) A chimeric transactivator allows tetracycline-responsive gene expression in whole plants. *Plant J* 5: 559–569
25 Bohner S, Lenk I, Rieping M, Herold M, Gatz C (1999) Technical advance: transcriptional activator TGV mediates dexamethasone-inducible and tetracycline-inactivatable gene expression. *Plant J* 19: 87–95

26 Zeidler M, Gatz C, Hartmann E, Hughes J (1996) Tetracycline-rcgulated reporter gene expression in the moss *Physcomitrella patens*. *Plant Mol Biol* 30: 199–205
27 Love J, Scott AC, Thompson WF (2000) Stringent control for transgene expression in *Arabidopsis thaliana* using the Top 10 promoter system. *Plant J* 21: 579–588
28. Gatz, C. personal communication
29 Corlett JE, Myatt SC, Thompson AJ (1996) Toxicity symptoms caused by high expression of Tet repressor in tomato are alleviated by tetracycline. *Plant Cell Environ* 447–454

Tetracyclines in Biology, Chemistry and Medicine
ed. by M. Nelson, W. Hillen and R.A. Greenwald

Tetracyclines in the control of gene expression in eukaryotes

Manfred Gossen and Hermann Bujard*

*Zentrum für Molekulare Biologie der Universität Heidelberg (ZMBH), Im Neuenheimer Feld 282, D-69120 Heidelberg, Germany

Introduction

The potential to conditionally alter the activity of individual genes in complex genetic systems and to observe accompanying phenotypic changes has provided new insights into numerous biological processes hitherto not amenable to genetic dissection. The most widely applied approach makes use of a transcriptional regulation system which allows stringent control of gene expression by tetracycline (Tc) or its derivative doxycycline (Dox). The various versions of the "Tet regulatory system" were successfully incorporated into a variety of cultured cells but, most interestingly, they were also shown to function in many biological model organisms such as *S. cerevisiae*, *Dictyostelium*, *Drosophila*, *Arabidopsis*, mice and rats. Thus they have developed into an efficient tool for the study of gene function *in vivo*. In particular for the mouse, which has become a widely used model for mammalian genetics thanks to transgenesis and embryonic stem cell technology, the advent of a generally applicable conditional gene expression system adds another level of sophistication for the dissection of gene function [1, 2]. Indeed, exploiting the Tet regulatory systems *in vivo* is beginning to provide fundamental insights into such complex biological processes as development, disease and behaviour [3–10]. Moreover, as tetracyclines are non-toxic compounds extensively used in human and animal medicine, the Tet regulatory systems may even hold promise in gene therapy.

The principle of using prokaryotic regulatory elements in a eukaryotic environment relies on the exceptional specificity to the Tet systems. Accordingly, their application is in general considered to be favourable compared to "homologous" gene expression systems which are based on control elements endogenous to the eukaryotic cell. The reasoning behind this is beyond the scope of this review, but has been dealt with previously [11]. More interesting in the context of this monograph are the reasons why Tc-regulated expression systems even outperformed other heterologous gene expression systems, which are based on DNA-binding proteins foreign to mammalian cells. As we will see, some unique virtues of the Tet-repressor as well as the pharmacological properties of tetracyclines as effector substances contribute to the success of the Tet system in its various modifications.

The principles

The initial and as of today still most widely applied version of Tc-controlled gene expression system [12] incorporates two crucial elements:

- the Tc-controlled transactivator tTA, a fusion protein comprising the N-terminally located wildtype Tet-repressor (TetR) – derived from the bacterial Tn*10* encoded Tc resistance operon – fused to the transcriptional transactivation domain of protein 16 of herpes simplex virus, VP16 (Fig. 1A, upper construct).
- a minimal (i.e., enhancerless) cytomegalovirus IE promotor fused to heptamerized tet operators (*tet*O), the binding sites of TetR. This promoter construct was originally designated P_{cmv*-1} [12] and more recently renamed P_{tet}-1 [13] (Fig. 1B, upper construct; the variants of tTA-responsive promoters are collectively referred to as P_{tet}).

When fused to a gene to be regulated, P_{tet}-1 is transcriptionally silent, provided it is inserted into a proper chromosomal locus. By contrast, in the presence of tTA which binds to the *tet*O sequences within P_{tet}-1, this hybrid promoter is strongly activated to levels often exceeding that of the intact enhancer-containing wildtype P_{cmv} [14]. This activation is abolished by Tc or various of its derivatives, in particular by Dox. These effector molecules bind to the TetR moiety of tTA and prevent its interaction with *tet*O. The resulting regulation factors measured are remarkable and reach up to 5 orders of magnitude. Comparisons between this system and other available so-called heterologous regulation systems have been discussed elsewhere [15]. Here, we would like to focus on parameters of the Tet regulatory system that make it a highly specific tool for the study of gene function in complex systems such as the mammalian cell. A first important aspect is the interaction between, e.g., Dox and TetR. The effector molecule Dox binds to the repressor protein highly efficiently (Ka ~ 10^{12} M^{-1}) allowing interference with the repressor/operator interaction at very low concentrations of the antibiotic. Most remarkably, the binding of Dox reduces the affinity of TetR by 6 orders of magnitude. The consequences for the Tet regulatory systems are that tTA activation of P_{tet}-1 is completely abolished in cell culture at Dox concentrations as low as 10 ng/ml. A second noteworthy feature is the uptake of tetracyclines by cells. It is not dependent on an active transport process; instead, tetracyclines appear to enter cells by diffusion. Consequently, their intracellular concentrations are reflected by the concentration in the medium surrounding the cell. Therefore, partial induction of P_{tet} is feasible at the single-cell level as confirmed by FACS analysis for mammalian cells [16], *S. cerevisiae* (J. H. Hegemann, personal com-

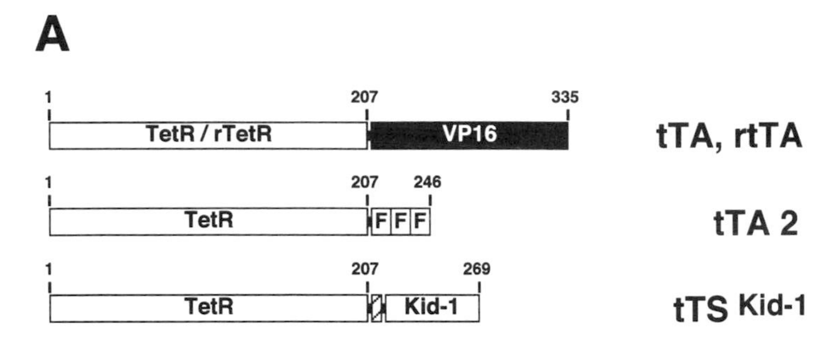

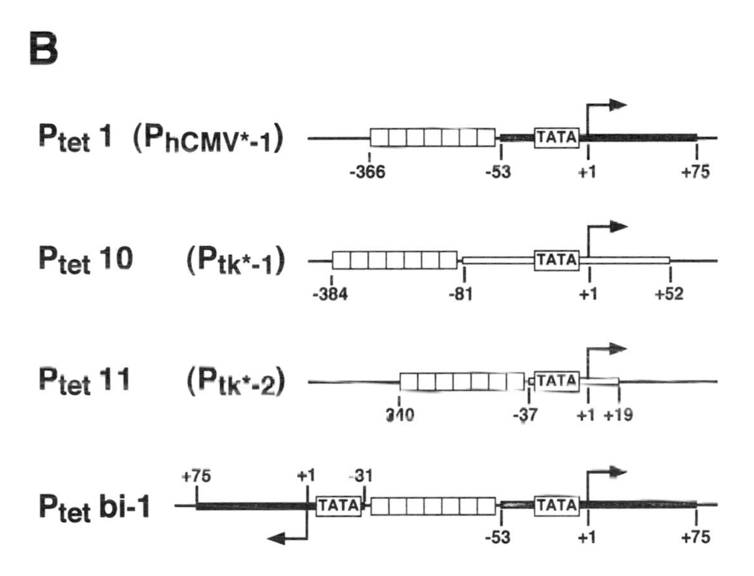

Figure 1. Tetracycline-controlled fusion proteins and their target promoters. (A) Fusions between TetR/rTetR with domains capable of either activating or silencing transcription, respectively. tTA is a fusion protein between the Tet repressor of the Tn*10* Tc operon of *E. coli* consisting of 207 amino acids and the 128 amino acids long carboxy-terminal portion of the transactivator protein VP16 from *Herpes simplex* virus. In tTA2, the VP16 moiety of tTA was replaced by three acidic minimal activation domains (F), each consisting of only 13 amino acids. tTS^{Kid-1} is a fusion between TetR and a 61 amino acid long KRAB domain, a transcriptional repression domain derived from the human kidney protein Kid-1. In tTS^{Kid-1}, the silencing moiety is connected to TetR via the nuclear localisation signal (nls) derived from the SV40 Tag. (B) tTA/rtTA-responsive promoters. These promoters are fusions between heptamerized *tet*O sequences (indicated as grey boxes) and minimal promoters derived from viral or cellular RNA polymerase II promoters. The original $P_{hCMV*-1}$ is derived from the human cytomegalovirus IE; the other two promoters (P_{tk*}–1 and P_{tk*}-2) are derivatives of the HSVtk promoter [12]. For simplicity, all tTA/rtTA-responsive promoters from our laboratory have been renamed P_{tet}-1 to x. The HSVtk derived promoter (P_{tet}-10, P_{tet}-11) show lower basal activity in transient expression experiments when compared to P_{tet}-1. However, they cannot really be activated to the level of the latter. In the bidirectional promoter P_{tet}bi-1 heptamerized *tet*O sequences are flanked by two divergently oriented hCMV derived minimal promoters. Positions spanning promoter regions and *tet* operator sequences are indicated with respect to the start site of transcription (+1). These bidirectional promoters permit coregulation of two genes.

munication) and *Dictyostelium discoideum* [17]. A heterogeneous induction of gene activity within cell populations as is sometimes observed has, thus, to be attributed to other parameters than a putative differential uptake of Dox. Third, many tetracyclines exhibit very favourable pharmacokinetic properties. While this has been known for a long time, it has recently been elegantly demonstrated, when the tTA system has been employed to measure mRNA half-life times in cell culture. These studies [18] show that transcription activation via tTA is terminated in less than 5 min upon addition of Dox to the culture medium. Finally, the specificity of TetR for its cognate operator sequence is unusually high as the binding constants for *tet*O and unspecific DNA differ by around 7 orders of magnitude. All these parameters contribute to the remarkable specificity of the Tet regulatory system which is also supported by recent expression-profiling experiments where neither Dox nor tTA had a detectable effect in cultured U2OS [19] and Raji cells (Weik, pers. communication). These favourable characteristics have led to the broad application of Tet regulation in a variety of systems.

Nevertheless, over the years many and in part significant modifications and improvements of the system have emerged. Most notable is the development of the so-called reverse Tc-controlled transactivator rtTA [20]. Compared to tTA it contains a few amino acid exchanges, which fundamentally change its binding characteristics towards the *tet*O. As rtTA requires Tc derivatives such as Dox for binding to the *tet*O sequences within P_{tet}-1, transcription from this promoter occurs only in the presence of Dox, as opposed to the initially described Tet system. This change in direction has its most striking advantages when the system is incorporated in transgenic animals as it substantially accelerates the induction of target genes (see below). A schematic outline of the action of the two basic regulatory systems that function in opposite fashion is given in Figure 2.

Only recently a new generation of rtTAs has been described [21]. They were selected in a functional screen in yeast. While showing the same basic properties as the original rtTA, they have lost any measurable residual binding affinity towards P_{tet}-1 in absence of Dox. Equally important, one of the novel transactivators, rtTA2^{S}-M2, senses Dox at considerably lower concentrations when compared to rtTA.

Several further modifications of the basic Tet systems have been described and will be discussed in the context of the paragraphs to follow.

Applying Tet regulation to cultured cells

The potential of the Tet system in stably transfected mammalian tissue culture cells is documented in hundreds of publications of which a compilation can be found in a commercial web site (http://www.clontech.com/tet/Refs/index.html). Despite its impact, this voluminous work will not be further discussed here except for some aspects related to the properties of tetracyclines in tissue cul-

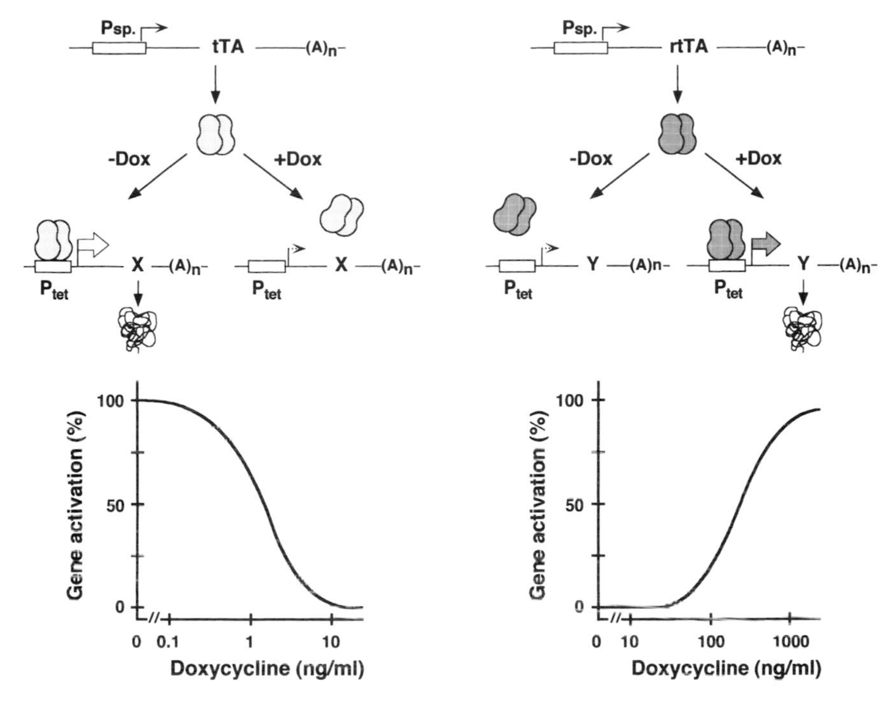

Figure 2. Schematic outline of the Tet regulatory systems. Left upper part shows the mode of action of the Tc-controlled transactivator (tTA). tTA binds in the absence of the effector molecule Dox to the *tet*O sequences within P_{tet} and activates transcription of gene x. Addition of Dox prevents tTA from binding and thus the initiation of transcription. Left lower part depicts the dose response of Dox on the tTA dependent gene expression. Gene activity is maximal in the absence of the antibiotic whereas increasing effector concentrations gradually decrease expression to background levels at concentrations >10 ng/ml. Right upper part illustrates the mechanism of action of the reverse Tc-controlled transactivator (rtTA). rtTA is identical to tTA with the exception of 4 amino acid substitutions in the TetR moiety of which 3 convey a reverse phenotype to the binding properties of TetR: rtTA requires Dox for binding to *tet*O sequences within P_{tet} in order to activate transcription of gene y. Right lower part outlines the dose response of Dox on the rtTA dependent transcription activation. By increasing the effector concentration beyond 50 ng/ml Dox, rtTA dependent gene expression is gradually stimulated.

tures. Toxicity is clearly a concern for substances commonly known as antibiotics. However, tetracyclines in general show a superb specificity for prokaryotes, where in most cases the ribosome is the target of their antibiotic action (see chapter by Berens) while higher concentrations are required for interference in eukaryotic cells. For example, only concentrations exceeding 10 µg/ml of Tc show an effect on the growth rate of HeLa cells, whereas 30 ng/ml are sufficient to completely downregulate the expression of a P_{tet}-1 controlled target gene in these cells [11]. For other tetracyclines the difference between the minimal effector concentration required to completely downregulate transcription and the concentration where this substance shows cyctotoxic effects is even more favourable [22]. By contrast, full induction via rtTA requires

1–2 μg/ml of Dox, which is only about five-fold below the concentration which causes side-effects in HeLa cells [20]. For most applications this "buffer" is sufficient. However, with the advent of a new generation of rtTAs [21] the maximal Dox concentration required for induction (80 ng/ml) is now comfortably below the critical threshold.

Establishing Tet control via viral vectors and episomes

The Tet system has been incorporated in a variety of viral vectors including retroviruses [23–26], adenoviruses [27–29], adeno-associated viruses [30–32] herpes simplex viruses [33] and lentiviruses [34]. In part, this work was aimed at the fast and efficient introduction of a transcription control system into tissue cultures or animals. However, a main goal of these efforts is to explore the feasibility of using the Tet system as a conditional expression system in gene therapy, where the stringent control of a gene's activity is an obvious concern (see below).

The approaches used in the development of the viral vectors can be divided into "one-virus" and "two-virus-strategies". Considering the problem of cotransduction, it is certainly an advantage when both the transactivator gene as well as the response unit are incorporated in a single vector. On the other hand, there might be a significant interference between the transactivator driving promoter and $P_{tet,}$ potentially resulting in a less stringent regulation (see below for discussion). However, strategies like the cointroduction of silencer proteins (see below) or the design of properly "insulated" response units might alleviate this problem in the future [35].

As for viral vectors, crosstalk between the regulatory elements of the Tet system is a concern whenever the components of the system are incorporated into a single episomal vector. The strategies to overcome such potential limitations are the same as discussed for viral vectors. However, it has been shown that, provided a proper design of the vectors is used, such systems can function efficiently even without the incorporation of insulator sequences or silencer proteins [36, 37]. Such vectors bear considerable potential for speeding up the establishment of the Tet system into cultured cells, whenever chromosomal integration of the regulatory system is not required.

Modifications of the system

The Tet systems were modified in various ways either to expand their applicability or to optimize their action. Some of the developments are summarized in Figure 1. A problem sometimes encountered is the so-called "leakiness" which refers to a basal level of expression in the uninduced state. Although this issue caught considerable attention, it has only rarely been addressed in a correct way. A deeper discussion of this topic can be found elsewhere [38], but in essence one has to discriminate between (i) promoter-dependent and (ii) integration-site dependent leakiness.

(i) P_{tet}-1 is not necessarily silent in all cellular environments, as it may have an intrinsically elevated activity in cells which contain trans-acting factors that are capable of binding to the promoter itself or to neighbouring sequences of the vector. In a first well-documented example, the binding motif of an αIFN-stimulated response element was identified within the linker region of the *tet*O sequences [39]. Accordingly, an αIFN-dependent activity was observed in respective cytokine studies. Recent promoter designs lack this as well as other potential binding motifs for transcription factors (Baron et al., unpublished). Nevertheless, as pointed out previously [12], an adaptation of P_{tet} to particular cellular environments may sometimes be advantageous but too little effort may have been directed towards this option in the past (for the few examples, see Fig. 1B and [40]).

(ii) The basal activity of a chromosomally integrated minimal promoter depends strongly on its integration site. When located close to a nearby enhancer, it will show an increased activity even in the absence of its cognate activator, in our case tTA/rtTA. This problem can be circumvented by increasing the number of clones (i.e., different integration events) to be screened, or – more actively – by reducing the activity of the P_{tet}-1 in the uninduced state, by shielding it via Tc-controlled transcriptional silencer (tTS) proteins (Fig. 1A, [41–43]. Ideally, tTS and rtTA bind mutually exclusively to the *tet*O sequences within P_{tet} depending on the presence and absence of Dox. This approach was shown to substantially reduce the basal activity of P_{tet}-1 in a number of cell lines [41].

As the latter approach allows active suppression of potential unregulated background activity of P_{tet}, even when caused by elements which are neighbours of the integration site of the minimal promoter, it seems to be more versatile than the adaptation of P_{tet} as proposed above. However, this principle requires one more component as compared to the originally described Tet systems and is therefore experimentally more demanding.

Another issue that has raised concern is the potential toxicity of tTA and rtTA. Like any other transcription factor, these transactivators will induce unwanted pleiotropic effects by "squelching" [44] that may even kill a cell. The consequence is that – as for any endogeneous transcription factors – a Tc-controlled transcription factor should not exceed a certain intracellular concentration. In cell culture as well as in transgenic animals, a selection for proper integration-site dependent expression levels of tTA/rtTA genes takes place and simple screening for proper clones is sufficient as demonstrated by dozens of tTA/rtTA cell lines and mouse strains.

The situation changes, however, when a tTA or rtTA gene has to be integrated at a specific locus, e.g., by homologous recombination following the so-called "knock-in" strategy. Here the result of integration is not subject to subsequent physiological preselection of a proper integration site. Instead, the expression level of the tTA/rtTA gene is locus-dependent and might be too high to be tolerated. In such cases, tTAs and rtTAs might be chosen which contain various minimal activation domains and therefore exhibit a graded activation potential [45]. Transactivators with a reduced activation potential are tolerated at higher intracellular concentrations.

Another attempt to overcome the squelching process is the creation of autoregulatory loops, where the transactivator not only controls the expression of the gene of interest, but also its own synthesis, i.e., the transactivator gene is under P_{tet}-1 control [46]. Results obtained with this modified Tet system should however be interpreted with caution. As the transactivator concentration will not be constant upon induction, unwanted pleiotropic effects will be readily elicited. This concern is highlighted by a recent publication showing the biological side-effects of tTA self-activation [47]. Nevertheless, whenever these limitations are not important as, e.g., in processes of protein production [48, 49] such autoregulatory systems might be useful.

A development that had considerable impact is the construction of bidirectional promoters (P_{tet}-bi) [45]. Here, the heptamerized *tet*O sequences are flanked on both sides by minimal promoters (Fig. 1B) which allows for the simultaneous regulation of two genes of interest, whereby the respective transcription units face in opposite directions. These constructs enjoy special popularity when one gene is an easily assayable reporter gene. Thus, the initial identification of positively transfected clones (or transgenic animals) is facilitated, particularly when expression of the actual gene of interest is difficult to detect. By using the luciferase gene as reporter, the range of regulation and the tightness of control can be readily assessed, whereas, e.g., the *lazZ* or GFP gene allows monitoring of the activity of the expression unit *in situ*.

Finally, two developments should be mentioned here that also indicate the surprisingly large sequence space of the Tet repressor. Making use of altered specificity of TetR for operator sequences [50, 51], two tTAs were generated that discriminate highly efficiently between two operator sequences [13]. By introducing into one of these tTA sequences the mutations that convey a reverse phenotype, a pair of transactivators was generated that would mutually exclusively control the activity of two genes: in the absence of Dox, tTA but not rtTA would be active, whereas in the presence of Dox, rtTA but not tTA would activate its respective promoter. Interestingly, with intermediate concentrations of Dox, both systems can be kept silent. To prevent heterodimerization of the two transactivators, when coexpressed in one cell, the dimerization surface of the TetR moiety of tTA was altered [13]. These developments suggest that further exploitations of the sequence space of TetR may be possible and that even transactivators that efficiently discriminate between different effector molecules may be feasible.

Two complications in using the Tet system should not go unmentioned here, as they might have a considerable impact on the performance of the system. It is frequently attempted to set up the Tet system in tissue cultures in a one-step procedure, i.e., by cotransfection of the two essential components of the system (or, in analogy, by coinjection of the respective DNAs when establishing transgenic mice). While such an experimental shortcut may appear attractive, one should be aware that in general the two DNAs cointegrate at the same chromosomal locus – possibly in multiple copies – which generally leads to considerable crosstalk between the enhancer driving the expression of the transactivator and the minimal promoter within P_{tet}, resulting in elevated background activity of the system ([38]; see also the paragraph dealing with viral vectors). Therefore this approach can only be recommended when efficient screening or selection systems are available.

A rather unexpected finding was the frequent Tc-contamination of bovine sera used in tissue culture media. The resulting complications are obvious, and great care has to be taken to appropriately prescreen the sera for their performance in combination with the Tet system [38]. As these latter issues have been addressed in greater detail in the references given, we will not further elaborate on them.

Tet regulation in the mouse

The prospect of possibly generating truly conditional mutants at the level of higher organisms has sparked numerous attempts to transfer the Tet regulatory systems also into the mouse via transgenesis. Obviously, using animals where the activity of a gene of interest may be reversibly controlled in a temporal, spatial and quantitative manner would open up new perspectives for the analysis of gene function *in vivo*. Early studies have shown that indeed individual genes can be tightly and cell type-specifically controlled in the mouse and that such gene activities can be quantitatively regulated in a non-invasive way by simply supplying the animals with Dox in the drinking water [52, 53].

Following the strategy most generally applied, two classes of mouse lines are generated of which one controls a tTA or rtTA gene by a tissue-specific promoter, whereas the other contains the gene of interest driven by P_{tet}-1. Crossing individuals of respective lines leads to double-transgenic animals where the activity of the target gene depends on the presence or absence of Dox in the water supply of the animal. A rapidly increasing number of mouse lines of the two classes has, in the meantime, been described, of which some are compiled in Table 1. As these mouse lines will become generally available, *in vivo* studies will be greatly facilitated and each new and well-characterized mouse strain of either class will be a precious addition to the system.

The long experience with tetracyclines in mammals was of great advantage for the application of the Tet system in the mouse. Thus, it was not unexpected that Tet regulation would even function in the developing embryo [3, 54, 55]. Actually, since Dox is also present in the milk of feeding mothers, genes can be kept inactive all through the development until adulthood of the animal before they may be turned on by Dox depletion [54]. Moreover, some tetracyclines cross the blood/brain barrier fairly well and thus allow regulation in the brain [9, 10]. Nevertheless, for regulating genes in different organs, it has to be

Table 1. Mouse lines expressing tTA or rtTA genes

Promoter Reference	Tissue specificity	tTA	rtTA	
MHCα	heart muscle	+		[70]
rat insulin	pancreatic β-cells	+		[68]
MMTV	salivary gland and various tissues	+		[71, 72]
hCMVIE	various tissues	+	+	[52]
LAP	hepatocytes	+		[52]
αCamKII	brain	+		[9]
			+	[10]
muscle creatine kinase	muscle	+		[73]
prion	brain	+		[5]
Tek/Tie	endothelial cells	+		[74]
NSE	brain	+		[75]
Ednrb	melanocytes, neural crest	+		[3]
keratin 6	hair follicles keratinocytes	+		[76]
keratin 14	epidermis		+	[8]
Fabp	small intestine colon, bladder		+	[77]
WAP	mammary gland		+	[78]
β-lactoglobulin	mammary gland		+	[79]
retinoblastoma	thalamus, eye, cerebellum muscle		+	[78]
IgHC-SRα	T-cells	+		[6]
tyrosinase	melanocytes		+	[7]
CD2	T-cells		+	[80]
CCSP	respiratory epithelial cells		+	[81]
SP-C	respiratory epithelial cells		+	
MHC class II	thymic epithelial cells	+		[82]

By placing tTA and rtTA genes un der the control of specific promoters in transgenic animals, tissue-specific regulation of P_{tet} controlled genes can be achieved. This compilation shows most of the published tTA/rtTA mouse lines that have been successfully used in various studies. In most cases, the specificity of the promoters driving tTA or rtTA expression depends also on the site of integration which, however, may also result in useful artificial specificities. Many more tTA and rtTA mouse lines have been generated and are under study in various laboratories. An even larger number of target mouse lines containing various genes under P_{tet} control exists, of which around 30 are published.

reconciled that the tissue distribution of tetracyclines can vary as does the biological half-life time. This is particularly important if partial induction in specific tissues is attempted [52]. Moreover, it should be pointed out that the kinetics of induction of tTA and rtTA differ dramatically, since in one case (tTA), the system has to be depleted of, whereas in the other (rtTA), it has to be saturated with Dox. The latter process is of course intrinsically more rapid. Therefore, in studies where a gene should be kept silent during development, the rtTA principle is the preferred one. Nevertheless, cycle times of 2–5 days can be achieved, e.g., in the liver for both systems (Fig. 3; Schönig and Bujard, unpublished) and partial induction in various organs is also feasible by proper adjustment of Dox in the drinking water [52]. Finally, it needs to be emphasized that even prolonged exposure of the animals to Dox, including breeding of offspring, at 2 mg/ml (in the drinking water), has not yielded any detectable adverse effects in the animals [56].

Clearly, the application of Tet regulation in transgenic mice has yielded some spectacular results in recent years. Several tumor models were generated which permit not only the study of tumor initiation but also of tumor progression and regression [6–8] (see also Chapter by Kudlow). Numerous other models for human diseases are at present being developed. Here, we would just like to point out some exciting findings obtained by applying Tet regulation to the central nervous system where the group of Eric Kandel has carried out pioneering work. By controlling the expression of mutant forms of the Ca^{2+}/calmoduline-dependent protein kinase II [9] and calcineurin [10] they could correlate conditionally altered synaptic plasticity with spatial memory storage and retrieval.

Furthermore, modelling of human neuropathologies in the rodent as a prerequisite for developing therapeutic treatments has recently also led to most remarkable results. Thus, the Tet systems were exploited to create mouse models for prion and Huntington's diseases [4, 5]. In these models, the power to turn a gene's transcription on or off with Dox has allowed assessment of not only the role of specific proteins thought to be involved in the etiology of the disorder, but also the consequence of their elimination when the respective gene is turned off after the manifestation of pathological symptoms. Thus, in the Huntington's disease model, a dominant negative mutant of the Huntington's gene was placed under Tet control and its expression led not only to neuroanatomical abnormalities characteristic of Huntington's disease, but also to pathologies resulting in motor dysfunction. Strikingly, in symptomatic adult animals the suppression of transgene expression by Dox led to full reversal of pathological symptoms and to normal behaviour. This example, together with several recently described tumor models, underline most impressively the potential of the Tet regulatory systems in elucidating gene functions *in vivo*.

Tet regulation in plants

The tTA system has meanwhile been shown to function in tobacco [57], moss [58] and most recently in *Arabidopsis* [59]. The latter organism is particularly interesting as it is the most widely used model system in plant development. The moss *Physcomitrella patens* might also gain special importance for the analysis of gene function in plants as its genome can be manipulated via homologous recombination. Whereas the application of tetracyclines seems to be unproblematic, to our knowledge no major efforts have been made to analyze the pharmacokinetics in plants, nor has the efficacy of different tetracycline derivatives in these organisms been investigated in any significant detail.

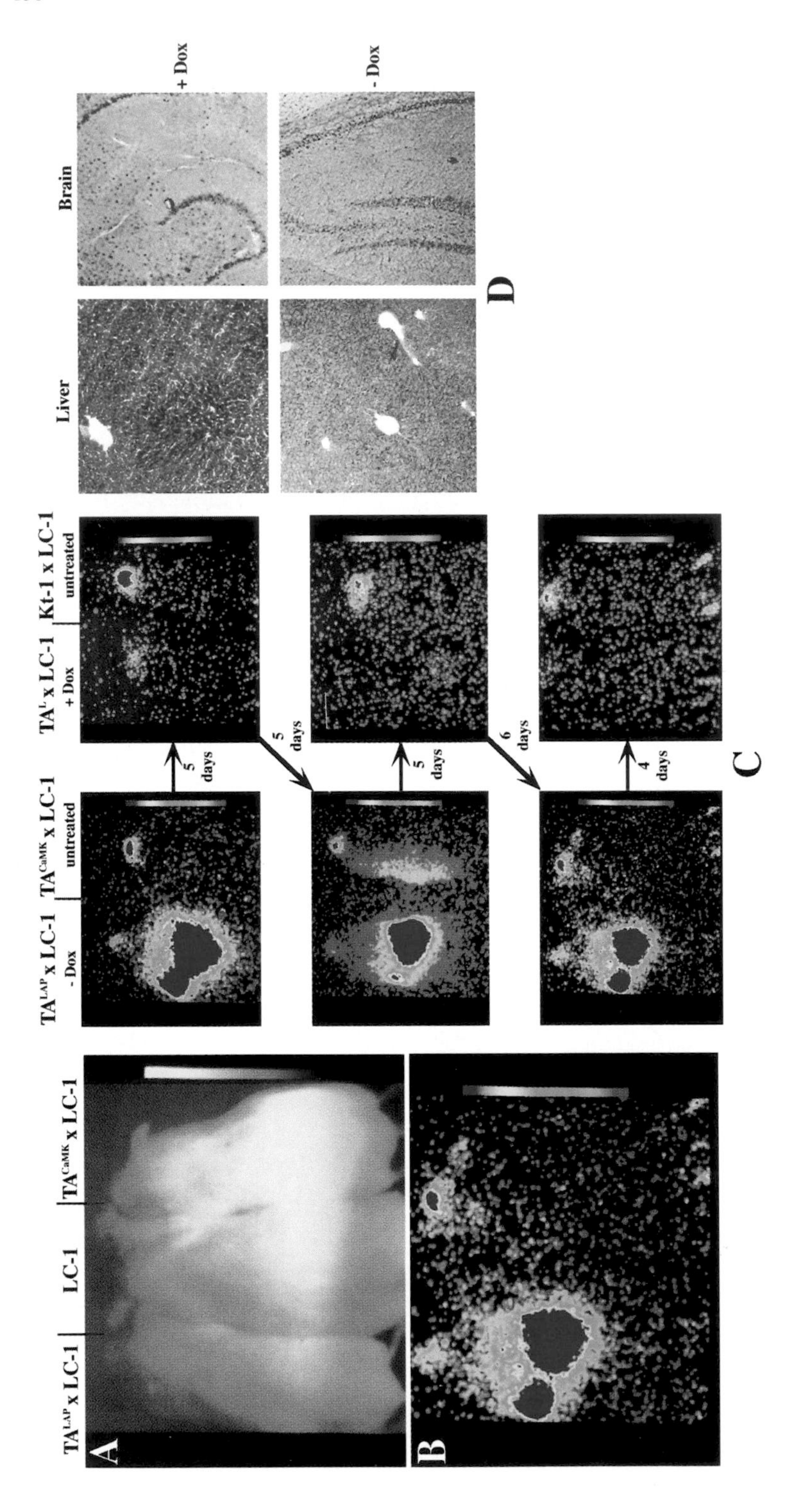
A
TA^{LAP} x LC-1
LC-1
TA^{CaMK} x LC-1
B
TA^{LAP} x LC-1
- Dox
TA^{CaMK} x LC-1
untreated
TA^{L} x LC-1
+ Dox
Kt-1 x LC-1
untreated
5 days
5 days
5 days
6 days
4 days
C
Liver
Brain
+ Dox
- Dox
D

Some applications in the plant field may, however, have to await the discovery of novel and more suitable effector molecules.

Tet regulation in unicellular organisms

The Tet systems were rather readily adapted to control genes in the yeast *Saccharomyces cerevisiae* [60, 61]. This was clearly in part due to the need for an efficient regulatory system in which the inducer is inert to the yeast's metabolism, contrary to, e.g., the widely used Gal-system. Meanwhile, an impressive effort is under way to exploit Tet regulation in yeast functional genomics by placing essential genes under Tc control, while their endogenous counterparts are knocked out by homologues recombination (see homepage of the EUROSCARF initiative (http://www.rz.uni-frankfurt.de/FB/fb16/mikro/euroscarf/data/ess.html) at the University of Frankfurt).

Only recently could the Tet system be adapted to *Dictyostelium discoideum* [17]. It is interesting to note here, that in order to establish Tc regulation in this amoeba, both the transcription control elements driving expression of tTA, as well as the tTA-responsive minimal promoter had to be adapted to meet the requirements for the transcription machinery in this organism.

Tc-controlled gene expression in *Drosophila*

In *Drosophila melanogaster* a highly efficient binary expression system is available, by which upon mating of individual transgenic flies, heterologous gene expression can be directed to specific tissues of the fly [62]. This system is based on the transgenic expression of yeast Gal4 transcription factor and its binding site containing response promoters. By driving expression of Gal4 via appropriate developmentally regulated promoters, the timing of expression can be controlled to a certain extent. However, the precise exogenous control as provided by the Tet system is not possible. Therefore the adaptation of the Tet system to *Drosophila* [63, 64] offers new prospects for analyzing gene functions in this important model organism. In two recent publications the Tet sys-

Figure 3. Turning gene expression on and off *in vivo*. (A) 3 transgenic mice anaesthesized and injected with luciferin placed in the open chamber of the Hamamatsu ICCD/Argus 20 device. Luciferase expression is controlled by tTA in hepatocytes (TA^{LAP}/LC-1) or in brain (TA^{CaMK}/LC-1). The LC-1 mouse in the middle is devoid of tTA and serves as a control. (B) The mice of (A) in the closed dark chamber; photons are collected and processed; exposure time 2 min, whole body images are displayed. (C) Time course of activation and deactivation of the luciferase gene in hepatocytes. Animals are fed via drinking water with Dox (200 µg/ml) for 5 days. Plain water reverses the situation within 5 days. The cycle was repeated several times. The TA^{CaMK}/LC-1 mouse serves as a control. The switching experiment was repeated 3 months later with the same animals yielding identical results. (D) Histological examination of TA^{LAP}/tgnZL2 (Z = β-galactosidase; L = luciferase) shows via β-gal staining that the enzyme is evenly expressed throughout the liver but not in other organs [52]. This expression is efficiently prevented by Dox.

tem has actually been used to generate conditional male-only transgenic *Drosophila* lines [65, 66]. This approach might break ground in establishing a new and convenient method to obtain large, exclusively male populations of other insect species to be used in sterile insect release programs.

Potential of gene regulation in gene therapy

It is generally agreed that in numerous future therapies where genes or cells harbouring therapeutic genes are transferred to patients, a stringent control of the respective gene activity is important and that even a total reversal of the therapeutic manipulation may be required for some regimens. For a number of reasons, the Tet regulatory systems hold promise in this context. First, Dox, at present the most powerful effector molecule for both tTA and rtTA, has been widely used as an antibiotic in human and veterinary medicine and its potential side-effects, also in long-term treatments, are well documented. Accordingly, this compound qualifies as a drug for which a wealth of pharmacological knowledge has been accumulated over the years. Second, experiments in transgenic mice but also in rats show that the Tet system functions well in various organs including the brain and that gene activity can be regulated over long periods of time (e.g., exceeding 1 year in transgenic mice). Third, in experiments where the Tet systems were transferred into adult mice in approaches which mimic *in vivo* or *ex vivo* therapeutic strategies, no immune response towards the transactivators was detected [67] indicating, at least in mice, a low immunogenicity of tTA and rtTA.

Here we would like to briefly discuss three therapeutic strategies in rodent models where Tet regulation has been included. In the first example, long-term control of erythropoietin (Epo) and accordingly of hematocrit has been achieved in mice. To this end mouse primary myogenic cells were generated which produce Epo under Dox control. Upon transplantation into the skeletal muscle of mice, the transduced cells differentiate into myotubes and express specifically muscle cell tTA which in turn activates the Epo gene under P_{tet-1} control. Synthesis and secretion into the circulation of Epo could be iteratively switched on and off during a 5-month period [67].

In a totally different approach of cellular therapy, Tet regulation was utilized to generate conditionally proliferating primary β-cells of the mouse pancreas [68]. By placing in transgenic mice the SV40 large T-antigen under P_{tet-1} control while driving tTA synthesis via the rat insulin promoter, β-cells could be harvested from the animals, which would proliferate in absence of Dox. Addition of Dox to the culture of the primary cells causes growth arrest and yields fully differentiated β-cells, which secrete insulin in a strictly glucose-dependent manner. Transplantation of such cells into diabetic mice and expansion of the population of β-cells until proper blood glucose levels were reached was followed by Dox supply via drinking water. The arrested and redifferentiated cells continued to control insulin secretion in dependence on blood glucose. One could envision that when using the rtTA system in an analogous way, but with properly encapsulated human β-cells, a regimen may become feasible that would be attractive for treating insulin-dependent diabetes mellitus [69].

Finally, an *in vivo* therapeutic approach for Parkinson's disease was modelled in rats with dopamine-depleted striatum. Direct injection into the lesioned striatum of an adenoviral vector which expresses the tryosine hydroxylase gene under Tet control resulted, after around 20 days, in expression of the enzyme which could be suppressed by treating the animals with Dox.

These model systems suggest that Tet regulation may indeed prove useful in gene- or cell-therapeutic approaches which, however, still require the solution of difficult problems before their broad application appears feasible.

Perspectives

Already at the present state of the technology, the application of Tet regulation in the study of gene function will rapidly expand due to synergisms which emerge from the work of different laboratories. Thus, an increasing number of cell lines harbouring tTA or rtTA is becoming available, many even on a commercial basis (Clontech Laboratories Inc., Palo Alto, USA). Similarly, the number of mouse lines that express tTA or rtTA genes via cell type-specific promoters is growing (Tab. 1) as are mouse lines that contain target genes under P_{tet} control. Again, some of these mouse lines are already commercially available (The Jackson Laboratory, Bar Harbor, USA) and it can be anticipated that many more will be accessible to the scientific community in the near future. An analogous development can be forseen for other model organisms, particularly for *S. cerevisiae* where the generation of respective strains is well underway, and in *Drosophila*. Together, these developments will increasingly permit a multitude of genetic studies without having to go through the costly and time-consuming generation of (all) required cell lines or strains of organisms.

Nevertheless, there is clearly room for further improvements of the Tet systems and particularly for their adaptation to special requirements. Among these are the design of improved transactivators and tTA/rtTA-responsive promoters. Whereas the latter has to be done in the specific context of the cellular system under study, the former can at least in part be achieved through functional screens in eukaryotic genetic systems. The sequence space of the Tet repressor can thus be thoroughly probed and recent results [13, 21] have indicated an exciting potential. Thus, besides specifying transactivators to discriminate between different DNA-binding sites, it may well be feasible to also adapt the protein to distinguish between different effector molecules. The independent regulation of two or more genes may thus become a possibility. Moreover, as pointed out in detail elsewhere in this section, the repressor moiety may be adapted to non-tetracycline compounds as inducers. Such inducers, when ecologically harmless and available at low cost, may even make crop plants amenable to Tet regulation, althouh some ground-breaking work will have to be accomplished before this ambitious goal may be achieved.

Acknowledgements
We would like to thank Ms. Sibylle Reinig and Ms. Gaik Wee for their help in producing this manuscript. Our work was supported by funds from the Volkswagen-Stiftung, the BMBF BioRegio Program, by an EC grant and by the Fonds der Chemischen Industrie Deutschlands.

References

1 [1] Mansuy I, Bujard H (2000) Tetracycline-regulated gene expression in the brain. *Curr Opin Neurobiol* 10: 593–596
2 Hasan MT, Schoenig K, Graewe W, Bujard H (2001) Long-term, non-invasive imaging of regulated gene expression in living mice. *Genesis* 29: 116–122
3 Shin MK, Levorse JM, Ingram RS, Tilghman SM (1999) The temporal requirement for endothelin receptor-B signaling during neural crest development. *Nature* 402: 496–501
4 Yamamoto A, Lucas JJ, Hen R (2000) Reversal of neuropathology and motor dysfunction in a conditional model of Huntington's disease [see comments]. *Cell* 101: 57–66
5 Tremblay P, Meiner Z, Galou M, Heinrich C, Petromilli C, Lisse T, Cayetano J, Torchia M, Mobley W, Bujard H et al (1998) Doxycycline control of prion protein transgene expression modulates prion disease in mice. *Proc Natl Acad Sci USA* 95: 12 580–12 585
6 Felsher DW, Bishop JM (1999) Reversible tumorigenesis by MYC in hematopoietic lineages. *Mol Cells* 4: 199–207
7 Chin L, Tam A, Pomerantz J, Wong M, Holash J, Bardeesy N, Shen Q, O'Hagan R, Pantginis J, Zhou H et al (1999) Essential role for oncogenic Ras in tumour maintenance. *Nature* 400: 468–472
8 Xie W, Chow LT, Paterson AJ, Chin E, Kudlow JE (1999) Conditional expression of the ErbB2 oncogene elicits reversible hyperplasia in stratified epithelia and up-regulation of TGFalpha expression in transgenic mice. *Oncogene* 18: 3593–3607
9 Mayford M, Bach ME, Huang YY, Wang L, Hawkins RD, Kandel ER (1996) Control of memory formation through regulated expression of a CaMKII transgene. *Science* 274: 1678–1683
10 Mansuy IM, Mayford M, Jacob B, Kandel ER, Bach ME (1998) Restricted and regulated overexpression reveals calcineurin as a key component in the transition from short-term to long-term memory. *Cell* 92: 39–49
11 Gossen M, Bonin AL, Bujard H (1993) Control of gene activity in higher eukaryotic cells by prokaryotic regulatory elements. *Trends Biochem Sci* 18: 471–475
12 Gossen M, Bujard H (1992) Tight control of gene expression in mammalian cells by tetracycline-responsive promoters. *Proc Natl Acad Sci USA* 89: 5547–5551
13 Baron U, Schnappinger D, Helbl V, Gossen M, Hillen W, Bujard H (1999) Generation of conditional mutants in higher eukaryotes by switching between the expression of two genes. *Proc Natl Acad Sci USA* 96: 1013–1018
14 Yin DX, Zhu L, Schimke RT (1996) Tetracycline-controlled gene expression system achieves high-level and quantitative control of gene expression. *Anal Biochem* 235: 195–201
15 Rossi FM, Blau HM (1998) Recent advances in inducible gene expression systems. *Curr Opin Biotechnol* 9: 451–456
16 Kringstein AM, Rossi FMV, Hofmann A, Blau HM (1998) Graded transcriptional response to different concentrations of a single transactivator. *Proc Natl Acad Sci USA* 95: 13 670–13 675
17 Blaauw M, Linskens MH, van Haastert PJ (2000) Efficient control of gene expression by a tetracycline-dependent transactivator in single *Dictyostelium discoideum* cells. *Gene* 252: 71–82
18 Clement MV, Ponton A, Pervaiz S (1998) Apoptosis induced by hydrogen peroxide is mediated by decreased superoxide anion concentration and reduction of intracellular milieu. *FEBS Lett* 440: 13–18
19 Lee SB, Huang K, Palmer R, Truong VB, Herzlinger D, Kolquist KA, Wong J, Paulding C, Yoon SK, Gerald W et al (1999) The Wilms tumor suppressor WT1 encodes a transcriptional activator of amphiregulin. *Cell* 98: 663–673

20 Gossen M, Freundlieb S, Bender G, Muller G, Hillen W, Bujard H (1995) Transcriptional activation by tetracyclines in mammalian cells. *Science* 268: 1766–1769
21 Urlinger S, Baron U, Thellmann M, Hasan MT, Bujard H, Hillen W (2000) Exploring the sequence space for tetracycline-dependent transcriptional activators: Novel mutations yield expanded range and sensitivity. *Proc Natl Acad Sci USA* 97: 7963–7968
22 Gossen M, Bujard H (1993) Anhydrotetracycline, a novel effector for tetracycline controlled gene expression systems in eukaryotic cells. *Nucl Acid Res* 21: 4411–4412
23 Lindemann D, Patriquin E, Feng S, Mulligan RC (1997) Versatile retrovirus vector systems for regulated gene expression *in vitro* and *in vivo*. *Mol Med* 3: 466–476
24 Hofmann A, Nolan GP, Blau HM (1996) Rapid retroviral delivery of tetracycline-inducible genes in a single autoregulatory cassette [see comments]. *Proc Natl Acad Sci USA* 93: 5185–5190
25 Hwang JJ, Scuric Z, Anderson WF (1996) Novel retroviral vector transferring a suicide gene and a selectable marker gene with enhanced gene expression by using a tetracycline-responsive expression system. *J Virol* 70: 8138–8141
26 Iida A, Chen ST, Friedmann T, Yee JK (1996) Inducible gene expression by retrovirus-mediated transfer of a modified tetracycline-regulated system. *J Virol* 70: 6054–6059
27 Neering SJ, Hardy SF, Minamoto D, Spratt SK, Jordan CT (1996) Transduction of primitive human hematopoietic cells with recombinant adenovirus vectors. *Blood* 88: 1147–1155
28 Yoshida Y, Hamada H (1997) Adenovirus-mediated inducible gene expression through tetracycline-controllable transactivator with nuclear localization signal. *Biochem Biophys Res Commun* 230: 426–430
29 Paulus W, Baur I, Boyce FM, Breakefield XO, Reeves SA (1996) Self-contained, tetracycline-regulated retroviral vector system for gene delivery to mammalian cells. *J Virol* 70: 62–67
30 Maxwell IH, Spitzer AL, Long CJ, Maxwell F (1996) Autonomous parvovirus transduction of a gene under control of tissue-specific or inducible promoters. *Gene Ther* 3: 28–36
31 Bertran J, Miller JL, Yang Y, Fenimore-Justman A, Rueda F, Vanin EF, Nienhuis AW (1996) Recombinant adeno-associated virus-mediated high-efficiency, transient expression of the murine cationic amino acid transporter (ecotropic retroviral receptor) permits stable transduction of human HeLa cells by ecotropic retroviral vectors. *J Virol* 70: 6759–6766
32 Haberman RP, McCown TJ, Samulski RJ (1998) Inducible long-term gene expression in brain with adeno-associated virus gene transfer. *Gene Ther* 5: 1604–1611
33 Ho DY, McLaughlin JR, Sapolsky RM (1996) Inducible gene expression from defective herpes simplex virus vectors using the tetracycline responsive promoter system. *Brain Res Mol Brain Res* 41: 200–209
34 Kafri T, van Praag H, Gage FH, Verma IM (2000) Lentiviral vectors: regulated gene expression. *Molec Ther* 1: 516–521
35 Bell AC, Felsenfeld G (1999) Stopped at the border: boundaries and insulators. *Curr Opin Genet Develop* 9: 191–198
36 Feuillard J, Schuhmacher M, Kohanna S, Asso-Bonnet M, Ledeur F, Joubert-Caron R, Bissieres P, Polack A, Bornkamm GW, Raphael M (2000) Inducible loss of NF-kappaB activity is associated with apoptosis and Bcl-2 down-regulation in Epstein-Barr virus transformed B lymphocytes. *Blood* 95: 2068–2075
37 Jost M, Kari C, Rodeck U (1997) An episomal vector for stable tetracycline-regulated gene expression. *Nucl Acid Res* 25: 3131–3134
38 Freundlieb S, Baron U, Bonin AL, Gossen M, Bujard H (1997) Use of tetracycline-controlled gene expression systems to study mammalian cell cycle. *Methods Enzymol* 283: 159–173
39 Rang A, Will H (2000) The tetracycline-responsive promoter contains functional interferon-inducible response elements. *Nucl Acid Res* 28: 1120–1125
40 Hoffmann A, Villalba M, Journot L, Spengler D (1997) A novel tetracycline-dependent expression vector with low basal expression and potent regulatory properties in various mammalian cell lines. *Nucl Acid Res* 25: 1078–1079
41 Freundlieb S, Schirra-Muller C, Bujard H (1999) A tetracycline controlled activation/repression system with increased potential for gene transfer into mammalian cells. *J Gene Med* 1: 4–12
42 Forster K, Helbl V, Lederer T, Urlinger S, Wittenburg N, Hillen W (1999) Tetracycline-inducible expression systems with reduced basal activity in mammalian cells. *Nucl Acid Res* 27: 708–710
43 Rossi FM, Guicherit OM, Spicher A, Kringstein AM, Fatyol K, Blakely BT, Blau HM (1998) Tetracycline-regulatable factors with distinct dimerization domains allow reversible growth inhibition by p16. *Nat Genet* 20: 389–393

44 Gill G, Ptashne M (1988) Negative effect of the transcriptional activator GAL4. *Nature* 334: 721–724
45 Baron U, Freundlieb S, Gossen M, Bujard H (1995) Co-regulation of two gene activities by tetracycline via a bidirectional promoter. *Nucl Acid Res* 23: 3605–3606
46 Shockett P, Difilippantonio M, Hellman N, Schatz DG (1995) A modified tetracycline-regulated system provides autoregulatory, inducible gene expression in cultured cells and transgenic mice. *Proc Natl Acad Sci USA* 92: 6522–6526
47 Gallia GL, Khalili K (1998) Evaluation of an autoregulatory tetracycline regulated system. *Oncogene* 16: 1879–1884
48 Zhang Y, Katakura Y, Ohashi H, Shirahata S (1997) An autocatalytic expression system for regulated production of recombinant protein in mammalian cells. *Anal Biochem* 252: 286–292
49 Fussenegger M, Moser S, Mazur X, Bailey JE (1997) Autoregulated multicistronic expression vectors provide one-step cloning of regulated product gene expression in mammalian cells. *Biotechnol Progr* 13: 733–740
50 Helbl V, Tiebel B, Hillen W (1998) Stepwise selection of TetR variants recognizing tet operator 6C with high affinity and specificity. *J Mol Biol* 276: 319–324
51 Helbl V, Hillen W (1998) Stepwise selection of TetR variants recognizing tet operator 4C with high affinity and specificity. *J Mol Biol* 276: 313–318
52 Kistner A, Gossen M, Zimmermann F, Jerecic J, Ullmer C, Lubbert H, Bujard H (1996) Doxycycline-mediated quantitative and tissue-specific control of gene expression in transgenic mice. *Proc Natl Acad Sci USA* 93: 10 933–10 938
53 Fishman GI, Kaplan ML, Buttrick PM (1994) Tetracycline-regulated cardiac gene expression *in vivo*. *J Clin Invest* 93: 1864–1868
54 Lee P, Morley G, Huang Q, Fischer A, Seiler S, Horner JW, Factor S, Vaidya D, Jalife J, Fishman GI (1998) Conditional lineage ablation to model human diseases. *Proc Natl Acad Sci USA* 95: 11 371–11 376
55 Ray P, Tang W, Wang P, Homer R, Kuhn C, Flavell RA, Elias JA (1997) Regulated Overexpression of Interleukin 11 in the lung. Use to dissociate development-dependent and -independent phenotypes. *J Clin Invest* 100: 2501–2511
56 Kistner A (1996) Stabilisierung Tetracyclin-kontrollierter Expressionssysteme in transgenen Mäusen. Ph.D. thesis, University of Heidelberg
57 Weinmann P, Gossen M, Hillen W, Bujard H, Gatz C (1994) A chimeric transactivator allows tetracycline-responsive gene expression in whole plants. *Plant J* 5: 559–569
58 Zeidler M, Gatz C, Hartmann E, Hughes J (1996) Tetracycline-regulated reporter gene expression in the moss *Physcomitrella patens*. *Plant Mol Biol* 30: 199–205
59 Love J, Scott AC, Thompson WF (2000) Technical Advance: Stringent control of transgene expression in *Arabidopsis thaliana* using the Top10 promoter system. *Plant J* 21: 579–588
60 Gari E, Piedrafita L, Aldea M, Herrero E (1997) A set of vectors with a tetracycline-regulatable promoter system for modulated gene expression in *Saccharomyces cerevisiae*. *Yeast* 13: 837–848
61 Nagahashi S, Nakayama H, Hamada K, Yang H, Arisawa M, Kitada K (1997) Regulation by tetracycline of gene expression in *Saccharomyces cerevisiae*. *Mol Gen Genet* 255: 372–375
62 Brand AH, Perrimon N (1993) Targeted gene expression as a means of altering cell fates and generating dominant phenotypes. *Development* 118: 401–415
63 Bello B, Resendez-Perez D, Gehring WJ (1998) Spatial and temporal targeting of gene expression in *Drosophila* by means of a tetracycline-dependent transactivator system. *Development* 125: 2193–2202
64 Bieschke ET, Wheeler JC, Tower J (1998) Doxycycline-induced transgene expression during *Drosophila* development and aging. *Mol Gen Genet* 258: 571–579
65 Heinrich JC, Scott MJ (2000) A repressible female-specific lethal genetic system for making transgenic insect strains suitable for a sterile-release program. *Proc Natl Acad Sci USA* 97: 8229–8232
66 Thomas DD, Donnelly CA, Wood RJ, Alphey LS (2000) Insect population control using a dominant, repressible, lethal genetic system. *Science* 287: 2474–2476
67 Bohl D, Naffakh N, Heard JM (1997) Long-term control of erythropoietin secretion by doxycycline in mice transplanted with engineered primary myoblasts [see comments]. *Nat Med* 3: 299–305
68 Efrat S, Fusco-DeMane D, Lemberg H, al Emran O, Wang X (1995) Conditional transformation of a pancreatic beta-cell line derived from transgenic mice expressing a tetracycline-regulated

oncogene. *Proc Natl Acad Sci USA* 92: 3576–3580
69 Efrat S (1998) Cell-based therapy for insulin-dependent diabetes mellitus. *Eur J Endocrinol* 138: 129–133
70 Passman RS, Fishman GI (1994) Regulated expression of foreign genes *in vivo* after germline transfer. *J Clin Invest* 94: 2421–2425
71 Hennighausen L, Wall RJ, Tillmann U, Li M, Furth PA (1995) Conditional gene expression in secretory tissues and skin of transgenic mice using the MMTV-LTR and the tetracycline responsive system. *J Cell Bioch* 59: 463–472
72 Huettner CS, Zhang P, Van Etten RA, Tenen DG (2000) Reversibility of acute B-cell leukaemia induced by BCR-ABL1. *Nat Genet* 24: 57–60
73 Ghersa P, Gobert RP, Sattonnet-Roche P, Richards CA, Pich EM, van Huijsduijnen RH (1998) Highly controlled gene expression using combinations of a tissue-specific promoter recombinant adenovirus and a tetracycline-regulatable transcription factor. *Gene Ther* 5: 1213–1220
74 Sarao R, Dumont DJ (1998) Conditional transgene expression in endothelial cells. *Transgenic Res* 7: 421–427
75 Kelz MB, Chen J, Carlezon WA Jr, Whisler K, Gilden L, Beckmann AM, Steffen C, Zhang YJ, Marotti L, Self DW, Tkatch T, Baranauskas G, Surmeier DJ, Neve RL, Duman RS, Picciotto MRNestler EJ (1999) Expression of the transcription factor deltaFosB in the brain controls sensitivity to cocaine. *Nature* 401: 272–276
76 Guo Y, Zhao J, Sawicki J, Peralta Soler A, O'Brien TG (1999) Conversion of C57Bl/6 mice from a tumor promotion-resistant to a -sensitive phenotype by enhanced ornithine decarboxylase expression. *Mol Carcinogen* 26: 32–36
77 Saam JR, Gordon JI (1999) Inducible gene knockouts in the small intestinal and colonic epithelium. *J Biol Chem* 274: 38 071–38 082
78 Utomo AR, Nikitin AY, Lee WH (1999) Temporal, spatial, and cell type-specific control of Cre-mediated DNA recombination in transgenic mice. *Nat Biotechnol* 17: 1091–1096
79 Soulier S, Stinnakre MG, Lepourry L, Mercier JC, Vilotte JL (1999) Use of doxycycline-controlled gene expression to reversibly alter milk-protein composition in transgenic mice. *Eur J Biochem* 260: 533–539
80 Legname G, Seddon B, Lovatt M, Tomlinson P, Sarner N, Tolaini M, Williams K, Norton T, Kioussis D, Zamoyska R (2000) Inducible expression of a p56Lck transgene reveals a central role for Lck in the differentiation of CD4 SP thymocytes. *Immunity* 12: 537–546
81 Tichelaar JW, Lu W, Whitsett JA (2000) Conditional expression of fibroblast growth factor-7 in the developing and mature lung. *J Biol Chem* 275: 11 858–16 649
82 Witherden D, van Oers N, Waltzinger C, Weiss A, Benoist C, Mathis D (2000) Tetracycline-controllable Selection of CD4(+) T Cells. Half-life and survival signals in the absence of major histocompatibility complex class ii molecules. *J Exp Med* 191: 355–364

Tetracyclines in Biology, Chemistry and Medicine
ed. by M. Nelson, W. Hillen and R.A. Greenwald

Tetracycline-regulated gene expression to study oncogenes *in vivo*

Jeffrey E. Kudlow

Department of Medicine/Endocrinology, University of Alabama at Birmingham, 1808 7th Avenue South, Rm 756, Birmingham, AL 35294-0012, USA

Introduction

In order to determine the role of signaling molecules, such as oncogenes, in cellular function, it is often necessary to express these proteins under the control of a heterologous promoter in the cells of interest and determine the downstream effects of this expression. While this strategy is often successful, there are circumstances when the constitutive expression of such proteins results in the death of the transfected cells or the entire organism if a transgenic model is used. Furthermore, constitutive expression of oncogenes can result in long-term secondary effects and genetic instability which may not relate in a proximal way to the mode of signal transduction of the oncogene under investigation. Because of these considerations, other means for inducible expression of oncogenes have been used. In the past, oncogenes were placed directly downstream of promoters such as the mouse mammary tumor virus (MMTV) LTR or the heavy metal-inducible metallothionein promoter. These promoters are useful in many cell culture lines because they can drive expression in a variably inducible manner. Using the MMTV promoter, 2- to 10-fold induction can be accomplished by provision of dexamethasone in the culture medium [1, 2]. Induction of the metallothionein promoter at similar levels can be obtained by exposure of the transfected cells to zinc or cadmium ions [3, 4]. However, these direct expression systems are of less use as inducible models in transgenic animals. The MMTV promoter drives expression to certain steroid-responsive tissues such as the mammary gland, seminal vesicles and salivary glands of mice [5–7], while the metallothionein promoter drives expression to multiple tissues [8]. Furthermore, MMTV-driven expression in transgenic mice becomes constitutive because of the endogenous steroids produced in the animal, whereas induction through the metallothionein promoter is difficult because of stimulatory levels of zinc in the diet, but zinc induction has been observed [9]. The major problems are the inability to completely control tissue expression by these direct promoter systems and the poor inducibility of expression. These difficulties were strong stimuli for the development of better methods for tissue-specific and inducible expression in transgenic animals.

To obtain both a broad selection of tissue-specific expression and inducibility, what was needed was a two-gene system that mimicked the physiologically inducible systems found in nature. The endocrine system of higher organisms is an example of a naturally evolved two-gene system in which a hormone and its receptor are expressed in specific tissues. In the endocrine system, hormone-dependent conditional expression in a specific tissue results from the tissue-specific expression of the receptor and changes in hormone concentration. A recent example of such an artificial transgene system that mimics this endocrine model is the ecdysone-inducible gene expression system [10, 11]. This system is literally a hormone-inducible system; however, the hormone is derived from insects where it is the signal to induce molting. The molting steroid hormone, muristerone, is not biologically active in mammalian cells, which lack the appropriate receptor for this hormone. The inducible transgenic system therefore provides this receptor to the tissue of choice. Thus one of the transgenes encodes the ecdysone receptor which can be expressed in the tissue of choice using an appropriate tissue-specific promoter. The second transgene contains the gene of choice placed downstream from a promoter whose activity can be driven by the liganded ecdysone receptor. Thus, while the gene to be expressed is present in all tissues, this gene will only be expressed at appreciable levels in the tissue in which the ecdysone receptor is expressed and this expression will depend on exposure of the cells or animals to the molting hormone, muristerone. Thus, this approach fulfills the need for both temporal and spatial control of gene expression. Evans and co-workers have reviewed ecdysone systems [10]. In summary, the ecdysone-dependent promoter as designed drives basal expression of the target gene in the absence of muristerone at very low levels, whereas induction can result in greater than 1000-fold increases in gene expression [11]. While this system was not tested with a highly toxic gene product to prove that unstimulated expression was sufficiently low to obviate toxicity, the luciferase assay was used in these studies, and this assay is very sensitive. This system has the advantage of rapid inducibility and rapid decay of the induction upon withdrawal of the insect steroid. The disadvantage is the expense of the muristerone which would be used in relatively large quantities for the study of transgenic models.

The tetracycline-inducible system

The tetracycline-inducible system is conceptually similar to the ecdysone-inducible system; however, considerable modification of the tet operon was required before it could be adapted for use in eukaryotic cells. The elements of this system are thoroughly reviewed elsewhere in this monograph. In summary, the major modification of the prokaryotic transcriptional regulator was to convert it to a eukaryotic activator by fusing the Tet repressor (TetR) DNA-binding domain to a strong mammalian transcriptional activation domain. In the first iteration, the portion of the wildtype TetR containing the DNA and tetracycline binding domains was fused to the herpes simplex VP16 activation domain to create a transcription factor, termed tTA, that binds to the minimal tet operator (*tetO*) sequence in the absence of tetracycline and is released from the *tetO* element in the presence of the antibiotic. The tTA fusion protein is directed to the nucleus by the VP16 domain and is capable of engaging and activating the mammalian transcriptional apparatus when the VP16 domain is directed to the DNA via the interaction of the TetR domain with the *tetO*. Usually in a separate construct, the *tetO* element is placed immediately upstream of a minimal mammalian promoter containing a TATA-box and transcriptional initiation site, allowing transcription activation by the VP16 domain. Thus, gene activation in mammalian cells that express tTA occurs when tetracycline is withheld, while repression of transcription is accomplished by provision of the antibiotic. The dynamic range for tetracycline conditional expression has also been around 3 orders of magnitude. This system is now termed the Tet-Off version. In the second iteration, a four amino acid mutation in the TetR results in a transcription factor with reversed responses to tetracycline. This rTetR mutant binds to DNA in a tetracycline-dependent manner. Thus, the rTetR-VP16 fusion protein, termed rtTA, activates transcription only in the presence of tetracycline. This system is appropriately named the Tet-On system.

Similar to the hormonal system described above, the tTA and rtTA transcription factors are analogous to a nuclear hormone receptor; however, the ligand is a tetracycline derivative. The advantages of tetracycline derivatives are that they are readily available, inexpensive and have no significant biological effects or teratogenic properties in mammals. The principal disadvantages are that this antibiotic can accumulate in bone and be released subsequent to the cessation of therapy and that the drug is present in the bovine sera used for the propagation of cells in culture. As described for the hormone-inducible system, these TetR-based transcription factors can be expressed in a wide variety of cultured cells using general promoters and in transgenic animals using tissue-specific promoters. The VP16 component of the tet-activators may be a problem if the transgene is integrated into a region of the genome that results in high-level expression of the activator. The VP16 activation domain is capable of 'squelching' the expression of unrelated genes, presumably by interacting with general transcription factors in a DNA-independent manner [12, 13]. This point is reviewed elsewhere [14] and appears to be overcome in stable cell transfection studies by natural selection against the toxic effects of VP16 overexpression. This selection results in the outgrowth of only those cell lines with moderate expression of the VP16 fusion protein. In transgenic animals, this problem has not been reported with the tet-activators, although high-level VP16 expression during early mouse embryogenesis results in impaired development [15]. Presumably, promoters that turn on later in development may not give rise to such problems. In our own experience with promoters that direct expression to the skin, pituitary and pancreatic β-cells, we have not seen developmental problems. Similarly, the integration site of the *tetO* containing promoter in stable cell lines and transgenic animals can have an effect on expression. In some cell culture studies, basal expression from the *tetO* may be independent of the tet-activator, presumably because of neighboring enhancer elements. This problem is generic to all inducible systems and does not appear to be a major problem in the reported transgenic mouse systems. At least half of our founder transgenic lines show good inducible expression (unpublished data) and the problem has been lack of expression rather than constitutive basal expression. In contrast, the problem of basal expression can be considerable in transient transfectants where basal expression is thought to not be suppressed by chromatin effects.

The use of the tetracycline-inducible system to express oncogenes in transgenic mice

For inducible oncogene expression, the oncogene cDNA of choice is placed downstream of the *tetO*-basal promoter. The transgenic animals with the transgene containing the oncogene under the control of the tetracycline-responsive element (TetRE), i.e., TetRE-oncogene transgene, are generally developed separately from the transgenic lines containing the tet-activator. This scheme requires that the TetRE-oncogene mice be crossed with the tet-activator mice to produce bitransgenic animals. This approach allows for the development of a combinatorial system for tissue-specific, inducible oncogene expression. For example, a series of mice can be developed with a variety of oncogenes or a set of mutant forms of these oncogenes placed downstream from the TetRE. This series of mice can then be crossed with another series of mice in which the transgene consists of the tet-activator placed downstream from a variety of tissue-specific promoters. From these two series of transgenics, bitransgenic animals can be produced to test the effect of the various oncogenes in a variety of selected tissues. Added to the spatial control is the temporal control provided by the tetracycline inducibility. This temporal control can be applied through all stages of development. As we and others have shown, the tetracycline derivative, doxycyline, can be provided in the drinking water of a mono-transgenic pregnant mouse carrying bitransgenic offspring. The maternal animal will not develop a phenotype because both components of the inducible system are not present. On the other hand, the bitransgenic embryonic mice receive doxycycline by transplacental transfer and thus the effect of oncogene expression can be tested during embyronic development. The mono-transgenic and non-transgenic littermates serve as controls. Similarly, doxycycline is also transmitted to neonatal offspring through lactation, thus allowing the initiation of gene expression at an early time point after birth.

This combinatorial power derived from the use of bitransgenic mice also provides an additional advantage. With conventional mono-transgenic mice, it is usually important to develop two or more lines with the same transgene in order to show that the phenotype is independent of the integration site in the genome. The combinatorial approach using bitransgenic mice reduces this requirement. If neither the mono-transgenic tet-activator nor the TetRE-oncogene mouse has a phenotype, while the bitransgenic mouse has a conditional phenotype that depends on exposure to tetracycline, then integration controls might not be necessary. Furthermore, if the phenotype can be shown to change by keeping the oncogene constant but varying the tissue in which the tet-activator is expressed, or by varying the oncogene while keeping the tissue constant, then the need for the integration control becomes even less important. Clearly, this property of the bitransgenic mouse system is leading to the development of a spectrum of tet-activator mice with varying tissue-specific promoters. Similarly, several TetRE-oncogene mice are being developed. The availability of these mouse lines should allow for collaborative exchange of these 'reagents' between investigators to exploit the combinatorial power of this system.

Examples that illustrate the use of the tet system *in vivo*

Keratin K14-erbB2 model

Our K14-erbB2 model illustrates many of the points made in the preceding general comments. It had been determined that erbB2 was normally expressed in the skin [16–18]; however, it was not clear what the effect of activation of this receptor-tyrosine kinase oncogene would have on skin development or maintenance and conversely, it has yet to be determined what would result from inactivation of this oncogene in the skin. Using the MMTV promoter to drive expression of an activated oncogenic form of erbB2 in transgenic mice [19, 20], previous investigators had shown that these animals would develop mammary carcinoma in a stochastic fashion, but this carcinoma would not be preceded by epithelial hyperplasia. In contrast, hyperplasia was observed in the salivary glands, Harderian glands, lungs [21] and kidney ducts [21] of these MMTV-erbB2 mice. This result suggested that different epithelial tissues respond variably to activation of the ErbB2 receptor. To test the skin response to ErbB2 activation, we expressed the oncogenic activated form of the receptor in the skin of transgenic mice using the K14 promoter [22]. The K14 promoter drives expression of the transgene to the basal cells of the epidermis and to the outer root sheath of the hair follicles, sites where erbB2 is normally expressed. The resulting K14-erbB2 transgenic mice developed a very dramatic phenotype, including epidermal hyperplasia, papilloma, hyperkeratosis, dyskeratosis and dermal hyperplasia. Most of the hair follicles were replaced by bizarre hyperproliferative intradermal invaginations resembling inverted papilloma, while the rest of the follicles exhibited severe hyperplasia and disorganization. Unfortunately, the non-chimeric founder mice died within the perinatal period and further study of this model was precluded. One of the mice was chimeric with a limited affected area of skin, mainly on the tail and central back. This animal did not have the transgene in its germ line and was unable to transmit the gene to its offspring.

Because the basal epidermis and hair follicles are exquisitely sensitive to activated ErbB2, expression of this oncogene in the skin results in a perinatal lethal phenotype. To allow further study of ErbB2 in the skin, we needed a model that avoided this lethality. This requirement could be fulfilled with an inducible system in which basal expression of the oncogene was sufficiently low to prevent the lethal outcome while induction of erbB2 expression could reproduce the original K14-erbB2 phenotype. The tetracycline/doxycycline-regulated systems appeared to provide a solution. At the time that we initiated our study with erbB2, all of the published applications had made use of the Tet-Off system [23–27]. None of these examples was applied to a transgene with embryonic or perinatal lethality. Furthermore, there was no evidence that the Tet-On system was sufficiently tetracycline-dependent to prevent leakiness of basal expression in the absence of doxycycline, such that the lethal phenotype might not be obviated. While our experiments were not designed primarily to test the Tet-On system for such a problem, our use of the lethal ErbB2/skin system suggests that gene expression with the Tet-On system does exhibit sufficient dependence on the presence of doxycycline, such that neither the lethal phenotype is expressed nor is there detectable expression of the erbB2 transgene in the absence of doxycycline [28].

The K14-erbB2 system was set up by the development of two mouse lines [28]. The first expressed the rtTA tet-on activator under the control of the K14 promoter. The second line had a transgene that consisted of the TetRE placed upstream from the activated erbB2 oncogene. In both transgenes, an SV40 intron and polyadenylation signal were included. Figure 1 indicates the schematic of this transgenic model.

Figure 2 demonstrates the gross phenotype of the K14-erbB2 mice at various stages of development. In panel A, two neonatal littermates are shown; the one on the left is monotransgenic for the TetRE-erbB2 transgene while the one on the right is bitransgenic. The mother had been administered doxycycline. The absence of phenotype in the monotransgenic mouse indicates that basal expression from the TetRE promoter is insufficient to drive biologically relevant expression of the erbB2 transgene. Panel B of this figure shows the phenotype of 11-day-old animals that had received doxycycline through maternal transfer via lactation starting at postnatal day 3. Although at day 3 the littermates were indistinguishable, over the ensuing 8 days, the bitransgenic mouse (top) developed the runted phenotype with thickened and wrinkled skin. The TetRE-erbB2 monotransgenic littermate maintained its normal development while exposed to doxycycline. Panel C is a closer view of the mice shown in panel B showing the severe skin hyperplasia in the snout area and eyelids. Panel D shows two bitransgenic mice at the age of 4 weeks. The mouse on the left was not administered doxycycline while the one on the right was adminis-

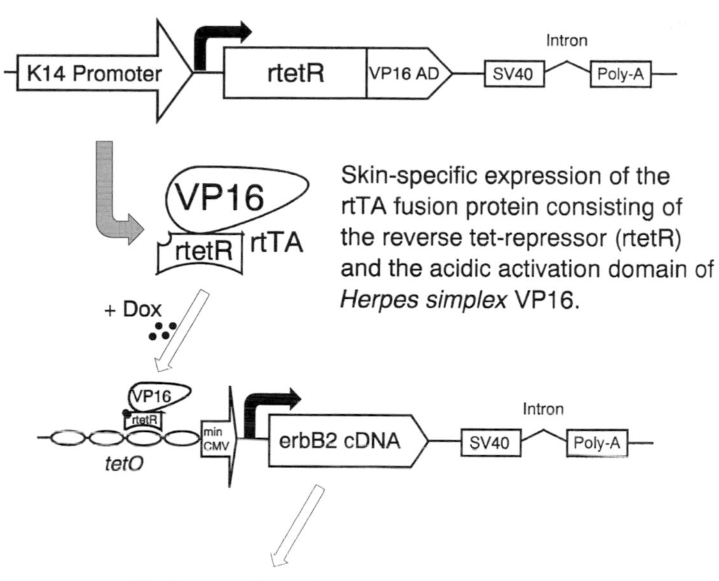

Figure 1. The Tet On inducible transgenic mouse system. A schematic outline of the K14-rtTA/TetRE-ErbB2 two-component Tet On transgenic mouse system. The K14-rtTA transgene directs expression of the reverse tetracycline transcriptional activator (rtTA) to the epidermis. The binding of rtTA to the tetracycline responsive element (*tetO*) and the induction of the transgene ErbB2 should only occur in the presence of Doxycycline (Dox).

tered the drug for 1 week prior to the photograph. The mouse on the right died 2 days later while the bitransgenic littermate that had not received doxycycline remained normal. These gross phenotypic features indicate that the lethal phenotype could be obviated by the conditional expression of erbB2 in the skin even in bitransgenic animals.

The inducibility of oncogene expression was confirmed by Northern blot analysis of the skin. Figure 3A is a Northern blot of skin from mono- and bitransgenic mice probed with the SV40 element that was present in both transgenes. Lanes 1, 3 and 4 indicate that the K14-rtTA gene was expressed constitutively in the skin of both mono- and bitransgenic animals. In contrast, the erbB2 transgene was not expressed at a detectable level in the TetRE-erbB2 monotransgenic mouse (lane 2) nor was it expressed in the bitransgenic mouse that was not treated with doxycycline. However, a biopsy sample of skin from this same bitransgenic mouse after being treated for 4 days with doxycycline shows very high expression of the TetRE-erbB2 transgene to a level even higher than seen in the salivary gland of an MMTV-erbB2 transgenic mouse.

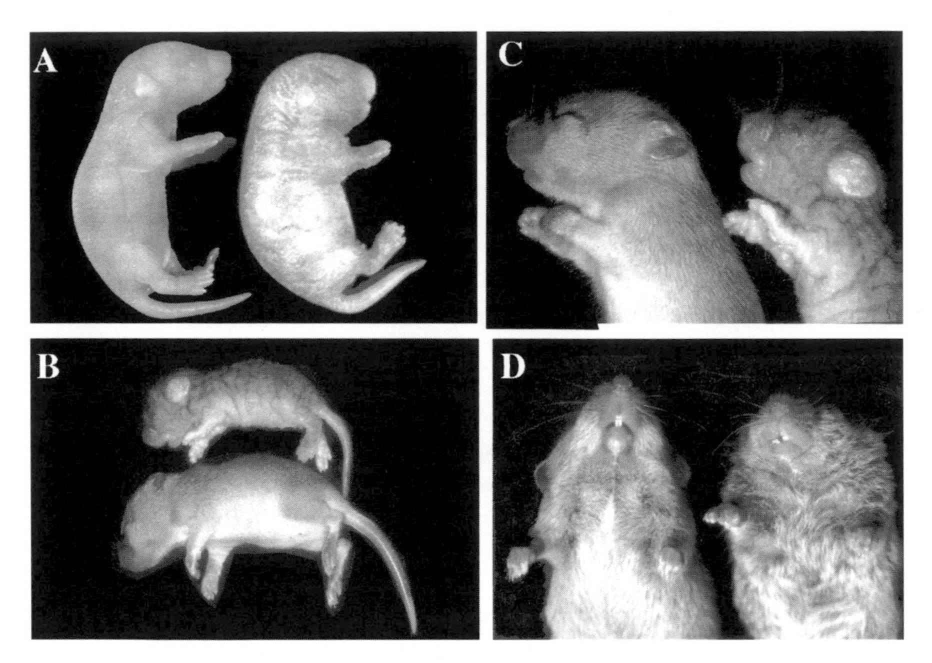

Figure 2. Gross skin phenotypes. (A) A newborn bitransgenic mouse with transplacental administration of Dox (right), as compared with its TetRE-ErbB2 single transgenic littermate (left). Note the bitransgenic skin is pale and thickened, but has no wrinkles. (B) An 11-day-old bitransgenic animal (top) under lactation from a female fed with Dox for 8 days (from day 3 to day 11), as compared with its TetRE-ErbB2 single transgenic littermate (bottom). Note the thickness of the skin of the ears, snout and paws, and the wrinkles in the bitransgenic skin. (C) Close-up view of the mice in B. (D) A weaned 4-week-old bitransgenic mouse after being fed with Dox-containing drinking water for 1 week (from week 3 to week 4) (right), as compared with its mock-treated bitransgenic littermate (left).

Figure 3. Northern blot analyses of the transgene expression. 10 µg of total cellular RNA (panels A, and C) or the poly (A) RNA (panel B) derived from 20 µg of total RNA from each skin sample was subjected to Northern blot analysis. The filters were hybridized with [^{32}P]-labeled SV40 probe (A, and C), or the ErbB2 cDNA probe (B). The filters were subsequently stripped and reprobed with the glyceraldehyde-3-phosphate dehydrogenase (GAPDH) cDNA as loading control (A, and C). (A) Dox-induced ErbB2 transgene expression in the epidermis. The rtTA transcripts were detected in the skin of all K14-rtTA bearing animals (lanes 1, and 3,4), while the expression of the 5.5 kb ErbB2 transgene was only observed in the bitransgenic mice in the presence of Dox (lane 4). The RNA in lane 5 was derived from a salivary tumor from a MMTV-ErbB2 transgenic mouse in which the ErbB2-SV40 poly (A) portion of the transgene is the same as in the TetRE-ErbB2 mice [22]. (B) A skin biopsy was obtained from a 5-week-old bitransgenic mouse (3-8-110) prior to the introduction of Dox in the drinking water (lane 1). The skin was sampled again after 4 days of Dox treatment (lane 2). The Northern blot analysis reveals that the expression of the ErbB2 transgene increased from undetectable to a level at least 20-fold higher than the endogenous ErbB2 mRNA. The induction of the ErbB2 transgene did not significantly alter the expression of the endogenous ErbB2 message (compare lanes 1 and 2). (C) Time course and the reversibility of ErbB2 induction in bitransgenic mice. Time following the initiation of Dox treatment is shown on the figure. Lane 6 shows RNA derived from liver. The induction was completely reversed after withdrawal of Dox for 7 days following a 2-day Dox treatment (lanes 7 and 8 show RNA from two mice derived from two independent experiments), and the ErbB2 transgene was re-induced 1 day after Dox was added to the drinking water after a 7-day withdrawal (lane 9). This figure is from Xie et al. [28].

Figure 3B confirms the inducibility of the erbB2 transgene expression in bitransgenic mice. This blot was probed with an erbB2 cDNA and shows RNA from the skin of the same mouse before (lane 1) and after (lane 2) 4 days of doxycycline treatment. The transgenic erbB2 gene was not expressed at a detectable level prior to antibiotic administration but after 4 days of antibiotic, it was expressed at a much higher level than the endogenous erbB2 gene. The

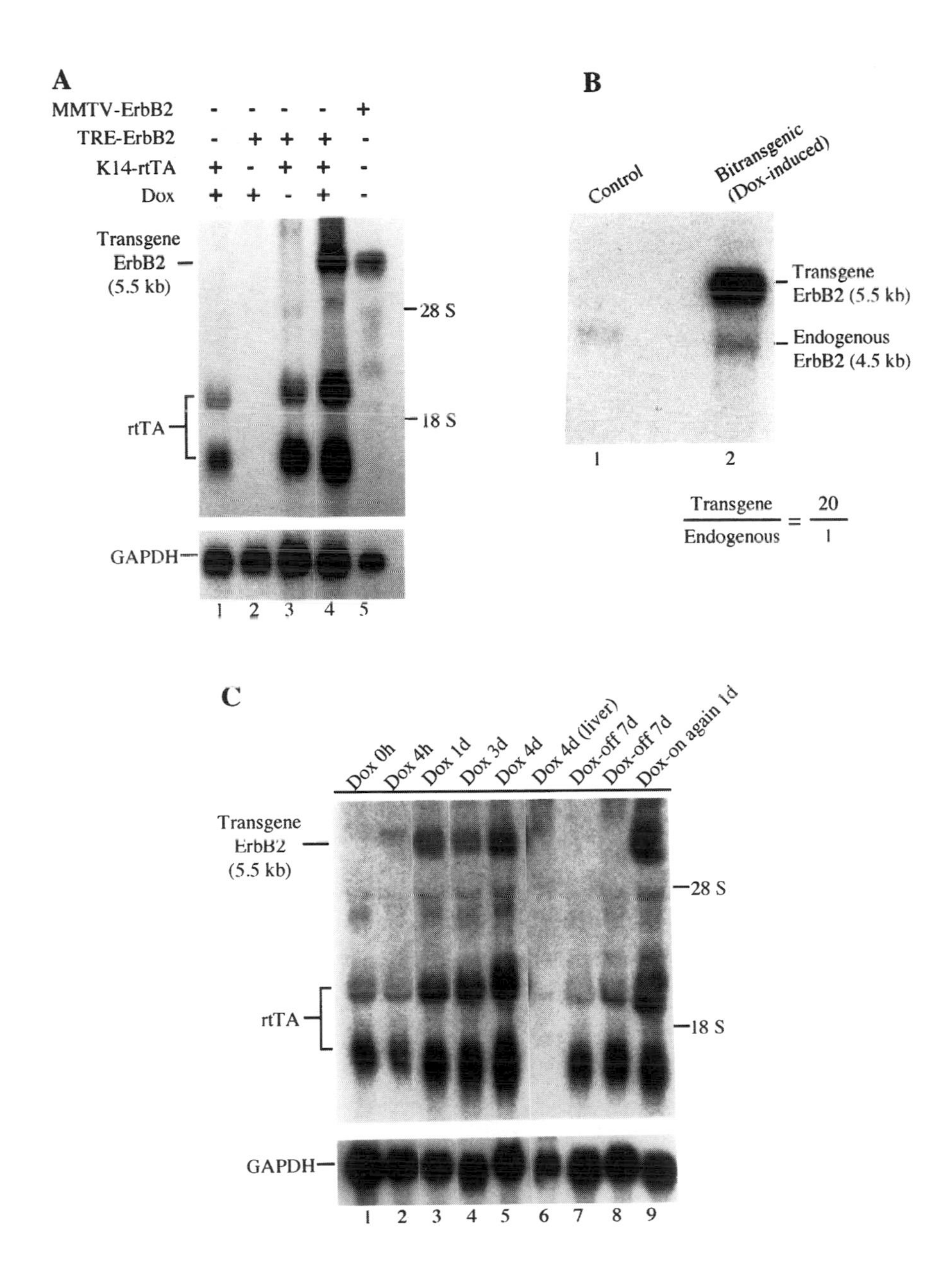

reversible inducibility and tissue-specificity of the erbB2 oncogene expression is shown in Figure 3C (see figure legend for explanation).

Microscopic examination of skin biopsies from a bitransgenic mouse also confirmed the inducibility and reversibility of the phenotype. Figure 4 shows a DNA synthesis experiment in which BrdU incorporation into DNA was monitored by immunohistochemistry. Figure 4A shows the microscopic skin morphology from a bitransgenic mouse prior to receiving doxycycline. The skin shows the normal layering and infrequent DNA synthesis confined to the basal layer. After 2 days of doxycycline treatment, there is a marked proliferation of the basal cells with frequent appearance of DNA synthesis in basal and suprabasal cells of this morphologically abnormal skin. Seven days later, after the doxycyline has been discontinued, the skin has returned to its normal morphology and rate of DNA synthesis. These morphological results correspond nicely with the observations on doxycyline-inducible transgene expression.

In addition to these proliferative effects of erbB2 expression, the differentiation pattern was also altered. In bitransgenic mice without induction, the normal progression of cytokeratin expression occurred as the cells matured from basal cells to superficial keratinocytes. However, induction of oncogene expression resulted in a maturation arrest as evidenced by persistent expression of K14/K5 keratins through all layers of the skin [28]. K14/K5 are markers of the basal cells of the epidermis and the expression of these keratins is normally turned off as the cells mature and are displaced to the more superficial layers of the skin.

Another observation made using the Tet-On K14-erbB2 model was that the induction of erbB2 expression in the skin was followed within 24 h by the induction of TGFα and EGF receptor gene expression. It had previously been observed in cell culture experiments that EGF receptor activation results in the increased expression of the EGF receptor gene [29–31] and the expression of its ligand, TGFα [32, 33]. However, this autocrine induction had not been shown *in vivo*. Because EGF receptor activation is thought to also involve the co-activation of ErbB2 [34, 35], it is possible that the tetracycline induction of ErbB2 signaling in this transgenic model activated the same pathway activated by the EGF receptor that leads to increased TGFα and EGF receptor gene transcription. This result indicates the utility of the Tet-On in studying early downstream signaling events that follow oncogene activation *in vivo*. Indeed, with the availability of state-specific antibodies that can detect phosphorylation of intracellular signaling molecules, the Tet-On system in transgenic animals will facilitate the dissection of the early downstream events following

Figure 4. Analysis of epidermal proliferation by BrdU labeling and immunostaining. Paraffin sections from bitransgenic animals without Dox-treatment (A), with 2 days of Dox-treatment (B), or 7 days of drug withdrawal following 2 days of Dox-treatment (C). Note the increased BrdU labeling in the basal cells, and the expansion of BrdU-positive nuclei into the suprabasal layer in the Dox-inducing skin (B). By 7 days after Dox withdrawal (C), the BrdU labeling shows no difference from that of the non-induced skin (A). Original magnification, A-D, 200×; E-K, 400×. From Xie et al. [28].

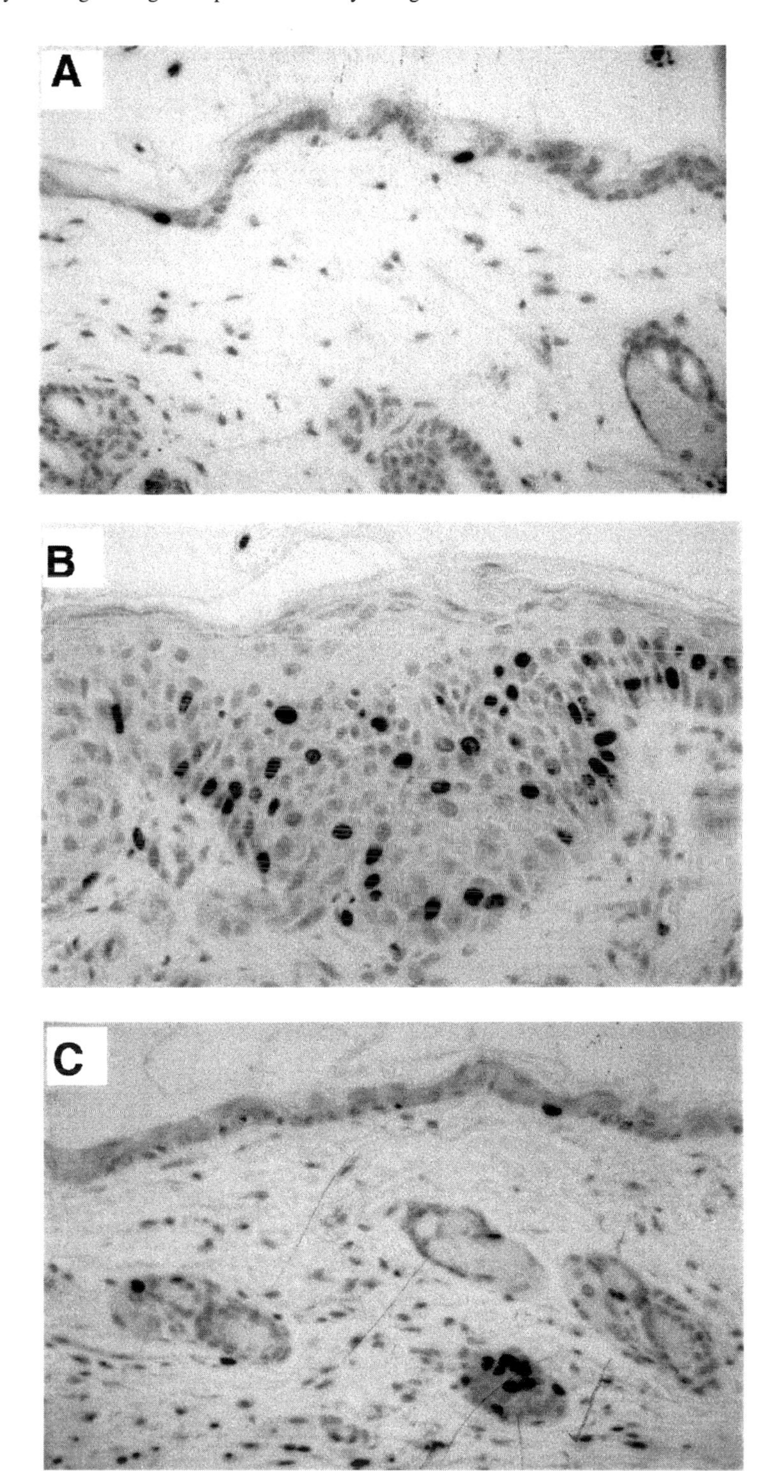
A
B
C

oncogene activation *in vivo*. Combined with dominant negative strategies, this system will allow the dissection of signal transduction pathways *in vivo*.

Tet-on and Tet-Off systems both take advantage of the pharmacological properties that doxycycline can cross the placental barrier, allowing the study of the effects of toxic genes during embryogenesis and that doxycycline is secreted into milk, allowing the induction of the transgene in neonates. But the Tet-On system has additional attractive attributes from the kinetics of induction relative to the Tet-Off system. Induction by doxycycline is rapid. Transgene expression was detected at 4 h, and reached a maximum at 24 h after exposure to doxycycline (Fig. 3). Induction is also rapidly reversible, requiring 2–3 days for reversal of the phenotype. The reversibility was observable in animals treated for as long as 7 days with doxycycline. In contrast, the Tet-Off system requires 4–7 days of the continuous presence of doxycycline [25, 36] to shut down tTA-dependent transcription and also requires the cessation of doxycycline for days before gene expression is fully induced, even in animals that have previously been fed with low doses of doxycycline to suppress gene expression [36]. In our own unpublished studies using the Tet-Off system in another transgenic mouse model, we had suppressed transgene expression from the time of conception until the mice had reached 8 weeks of age. Even after 2 weeks of withdrawal of doxycycline, we were unable to observe transgene expression. Mice from the same transgenic lines not treated with doxycycline showed robust expression of the transgene. It is likely that the doxycycline accumulated in the tissues such as bone in the developing mice, and despite withdrawal of the drug, sufficient concentrations of the antibiotic were achieved to suppress transgene expression in this Tet-Off model. Another advantage of the Tet-On system is that it is convenient and cost-effective, especially if a transgene is embryonically or perinatally toxic. While the Tet-Off system requires maternal supplementation of doxycycline throughout gestation and postnatally in order to keep the transgene off, the Tet-On system does not require this effort and expense. Our demonstration of the usefulness of the Tet-On system for the conditional expression of a lethal transgene should encourage the use of this system for the generation of other models.

SV40 T-antigen in pancreatic β-cells

Efrat and coworkers made use of an early observation by Hanahan [37] that expression of the SV40 T-antigen in pancreatic β-cells leads to the transformation of these cells but the maintenance of a differentiated phenotype. Using the Tet-Off system, Efrat's group developed transgenic mouse lines that allowed the tetracycline-inducible expression of the SV40 T-antigen in β-cells [23]. In the absence of tetracycline suppression, β-cell tumors developed and cell lines were derived from the tumors. These cell lines proliferated actively in the absence of tetracycline, but the growth was suppressed in the presence of the antibiotic. In a subsequent study [38], it was shown that the growth-arrested cells could survive in culture for an extended period of time, secrete insulin normally in response to glucose and could secrete insulin *in vivo* when transplanted into syngeneic animals. This group has proposed using a similar system for the expansion of a β-cell population for subsequent transplant therapy of patients with type 1 diabetes.

Myc in hemopoietic cells

Depending on the conditionally expressed oncogene, reversibility of the transformed phenotype may not consistently occur. Felsher and Bishop [39] investigated the myc oncogene in hematopoietic cells of transgenic mice. They wanted to determine if the transformation of these cells was reversible upon cessation of oncogene expression. They found that sustained induction of myc expression resulted in the formation of malignant T cell lymphomas and acute myeloid leukemias. With cessation of expression of the oncogene, the established tumors regressed. Tumor regression was associated with a rapid cessation of proliferation, hematopoietic differentiation, apoptosis of tumor cells and resumption of normal host hematopoiesis. The conditional expression of the myc oncogene made it possible to draw the conclusion that the entire malignant phenotype was dependent on continued expression of this single oncogene.

Bcr-Abl in hemopoietic stem cells

In another study, Era and Witte [40] investigated a similar question of the role of a single oncogene, in this case, p210 Bcr-Abl, the activated tyrosine kinase oncogene encoded by the Philadelphia chromosome associated with human chronic myelogenous leukemia. Using the Tet-Off system, they expressed the Bcr-Abl oncogene in stem cells in a conditional manner. They showed that Bcr-Abl expression alone was sufficient to shift differentiation of the stem cells towards the multipotent and myeloid lineages and away from the erythroid progenitors. This effect on differentiation was reversible upon suppression of oncogene expression. The use of this conditional system established the notion that Bcr-Abl is the sole genetic change needed for the establishment of the chronic myelogenous leukemia and that this genetic alteration was not the result of a secondary effect in this form of leukemia. This result, like the myc result, suggest that even though tumorigenesis is a multistep process, reversal of a single genetic lesion may be sufficient to reverse malignancy.

H-Ras in a melanoma model

Dependence of a transformed line on a single oncogene was also demonstrated by Chin et al. [41] using the Tet-On system. In a mouse melanoma model null for the INK4a tumor suppressor, expression of oncogenic H-RasV12G was found to be necessary for tumor maintenance. When expression of the oncogene was extinguished upon withdrawal of the doxycycline, not only did the tumors regress with marked apoptosis, but the host-derived endothelial cells also underwent apoptosis. The endothelial apoptosis could not be prevented by provision of exogenous vascular endothelial growth factor, indicating that activated Ras in the tumor cells was required for tumor maintenance and vascular supply in a manner that could not be substituted by the growth factor. The apoptosis of these tumor cells upon removal of the oncogene stimulus differs from the observation of Efrat with SV40 T antigen in β-cells [23, 38]. In the β-cells, growth arrest was not associated with apoptosis.

SV40 T-antigen in submandibular gland

Reversibility of oncogene effects also appears to be tissue-dependent and time dependent. When the SV40 T-antigen was expressed in the submandibular gland of transgenic mice, using the Tet-Off system, ductal hyperplasia and cellular transformation were observed [24]. If induction of oncogene expression was silenced for 3 weeks with tetracycline in an animal that had expressed the oncogene from birth until 4 months of age, the entire phenotype was reversed. However, if induction of the oncogene was maintained for 7 months and then silenced, the hyperplasia and ductal cell polyploidy persisted. The authors concluded that a secondary event occurred in a time-dependent manner that allowed transformation to continue even in the absence of the original inciting oncogene. This point emphasizes the power of an inducible system over a constitutive one.

Matrix metalloproteinases in mammary cancer

Matrix metalloproteinases (MMPs) are generally thought to play a late role in cancer development. The MMPs are upregulated in the stromal compartment of epithelial cancers and appear to promote invasion and metastasis. In a study in which MMP3/stromelysin-1 was expressed conditionally in mammary epithelia of transgenic mice using tetracycline regulation, it was shown that induction of MMP expression in the epithelial cells caused the development of invasive mesenchymal-like tumors [42]. Interestingly, once these changes were initiated, cessation of MMP expression did not reverse the phenotype. While MMPs are not thought of as classic oncogenes, this study suggests that these proteases play an earlier role in carcinogenesis than had been appreciated. The conditional expression of the protease was vital for this observation.

Conclusions

In this review, I have not included every *in vivo* study of inducible oncogene expression using a tetracycline system. Rather, I have chosen to illustrate, using the published studies, the power of this system to dissect the role of oncogenes in cancer and physiological regulation. The major advantages of this inducible system are as follows: 1. Combinatorial selection of the oncogene and tissue of choice; 2. Obviates a lethal phenotype; 3. Once a tet-activator or TetRE-oncogene transgenic line is established, the line can be used for further studies without the need for integration site controls; 4. Inducibility allows the determination of early consequences of oncogene activation and avoids the long-term secondary effects or allows the systematic study of these secondary effects. These properties are absolutely vital for the study of oncogenes in transgenic animals or other *in vivo* models because the consequences of oncogene activation depend strongly on the stage of development of the animal, the time period over which activation has occurred and the tissue in which the oncogene is expressed.

References

1 Strauss M, Hering S, Lubbe L, Griffin BE (1990) Immortalization and transformation of human fibroblasts by regulated expression of polyoma virus T antigens. *Oncogene* 5: 1223–1229

2 Mercer WE, Shields MT, Amin M, Sauve GJ, Appella E, Romano JW, Ullrich SJ (1990) Negative growth regulation in a glioblastoma tumor cell line that conditionally expresses human wild-type p53. *Proc Natl Acad Sci USA* 87: 6166–6170

3 Mueller SG, Paterson AJ, Kudlow JE (1990) Transforming growth factor-alpha in arterioles, cell surface processing of its precursor by elastases. *Mol Cell Biol* 10: 4596–602, 1990

4 Musgrove EA, Lee CS, Buckley MF, Sutherland RL (1994) Cyclin D1 induction in breast cancer cells shortens G1 and is sufficient for cells arrested in G1 to complete the cell cycle. *Proc Natl Acad Sci USA* 91: 8022–8026

5 Li Y, Hively WP, Varmus HE (2000) Use of MMTV-Wnt-1 transgenic mice for studying the genetic basis of breast cancer. *Oncogene* 19: 1002–1009

6 Yao Y, Slosberg ED, Wang L, Hibshoosh H, Zhang YJ, Xing WQ, Santella RM, Weinstein IB (1999) Increased susceptibility to carcinogen-induced mammary tumors in MMTV-Cdc25B transgenic mice. *Oncogene* 18: 5159–5166

7 Xie W, Paterson AJ, Chin E, Nabell LM, Kudlow JE (1997) Targeted expression of a dominant negative epidermal growth factor receptor in the mammary gland of transgenic mice inhibits pubertal mammary duct development. *Mol Endocrinol* 11: 1766–1781

8 Sandgren EP, Luetteke NC, Palmiter RD, Brinster RL, Lee DC (1990) Overexpression of TGFα in transgenic mice: induction of epithelial hyperplasia, pancreatic metaplasia, and carcinoma of the breast. *Cell* 61: 1121–1135

9 Joseph H, Gorska AE, Sohn P, Moses HL, Serra R (1999) Overexpression of a kinase-deficient transforming growth factor-beta type II receptor in mouse mammary stroma results in increased epithelial branching. *Mol Biol Cell* 10: 1221–1234

10 Saez E, No D, West A, Evans RM (1997) Inducible gene expression in mammalian cells and transgenic mice. *Curr Opin Biotechnol* 8: 608–616

11 No D, Yao TP, Evans RM (1996) Ecdysone-inducible gene expression in mammalian cells and transgenic mice. *Proc Natl Acad Sci USA* 93: 3346–3351

12 Gill G, Ptashne M (1988) Negative effect of the transcriptional activator GAL4. *Nature* 334: 721–724

13 Min S, Crider-Miller SJ, Taparowsky EJ (1994) The transcription activation domains of v-Myc and VP16 interact with common factors required for cellular transformation and proliferation. *Cell Growth Differ* 5: 563–573
14 Freundlieb S, Baron U, Bonin AL, Gossen M, Bujard H (1997) Use of tetracycline-controlled gene expression systems to study mammalian cell cycle. *Methods Enzymol* 283: 159–173
15 Yueh YG, Yaworsky PJ, Kappen C (2000) *Herpes simplex* virus transcriptional activator VP16 is detrimental to preimplantation development in mice. *Mol Reprod Dev* 55: 37–46
16 Kokai Y, Cohen JA, Drebin JA, Greene MI (1987) Stage- and tissue-specific expression of the neu oncogene in rat development. *Proc Natl Acad Sci USA* 84: 8498–8501
17 Quirke P, Pickles A, Tuzi NL, Mohamdee O, Gullick WJ (1989) Pattern of expression of c-erbB-2 oncoprotein in human fetuses. *Brit J Cancer* 60: 64–69
18 Maguire HC Jr, Jaworsky C, Cohen JA, Hellman M, Weiner DB, Green MI (1989) Distribution of neu (c-erbB-2) protein in human skin. *J Invest Dermatol* 89: 786–790
19 Bouchard L, Lamarre L, Tremblay PJ, Jolicoeur P (1989) Stochastic appearance of mammary tumors in transgenic mice carrying the MMTV/c-neu oncogene. *Cell* 57: 931–936
20 Muller WJ, Sinn E, Pattengale PK, Wallace R, Leder P (1988) Single-step induction of mammary adenocarcinoma in transgenic mice bearing the activated c-neu oncogene. *Cell* 54: 105–115
21 Stocklin E, Botteri F, Groner B (1993) An activated allele of the c-erbB-2 oncogene impairs kidney and lung function and causes early death of transgenic mice. *J Cell Biol* 122: 199–208
22 Xie W, Chow LT, Wu X, Chin E, Paterson AJ, Kudlow JE (1998) Targeted expression of activated erbB2 to the epidermis of transgenic mice elicits striking developmental abnormalities in the epidermis and hair follicles. *Cell Growth Differ* 9: 313–325
23 Efrat S, Fusco-DeMane D, Lemberg H, al Emran O, Wang X (1995) Conditional transformation of a pancreatic beta-cell line derived from transgenic mice expressing a tetracycline-regulated oncogene. *Proc Natl Acad Sci USA* 92: 3576–80
24 Ewald D, Li M, Efrat S, Auer G, Wall RJ, Furth PA, Hennighausen L (1996) Time-sensitive reversal of hyperplasia in transgenic mice expressing SV40 T antigen. *Science* 273: 1384–1386
25 Mayford M, Bach ME, Huang Y-Y, Wang L, Hawkins RD, Kandel ER (1996) Control of memory formation through regulated expression of a CaMKII transgene. *Science* 274: 1678–1683
26 Yu Z, Redfern CS, Fishman GI (1996) Conditional transgene expression in the heart. *Circ Res* 79: 691–697
27 Lee P, Morley G, Huang Q, Fischer A, Seiler S, Horner JW, Factor S, Vaidya D, Jalife J, Fishman GI (1998) Conditional lineage ablation to model human diseases. *Proc Natl Acad Sci USA* 95: 11 371–11 376
28 Xie W, Chow LT, Paterson AJ, Chin E, Kudlow JE (1999) Conditional expression of the erbB2 oncogene elicits reversible hyperplasia in stratified epithelia and up-regulation of TGFα expression in transgenic mice. *Oncogene* 18: 3593–3607
29 Clark AJ, Ishii S, Richert N, Merlino GT, Pastan I (1985) Epidermal growth factor regulates the expression of its own receptor. *Proc Natl Acad Sci USA* 82: 8374–8378
30 Kudlow JE, Cheung M, Bjorge JD (1986) Epidermal growth factor stimulates the synthesis of its own receptor in a human breast cancer cell line. *J Biol Chem* 261: 4134–4138
31 Earp HS, Austin KS, Blaisdell J, Rubin RA, Nelson KG, Lee LW, Grisham JW (1986) Epidermal growth factor (EGF) stimulates EGF receptor synthesis. *J Biol Chem* 261: 4777–4780
32 Coffey RJ Jr, Derynck R, Wilcox JN, Bringman TS, Goustin AS, Moses HL, Pittelkow MR (1987) Production and auto-induction of transforming growth factor-α in human keratinocytes. *Nature* 328: 817–820
33 Bjorge JD, Paterson AJ, Kudlow JE (1989) Phorbol ester or epidermal growth factor (EGF) stimulate the concurrent accumulation of mRNA for the EGF receptor and its ligand transforming growth factor-alpha in a breast cancer cell line. *J Biol Chem* 264: 3880–3883
34 Carraway KL, III, Cantley LC (1994) A Neu acquaintance for erbB3 and erbB4: a role for receptor heterodimerization in growth signaling. *Cell 78*: 5–8
35 Dougall WC, Qian X, Paterson NC, Miller MJ, Samanta A, Green MI (1994) The neu-oncogene: signal transduction pathways, transformation mechanism and evolving therapies. *Oncogene* 9: 2109–2123
36 Kistner A, Gossen M, Zimmermann F, Jerecic J, Ullmer C, Lubbert H, Bujard H (1996) Doxycycline-mediated quantitative and tissue-specific control of gene expression in transgenic mice. *Proc Natl Acad Sci USA* 93: 10 933–10 938
37 Hanahan D (1985) Heritable formation of pancreatic beta-cell tumours in transgenic mice express-

ing recombinant insulin/simian virus 40 oncogenes. *Nature* 315: 115–122
38 Fleischer N, Chen C, Surana M, Leiser M, Rossetti L, Pralong W, Efrat S (1998) Functional analysis of a conditionally transformed pancreatic beta-cell line. *Diabetes* 47: 1419–1425
39 Felsher DW, Bishop JM (1999) Reversible tumorigenesis by MYC in hematopoietic lineages. *Mol Cells* 4: 199–207
40 Era T, Witte ON (2000) Regulated expression of p210 Bcr-Abl during embryonic stem cell differentiation stimulates multipotential progenitor expansion and myeloid cell fate. *Proc Natl Acad Sci USA* 97: 1737–1742
41 Chin L, Tam A, Pomerantz J, Wong M, Holash J, Bardeesy N, Shen Q, O'Hagan R, Pantginis J, Zhou H et al (1999) Essential role for oncogenic Ras in tumour maintenance. *Nature* 400: 468–472
42 Sternlicht MD, Lochter A, Sympson CJ, Huey B, Rougier JP, Gray JW, Pinkel D, Bissell MJ, Werb Z (1999) The stromal proteinase MMP3/stromelysin-1 promotes mammary carcinogenesis. *Cell* 98: 137–146

Tetracyclines in Biology, Chemistry and Medicine
ed. by M. Nelson, W. Hillen and R.A. Greenwald

Interactions of tetracyclines with RNA

Christian Berens

Lehrstuhl für Mikrobiologie, Institut für Mikrobiologie, Biochemie und Genetik, Friedrich-Alexander Universität Erlangen-Nürnberg, Staudtstr. 5, D-91058 Erlangen, Germany

Introduction

The tetracyclines (Fig. 1) are secondary metabolites produced by Gram-positive bacteria from the genera *Streptomyces* or *Nocardia*. They act as antibiotics and inhibit protein biosynthesis by interfering with binding of the aminoacyl-tRNA-EF-Tu-GTP ternary complex to the ribosomal A-site [1]. Tetracyclines show broad-spectrum activity against a wide range of Gram-negative and

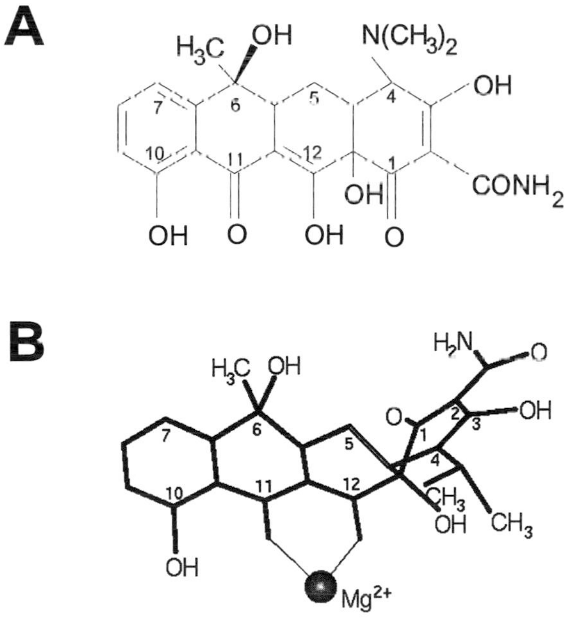

Figure 1. Structure of tetracycline. (A) Structure of tetracycline. (B) Three-dimensional structure of the tetracycline-Mg^{2+} complex. Tetracycline is shown as a stick model with chemical groups indicated, Mg^{2+} is represented by a gray sphere. Thin gray lines indicate direct contacts between Mg^{2+} and the ketoenolate group O11/O12 of tetracycline. The coordinates were taken from the crystal structure of the Tet repressor-tetracycline complex 2TRT [107].

Gram-positive bacteria, cell wall free mycoplasmas, rickettsiae, chlamydiae and protozoan parasites.

Over the past 45 years, the tetracyclines have been the subject of numerous reviews concerning their mode of antibiotic action [2–4], the biosynthetic pathways [5], chemistry [6–9], structure-function relationships [10, 11] and their uptake into cells [12, 13], as well as the emergence, spread and mechanisms of action of resistance determinants [14–20].

This review, however, covers a different aspect of tetracycline biochemistry: the interactions of tetracyclines with ribonucleic acids. About 30 years ago, tetracycline was proposed to bind to the ribosome by chelating RNA-bound Mg^{2+} ions [21] and the first model of a tetracycline binding site on the ribosome consisting solely of RNA was published [22]. Since then, only little experimental attention was devoted to studying tetracycline-RNA interactions. The excitement following the discovery of catalytic RNA [23, 24], the development of SELEX, a powerful method to isolate RNA molecules as high-affinity ligands for an incredible variety of molecular targets [25], as well as the observation that inhibitors of protein biosynthesis bind RNA (for reviews, see Cundliffe [26] and Wallis and Schroeder [27]) and can also inhibit ribozymes [28], have sparked great interest in essentially all aspects of RNA molecular biology (see both editions of The RNA World [29, 30]). This has also rekindled the interest in tetracycline-RNA interactions, leading to the publication of quite a few papers over the last 10–15 years, which will be the subject of this review.

Typical minimum inhibitory concentrations of sensitive *E. coli* cells are in the range of 1–2 μM tetracycline in the growth medium (see, e.g., Rasmussen et al. [31]). pH differences between the growth medium and the bacterial cytoplasm can lead to up to 25-fold accumulation of tetracycline within the cells [12, 32, 33]. So, tetracycline concentrations below 50 μM should be sufficient for antibiotic activity. I would therefore like to emphasize that data from *in vitro* experiments utilizing tetracycline concentrations higher than 100 μM must be treated with caution, as these concentrations are close to or more than one order of magnitude higher than the tetracycline concentrations needed *in vivo* for antibiotic activity.

Interaction of tetracyclines with bulk RNA

Binding of tetracycline to total RNA from yeast [34], to poly-uridinylic acid [35, 36], poly-adenylic acid and tRNA [36] has been demonstrated by equilibrium dialysis, ultracentrifugation, and filtration through Sephadex G25 columns. Equilibrium binding constants and stoichiometries were not determined, but this binding appears to be weak and nonspecific. Hydrophobic interactions might control binding of tetracyclines to bulk RNA, since 7-chlortetracycline (see Fig. 2 for the structures of tetracycline derivatives) is more hydrophobic than 6-demethyl-7-chlortetracycline, due to the additional methyl group, and also binds more strongly to bulk RNA [34]. However, this cannot

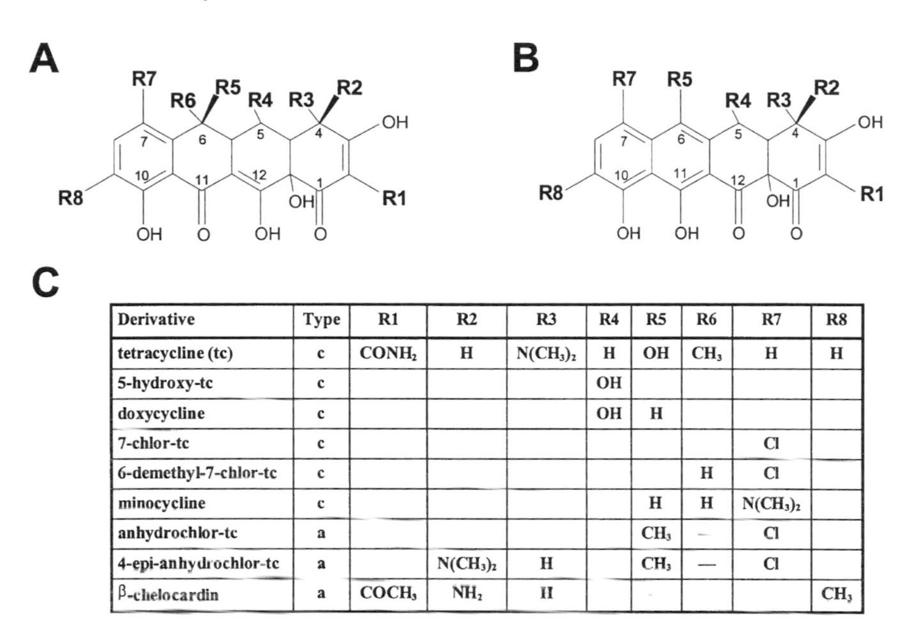

Derivative	Type	R1	R2	R3	R4	R5	R6	R7	R8
tetracycline (tc)	c	$CONH_2$	H	$N(CH_3)_2$	H	OH	CH_3	H	H
5-hydroxy-tc	c				OH				
doxycycline	c				OH	H			
7-chlor-tc	c							Cl	
6-demethyl-7-chlor-tc	c						H	Cl	
minocycline	c					H	H	$N(CH_3)_2$	
anhydrochlor-tc	a					CH_3	–	Cl	
4-epi-anhydrochlor-tc	a		$N(CH_3)_2$	H		CH_3	—	Cl	
β-chelocardin	a	$COCH_3$	NH_2	H					CH_3

Figure 2. Tetracycline derivatives. (A) Structure of tetracycline. (B) Structure of anhydrotetracycline. (C) Substitutions in the tetracycline and anhydrotetracycline derivatives tested. Classical tetracyclines are designated with "c", atypical tetracyclines by "a" [3]. Only the respective differences to tetracycline or anhydrotetracycline are shown.

be generalized from the analysis of only two tetracyclines. In any case, this binding is not physiologically important, as pretreatment of poly adenylic acid and tRNA with tetracycline did not interfere with protein synthesis *in vitro* [37].

Recently, experiments in tissue culture or in whole organisms have shown that the administration of tetracycline derivatives like doxycycline and minocycline can upregulate [38] or downregulate [39–42] the expression of specific mRNAs. Unfortunately, it is not known if the changes in the mRNA levels observed in all these cases are caused by direct interaction of the tetracycline derivative with the respective mRNA or if the tetracyclines act indirectly via one or more unknown target molecules. Further experiments are clearly needed to clarify this issue.

Binding of tetracyclines to the ribosome

A large number of studies have shown binding of tetracycline to the *E. coli* ribosome [36, 43–48]. The general consensus has emerged that there is one strong binding site on the 30S subunit with an equilibrium binding constant K_d in the range of 1–20 μM and that binding to this site is necessary and sufficient to interfere with protein synthesis. In addition, there are many low-affinity binding sites on both subunits with K_ds in the high μM to low mM range. These low-affinity binding sites can account for up to 30–40% of the overall tetracycline bound to the ribosome.

In order to identify the tetracycline binding site on the ribosome, empty *E. coli* ribosomes were UV-irradiated in the presence of [^{3}H]-tetracycline [49]. Under the reaction conditions, radioactivity was almost exclusively incorporated into protein. The following evidence supports the notion that this label originates from a functional binding site for tetracycline: (i) the protein predominantly labeled by [^{3}H]-tetracycline, the S7 protein, is located in the 30S subunit which contains the high-affinity binding site for tetracycline. (ii) Not only inhibition of bacterial translation, but also binding to the 30S subunit and labeling of S7 by 4-epi-tetracycline is significantly weaker than by tetracycline. (iii) Labeling of S7 is competed for by nonradioactive tetracycline, indicating a saturable site. This would not be expected for binding to nonspecific sites.

Data from a single protein omission reconstitution study (SPORE) complement these results [50]. Four proteins, S3, S7, S14 and S19, that interact with rRNA in the 3' major domain of 16S rRNA [51], are important for tetracycline binding. SPORE particles lacking one of these proteins bound less than 30% of the tetracycline bound by a native 30S subunit. In footprinting studies [51], three of the four proteins (S3, S7 and S19) protected bases close to other sites of tetracycline-RNA interaction (see below and in Fig. 3) from cleavage by Fe^{2+}/EDTA.

Interaction of tetracyclines with ribosomal RNA

Peptidyl transferase activity is unusually resistant to protein extraction procedures [52, 53], and peptide bond formation can be catalyzed by ribozymes selected *in vitro* [54], which indicates that the rRNA is the component responsible for ribosomal activity. It is, therefore, a sensible working hypothesis that tetracycline acts either by directly interacting with, or by otherwise perturbing, the structure of rRNA.

Chemical probing of tetracycline–RNA interactions

Chemical probing determines the accessibility of every nucleotide in an RNA towards chemicals that attack atoms in bases, ribose, or phosphate groups. It is a powerful tool to study RNA structure or to identify ligand interaction sites. Upon ligand binding, the modification pattern of an RNA under study can change either due to the close proximity of the ligand to a nucleotide or by direct ligand-residue interactions. Alternatively, a ligand-induced conformational change in the RNA might modify the accessibility to the chemical [55, 56]. 70S ribosomes from *E. coli* that were not charged with tRNA were probed

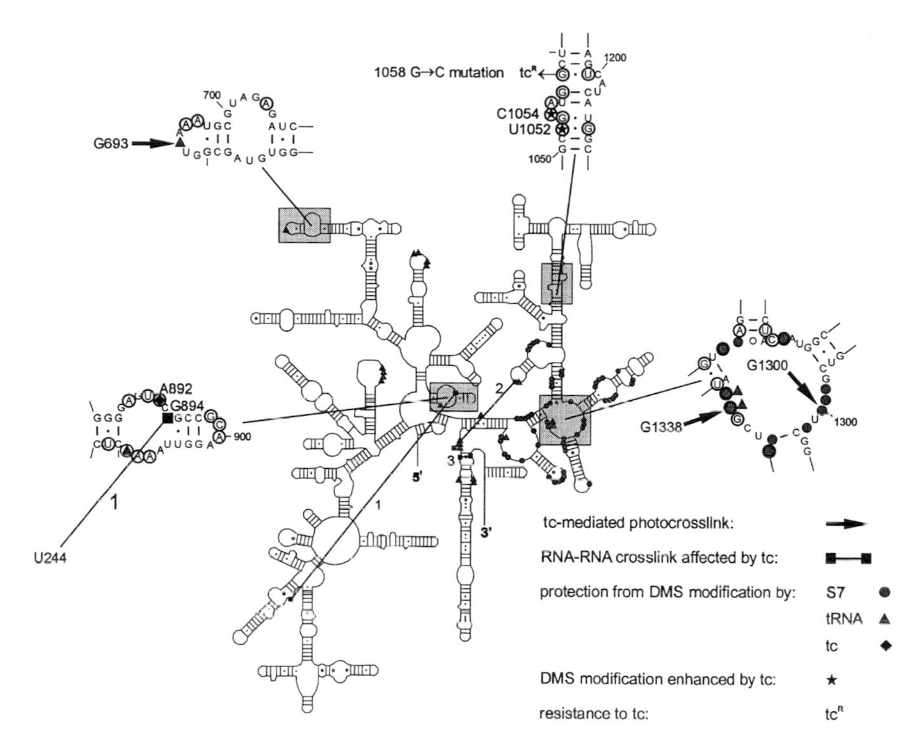

Figure 3. Sites of interaction of tetracycline with the 16S rRNA. The secondary structure of the *Escherichia coli* 16S rRNA [108] is shown schematically. Located within the gray boxes and shown in more detail in the enlarged sections are bases that (i) become photocrosslinked to tetracycline [61], (ii) display altered reactivity towards chemical probing in the presence of tetracycline [57], or (iii) when mutated lead to resistance against tetracycline [64]. RNA-RNA crosslinks whose intensity is altered in the presence of tetracycline [63] are numbered 1 through 3. Universally conserved bases [109] within the enlarged sections are encircled. In addition, filled circles highlight bases protected from chemical modification by binding of the S7 protein [110], while filled triangles indicate bases protected from chemical probing in the presence of tRNA [111].

with DMS in the presence of 250 μM tetracycline [57]. In the 16S rRNA, strong protection against methylation by DMS was observed at position A892 in the internal loop of helix 27, while methylation by DMS was enhanced at positions U1052 and C1054 in helix 34 (Fig. 3). The latter sites are close to bases footprinted by the S3 protein [51]. Analysis of the DMS probing pattern of tetracycline derivatives revealed that protection of A892 is confined to derivatives that directly inhibit protein synthesis and additionally contain a pseudoaxial OH group at the C-6 position (Fig. 2C) [31]. This protection most likely represents a close proximity or direct contact between drug and base. The atypical analogs chelocardin and 4-epi-anhydrochlortetracycline are neither ribosomal inhibitors, nor do they alter reactivity at the three bases to DMS. This probably relates to the amino- and dimethylamino groups at C-4 which are in the β-epimer configuration, as opposed to the α-epimer configuration present in the other derivatives (Fig. 2C). A third group of derivatives containing both inhibitors (minocycline, doxycycline, 6-demethyl-6-deoxytetracycline) and non-inhibitors (anhydrochlortetracycline, 6-thiatetracycline) of ribosomal translation only altered reactivity to DMS at U1052 and C1054. These analogs apparently bind to the ribosome, but not all inhibit protein synthesis.

In the three-dimensional structures of the 16S rRNA, the modified bases are too remote for a single drug molecule to be able to contact both [58, 59]. Due to the high tetracycline concentrations used (250 μM [57] and 500 μM [31]), nonspecific binding of tetracyclines to the ribosome cannot be ruled out, so that more than one tetracycline molecule might participate in the protection and stimulation observed. It is also not known if the changes in base reactivity to DMS are caused by a tetracycline molecule bound to the inhibitory site.

Crosslinking of tetracycline to the ribosome

The observation that tetracycline crosslinked efficiently to loop V of 23S rRNA [60] led to a reinvestigation of tetracycline photoincorporation into rRNA. Under different experimental conditions than in the previous study [49], tetracycline photocrosslinked to ribosomal proteins and rRNA at an approximately equal ratio [61]. In agreement with Goldman's results [49], the major protein labeled was S7. In addition, tetracycline crosslinked to sites in the 16S and 23S rRNAs. The sites in the 23S rRNA are not discussed, because they do not affect ribosomal function [61]. The positions of the three sites in the 16S rRNA (G693, G1300, G1338) are shown in Figure 3. They overlap with or are close to footprint sites of protein S7 or tRNA. Chimeric ribosomes consisting of photoaffinity-labeled 30S and non-irradiated 50S subunits were impaired in tRNA binding to the A-site, while P-site tRNA binding and peptidyl transferase activity were not affected. This indicates that the crosslinks originate from tetracycline bound to a functional site since tetracycline interferes with tRNA binding to the ribosomal A-site *in vitro* [4].

Tetracycline affects intramolecular 16s rRNA crosslinks

Fourteen intramolecular crosslinks induced by UV irradiation have been identified in the 16S rRNA of *Escherichia coli* [62]. Monitoring changes in their intensity provides an opportunity to detect conformational changes in the ribosome structure. Irradiation of empty 70S ribosomes in the presence of 25 μM tetracycline affected 3 of the 14 crosslinks [63]. They are shown in Figure 3. Formation of the C967×C1400 crosslink ("1") was completely inhibited. That of U244×G894 ("2") decreased 2-fold, while that of C1402×C1501 ("3") increased 2-fold. Five of the six bases crosslinked are close to footprint sites of tetracycline and tRNA, and the proteins S3 and S19 [51]. In a three-dimensional model of the 16S rRNA [63], the sites of tetracycline interaction with the 16S rRNA (rRNA-rRNA crosslinks, tetracycline-rRNA crosslinks, tetracycline footprints) are significantly separated from each other, suggesting that tetracycline binding affects the structure of the 30S subunit regionally or globally, rather than just locally.

A mutation in helix 34 of 16 S rRNA leads to tetracycline resistance

Treatment of acne vulgaris with tetracyclines has led to the development of resistant *Propionibacterium acnes* strains. The majority of these strains contained a single base change (G $\rightarrow$ C) at position 1058 (*E. coli* numbering) in helix 34 of the 16S rRNA [64]. This mutation disrupts a Watson-Crick base pair and is in close proximity to bases U1052 and C1054, which show enhanced reactivity towards DMS in the presence of tetracycline [57]. To check if this mutation was responsible for resistance, it was introduced into *E. coli* on a plasmid (pKK3535) containing the ribosomal *rrnB* operon, since propionibacteria are currently not genetically manipulatable. Transformants containing mutant plasmids were 2-fold less sensitive to tetracycline than cells containing the wildtype plasmid. This low level of resistance is somewhat misleading, because the plasmid-encoded rRNA is expressed in the background of expression of the seven chromosomally encoded *rrn* operons. 16S rRNA mutations conferring resistance to streptomycin and to neomycin showed no or only little resistance to the antibiotic when introduced into pKK3535 [65, 66]. It was then demonstrated that high levels of resistance to these antibiotics could be achieved when a mutation conferring resistance to spectinomycin was additionally introduced into the plasmid [65, 66]. In this case, the presence of spectinomycin eliminates the chromosome-borne and spectinomycin-sensitive ribosomes from the population of actively translating ribosomes [65, 66]. Thus, the bacteria become solely dependent on the plasmid-encoded rRNA for translation. The introduction of the G1058C mutation into a spectinomycin-resistant 16S rRNA and its subsequent expression in *E. coli* would allow for a more accurate evaluation of the resistance to tetracycline mediated by this base transversion.

Conclusions

As shown in Figure 3, tetracycline affects bases at many different positions in the 16S rRNA. In the recently published three-dimensional structures of the 30 S subunit from *Thermus thermophilus* [58, 59], the distance between the S7 binding site and the decoding site is clearly much larger than the size of a single tetracycline molecule (see Fig. 7 in Clemons et al. [58] and Figure 3 from Tocilj et al. [59]). So, do the interactions originate from several nonspecifically bound molecules or from just one which induces conformational changes in the 30S subunit? While the former possibility cannot formally be excluded, several lines of evidence suggest that tetracycline acts by inducing conformational changes in the 30S subunit. (i) Tetracycline does not directly compete with tRNA at the decoding site and the peptidyl transferase center, but rather weakens tRNA binding allosterically. (ii) The sites in helices 27 and 34 in the 16S rRNA that are affected by tetracycline are close to the ribosomal decoding site, but about 50 Å away from the S7 binding site. (iii) Enhancement of modification to DMS, as observed at positions 1052 and 1054, is mostly due to conformational changes which alter accessibility of the base to the modifying agent. Close proximity of or direct contacts between a ligand and a base generally lead to protection from modification. Taken together, this suggests that tetracycline binds to the 30S subunit and induces conformational changes that prevent tRNA from binding to the A-site of the ribosome.

The high-affinity binding site for tetracycline seems to be located at the base of the 3' domain of 16S rRNA. Photocrosslinking of tetracycline to the 16S rRNA at concentrations close to the K_ds for binding to the high-affinity site and for *in vivo* antibiotic activity yielded two interaction sites that overlap with footprint sites of S7, the major protein crosslinked to tetracycline. This site is also a functional site and most likely the site of antibiotic action, as chimeric 70S ribosomes containing photocrosslinked 30S subunits are impaired in tRNA binding to the A-site, reflecting the drug's activity *in vitro*. Tetracycline is incorporated to an approximately equal extent into RNA and S7, but it does not photocrosslink to isolated proteins from the 30S subunit [50]. Therefore, tetracycline might bind to a mixed RNA-protein site.

Inhibition of ribozymes by tetracyclines

Except for the hairpin ribozyme [67] and RNase P RNA [68], at least one member from each of the ribozyme families known to occur naturally has been checked for inhibition by either tetracycline or by one of its derivatives. The secondary structures of ribozymes that are inhibited by tetracyclines are shown in Figures 4–7. As the experimental data set is very heterogenous, I will first present the published data on each of the different ribozymes in the following section and then try to identify trends.

Group I introns

Group I introns are autocatalytic genetic elements that have been found in the nucleus and in the organelles of eucaryotes, in their viruses, in eubacteria and bacteriophages. They mediate their own removal from transcripts containing them via a characteristic structure which distinguishes them from other introns (for reviews, see [69] and [70]) and a splicing pathway which utilizes an external guanosine as nucleophile in the first step of splicing. *In vitro* self-splicing

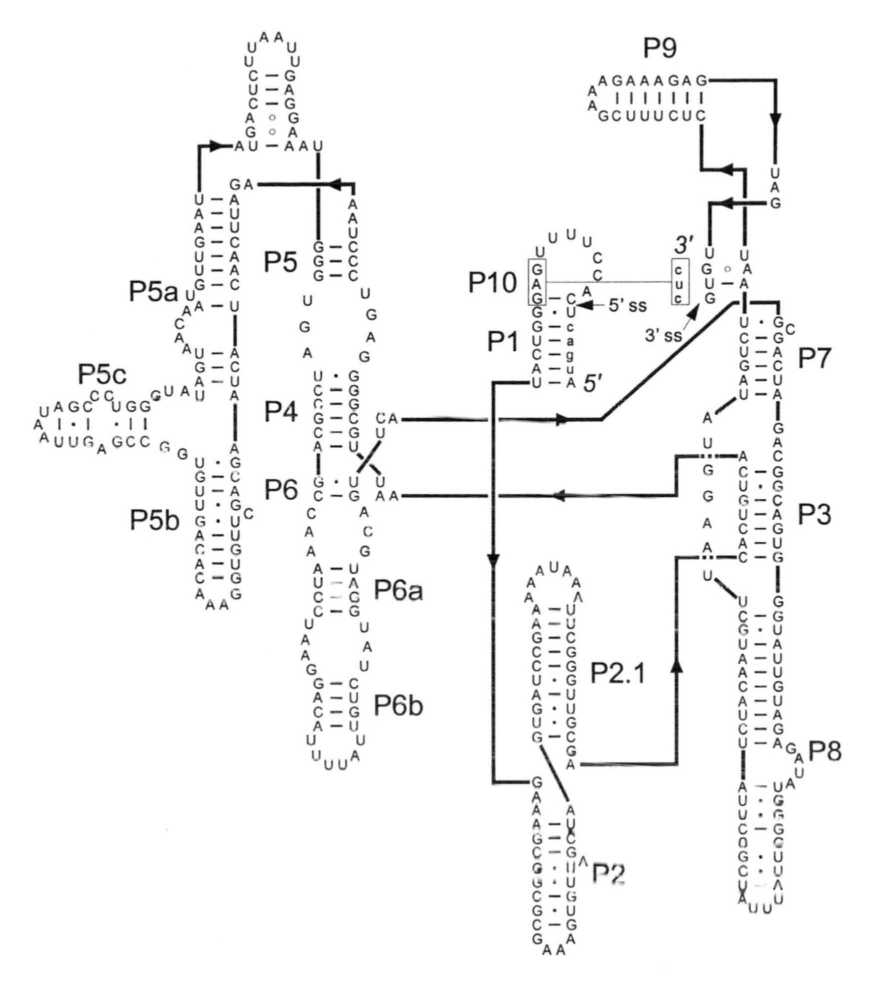

Figure 4. Group I intron secondary structure. The secondary structure from the mitochondrial LSU intron of *Pneumocystis carinii* (Pc1) [112] was drawn according to Cech et al. [113] and taken from Damberger et al. [114]. Intron sequences are in upper case, exon sequences in lower case letters. Splice sites are indicated by closed arrows and helical stems are numbered P1 through P10. Thick lines show the continuity of the strand with arrows indicating 5' → 3' polarity.

of the *Pneumocystis carinii* LSU group I intron (Fig. 4) was found to be inhibited by tetracycline [71]. The inhibition was noncompetitive with respect to the external cofactor guanosine and occurred with an observed K_i of 27 μM (Tab. 1) [71]. In contrast, a 20-fold higher tetracycline concentration was needed for 50% inhibition of splicing of the *sunY* intron, and splicing of the *td* and *Tetrahymena thermophila* LSU introns was not inhibited visibly in the presence of 100 μM tetracycline (Tab. 1) [72]. Both the *Pneumocystis* and the *Tetrahymena* LSU introns belong to the IC1 subgroup [73]. The nucleotides in their core regions are 68% identical and the peripheral extensions are very sim-

Table 1. Inhibition of ribozymes by tetracycline and by tetracycline derivatives

Ribozyme[a]	Tetracycline derivative	Inhibiting concentration [μM]	Reference
Group I introns			
T4 *sunY*	tetracycline	500[b]	[88]
T4 *td*	tetracycline	n. d. at 100[c]	[72]
T.th. LSU	tetracycline	n. d. at 100[c]	[72]
P.c. LSU 1	tetracycline	27[d]	[71]
Group II introns			
S.c. aI5γ	tetracycline	50[e]	[71]
S.c. bI1	tetracycline	500[b]	[78, 79]
S.c. bI1	7-chlor-tc	300[f]	[79]
S.c. bI1	7-chlor-tc	150[g]	[79]
Hammerhead[h]	tetracycline	420[b]	[83]
	7-chlor-tc	300[b]	[83]
HDV, genomic	tetracycline	500[b]	[88]
	5-hydroxy-tc	500[b]	[88]
	minocycline	100[b]	[88]
	anhydrochlor-tc	100[b]	[88]
	4-epi-anhydrochlor-tc	20[b]	[88]
	chelocardin	10[b]	[88]
HDV, antigenomic	tetracycline	500[b]	[88]
	chelocardin	50[b]	[88]
N.c. VS	tetracycline	n. d. at 100[c]	[91]

[a] *T.th.*, *Tetrahymena thermophila*; *P.c.*, *Pneumocystis carinii*; *S.c.*, *Saccharomyces cerevisiae*; HDV, Hepatitis Delta Virus; *N.c.*, *Neurospora crassa*
[b] concentration at which approximately 50% of the ribozyme reaction is inhibited
[c] no inhibition detected at 100 μM
[d] K_i
[e] indicates the minimum concentration that completely inhibits splicing
[f] under "standard" splicing conditions (40 mM Hepes pH 7.5; 60 mM $MgCl_2$; 1 M NH_4Cl)
[g] under "sensitive" splicing conditions (40 mM Hepes pH 7.5; 10 mM $MgCl_2$; 0.5 M NH_4Cl)
[h] from Uhlenbeck [106]; derived from the ABSV(-) strand

ilar, except that the *Pneumocystis* intron lacks P9.1 and P9.2 extensions. In the *Tetrahymena* intron, the L9.1a loop forms a tertiary interaction with L2.1, termed P13. Disruption of this interaction reduces intron splicing activity [74]. Chemical probing [75] and Fe^{2+}/EDTA cleavage [76] show that deletion of the P9.1 and P9.2 extensions from the *Tetrahymena* intron destabilize it. It is attractive to speculate that the *Pneumocystis* intron is more sensitive towards tetracycline than the *Tetrahymena* intron because it is less stable, an assumption that can easily be tested experimentally. However, it cannot be ruled out that the remaining sequence differences between the two introns are responsible for the altered sensitivity to tetracycline.

Group II introns

Group II introns, which are not related to group I introns, have been found in organellar genomes of eucaryotes (mostly plants and fungi) and in eubacteria. Their splicing pathway involves formation of a lariat. Based on this and other mechanistic similarities, an evolutionary relationship with spliceosomal introns has been proposed (for a review see Michel & Ferat [77]). Splicing of the aI5γ intron (Fig. 5) of the *coxI* gene from yeast mitochondrial DNA was completely inhibited in the presence of 50 μM tetracycline [71], while 300 μM chlortetracycline and more than 500 μM tetracycline were needed to achieve 50% splicing inhibition of the bI1 intron from the yeast mitochondrial *cob* gene under "standard" splicing conditions (60 mM $MgCl_2$) [78, 79]. Under "sensitive" splicing conditions (10 mM $MgCl_2$), 150 μM chlortetracycline were sufficient for 50% inhibition of the bI1 intron [79], indicating competition between chlortetracycline and Mg^{2+} ions. Although both introns belong to the bI1 subgroup and seem to be very similar in structure and activity [77],

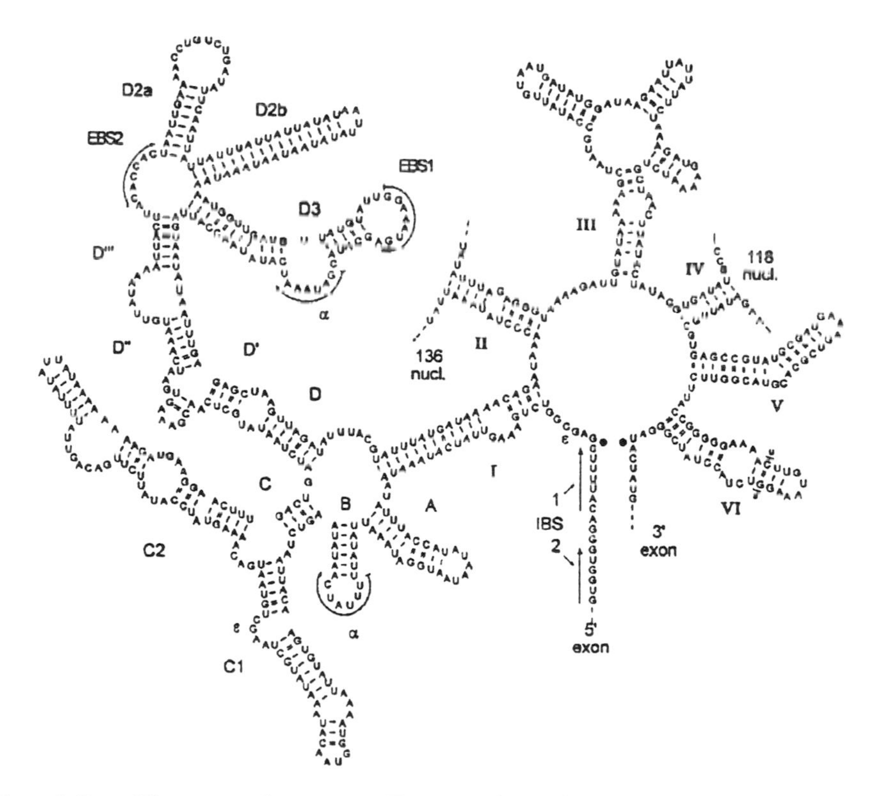

Figure 5. Group II intron secondary structure. The group II intron displayed is aI5γ from the *coxI* gene of yeast mitochondria [115, 116]. Domains I through VI are indicated, as are the helical stems of domain I. Tertiary interactions are indicated by curved arrows (EBS and IBS stand for exon- and intron-binding sites) and Greek letters (α, ε). The splice sites are highlighted by closed circles.

they have not been characterized well enough to allow speculation on the reason(s) for their different sensitivity to the drug.

Hammerhead

The hammerhead ribozyme was originally found in single-stranded plant pathogenic RNAs where it is most likely involved in processing the concatemeric replication products into monomers. It folds into a wishbone-shaped conformation and self-cleaves at a specific phosphodiester bond yielding 5'-hydroxyl and 2',3'-cyclic phosphate termini (for reviews, see Birikh et al. [80], Symons [81] and Stage-Zimmermann and Uhlenbeck [82]). The *trans*-cleaving ribozyme shown in Figure 6 was assayed for cleavage inhibition by tetracycline and 7-chlortetracycline (Fig. 2) and the respective 50% inhibition values determined were 420 µM and 300 µM (Tab. 1) [83]. A plot of the calculated cleavage rate *versus* the antibiotic concentration did not fit to a simple hyperbola, suggesting more than one interaction site of the drug. These binding sites do not overlap with the inhibitory binding site of neomycin [83] which is proposed to be the catalytic site of the hammerhead [84]. Under certain experimental conditions, low concentrations of chlortetracycline enhanced hammerhead activity, an effect that was not observed for any of the other ribozymes. The authors interpreted this as denaturation induced by the drug, analogous to the rate enhancement they observed in the presence of 2.5% (v/v) dimethylsulfoxide [83]. But, at the low concentrations of the drug used (about 50 µM), tetracycline might also act by stabilizing an active conformation of the ribozyme before the second site is bound at higher concentrations leading to inhibition. Such an effect has been observed for lanthanide ion binding to the hammerhead [85], and binding of tetracycline to the ribosome stabilizes it against thermal denaturation [46]. In agreement with the bI1 intron, the

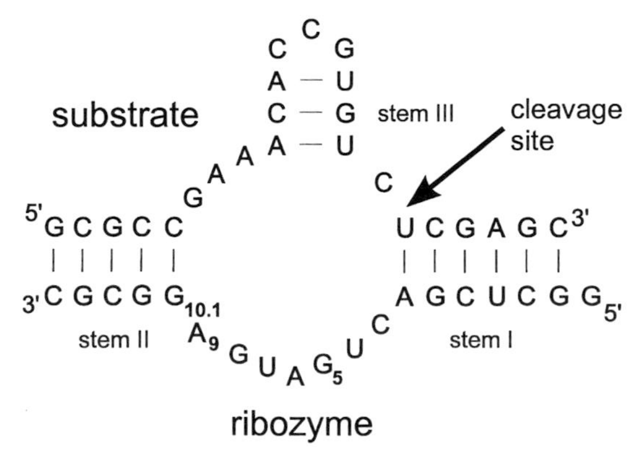

Figure 6. Hammerhead secondary structure. Shown is a *trans*-cleaving hammerhead derived from the ABSV(-) strand [106]. The substrate and ribozyme strands, the cleavage site (closed arrow) and the three helical stems are indicated. The nucleotides G5, A9 and G10.1 (numbering according to Hertel et al. [117]) are highlighted.

inhibitory activity of 7-chlortetracycline can be suppressed by increasing the Mg^{2+} concentration [83].

Hepatitis Delta virus

The Hepatitis Delta virus (HDV) is a satellite virus of hepatitis B. It is a single-stranded RNA virus and resembles plant pathogenic RNAs. The HDV RNA contains a ribozyme (Fig. 7) that folds into a nested double pseudoknot structure [86] which is capable of autocatalytic cleavage *in vitro*, and which is thought to process the replication products *in vivo* [87]. It is also the sole ribozyme for which a structure-function analysis of the determinants for tetracycline activity has been attempted. Rogers et al. [88] tested six tetracycline derivatives shown in Figure 2 for their ability to inhibit cleavage of a genomic HDV ribozyme (Tab. 1). Inhibition of cleavage clearly correlates with the lipophilicity of the tetracycline derivative assayed. Of all derivatives, tetracycline and 5-hydroxy-tetracycline are the least lipophilic and the least active. Minocycline which is more lipophilic requires a five-fold lower concentration for 50% inhibition. The atypical tetracyclines are also very lipophilic [8]. Of these, anhydrochlortetracycline is as active as minocycline, 4-epi-anhydrochlortetracycline and chelocardin require 5- and 10-fold lower concentra-

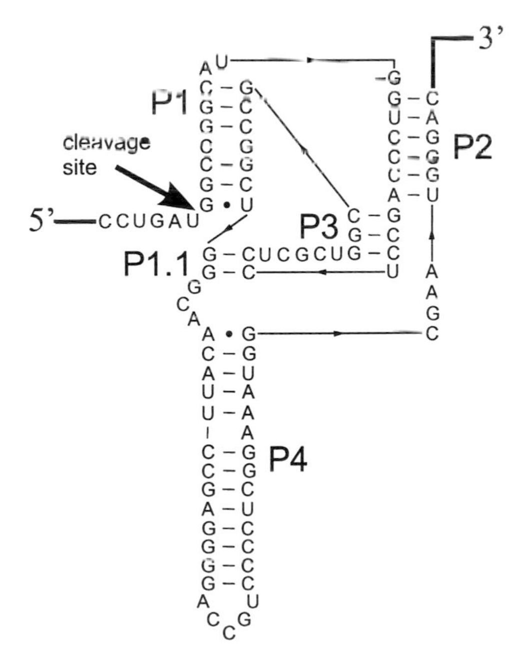

Figure 7. Hepatitis Delta Virus secondary structure. The genomic sequence of hepatitis delta virus [118] is shown. The cleavage site is marked by a closed arrow and the helical stems [86, 119] are numbered P1 through P4.

tions. In addition, having a large group in the β-epimer at position C-4 (Fig. 2), as in the last two derivatives, seems to promote activity.

Neurospora crassa VS

Certain natural isolates of *Neurospora crassa* contain a single-stranded, abundant RNA termed Varkud satellite (VS) in their mitochondria. VS RNA is capable of self-cleavage and self-ligation, presumably as part of its replication cycle [89]. Although self-cleavage of the VS RNA generates 5'-hydroxyl and 2',3'-cyclic phosphate ends like the hammerhead, hairpin, and HDV ribozymes, activity is mediated by an RNA motif distinct in sequence and secondary structure from these ribozymes [90]. Tetracycline failed to inhibit cleavage of the *Neurospora crassa* VS ribozyme G11 at a concentration of 100 μM (Tab. 1) [91]. This concentration, however, is below the range of 400–500 μM normally needed to inhibit ribozyme cleavage (Tab. 1). At higher concentrations, tetracycline might still inhibit the VS ribozyme. Tetracycline analogs were not tested.

Trends

Although the data set is incomplete and very diverse, several trends do emerge: (i) Compared to other inhibitors of ribozyme activity, like the aminoglycosides [88, 92–94], the tuberactinomycins [95] or the pseudodisaccharides [96], tetracycline is a rather weak inhibitor. The approximate concentration of 500 μM tetracycline needed to inhibit cleavage of most ribozymes by 50% is only about an order of magnitude lower than the binding constants determined for nonspecific binding of tetracycline to the ribosome [47]. It therefore cannot be ruled out that ribozymes are inhibited nonspecifically by tetracycline. This must not necessarily be the case for other tetracycline derivatives. Chelocardin and 4-epi-anhydrochlortetracycline show 50% inhibition of the genomic HDV ribozyme in the low μM range, a concentration at which several tuberactinomycins specifically inhibit group I intron splicing [95]. (ii) More hydrophobic tetracyclines are more active in inhibiting ribozymes. Among the classical tetracyclines, the hydrophobicity decreases in the following order: minocycline > 7-chlor-tetracycline > tetracycline > 5-hydroxy-tetracycline [97]. This is exactly the same order of activity observed for inhibition of the bI1 group II intron [79], the hammerhead [83] and the HDV [88] ribozymes. The lipophilic atypical tetracyclines are the strongest inhibitors of the HDV ribozyme. (iii) Tetracyclines may inhibit translation and ribozymes differently. However, this has only tentatively been shown for the HDV ribozyme. It is inhibited most strongly by chelocardin and 4-epi-anhydrochlortetracycline [88], two atypical tetracyclines which are poor inhibitors of protein synthesis [31], while the strong translation inhibitors tetracycline and 5-hydroxy-tetracycline are rather ineffective in inhibiting ribozyme cleavage. This is not without precedent. Neomycin inhibits group I intron and HDV activity effectively *in vitro* [93], but fails to do so *in vivo* [98, 99]. In ribozymes, neomycin acts by displacing metal ions from the catalytic site [84, 88, 100]. In contrast, neomycin binds in the absence of divalent metal ions to the major groove of an oligonucleotide representing the decoding site on the 16S rRNA and induces a conformational change [101]. This is thought to reflect the action of the drug on the ribosome.

Perspectives

The recently published three-dimensional structures of the ribosome [102] and ribosomal subunits [58, 59, 103] will provide a structural framework to evaluate the biochemical data on tetracycline-rRNA interactions. The structures will help identify the tetracycline binding site with possible aid coming from SERF (SElection of Random Fragments; [104]), a method used to identify minimal binding regions within ribonucleo-protein complexes and the recently published S7 mutants [105] which could be analyzed for altered tetracycline sensitivity. The crystal structures will also help to suggest experiments to determine the conformational changes associated with tetracycline binding to the inhibitory binding site. For the ribozymes, a thorough structure-function analysis has to be performed for several different ribozymes to identify the tetracycline functions necessary for inhibition of ribozyme activity and to elucidate the mode and site-specificity of tetracycline binding.

Acknowledgments
I thank Jennifer Swisher for providing Figure 5. I would also like to thank Mark Nelson from Paratek Pharmaceuticals, Albrecht Sigler and Beatrix Süß in Erlangen, as well as Renée Schroeder, Dolly Wittberger, Norbert Polacek, Alison Sleigh, Christina Waldsich and Uwe von Ahsen at the Vienna Biocenter for many stimulating discussions. Work on tetracycline-RNA interactions was funded by the EU TMR program, grant no. FMRX-CT97-0154 to Renée Schroeder.

Note added in proof
Two crystal structures of 30S ribosomal subunits from *Thermus thermophilus* complexed with tetracycline [1, 2] show that several tetracycline molecules can bind independently to the ribosomal RNA. No large-scale conformational changes due to tetracycline binding were observed in the 30S subunits. This suggests that the biochemical interaction data is caused by binding of more than one tetracycline molecule to the ribosome, rather than by only one tetracycline binding and inducing conformational changes. The inhibitory tetracycline is proposed to bind to helix 34 just touching the A-site tRNA, so it could prevent tRNA binding. This is supported by a 16 Å cryo-electron microscopic reconstruction of a complex of Tet(O) with the *E. coli* 70S ribosome [3]. While the tip of domain IV from Tet(O) also interacts with helix 34 and comes within 6 Å of the tetracycline binding site, domain IV in EF-G is positioned differently and does not come close to this tetracycline binding site.

In the ribozyme section, trends (ii) and (iii) are strengthened by data from a fluorescence-based assay screening for modulators of hammerhead ribozyme cleavage [4]. Here, chelocardin was most active in inhibiting hammerhead cleavage, followed by tetracycline, doxycycline and oxytetracycline, thus showing a similar structure-activity profile as inhibition of HDV ribozyme cleavage.

1 Brodersen DE, Clemons WM, Carter AP, Morgan-Warren RJ, Wimberly BT, Ramakrishnan V (2000) The structural basis for the action of the antibiotics tetracycline, pactamycin, and

hygromycin B on the 30S ribosomal subunit. *Cell* 103: 1143–1154
2 Pioletti M, Schlünzen F, Harms J, Zarivach R, Gluhmann M, Avila H, Bashan A, Bartels H, Auerbach T, Jacobi C, Hartsch T, Yonath A, Franceschi F (2001) Crystal structures of complexes of the small ribosomal subunit with tetracycline, edeine and IF3. *EMBO J* 20: 1829–1839
3 Spahn CM, Blaha G, Agrawal RK, Penczek P, Grassucci RA, Trieber CA, Connell SR, Taylor DE, Nierhaus KH, Frank J (2001) Localization of the ribosomal protection protein Tet(O) on the ribosome and the mechanism of tetracycline resistance. *Mol Cell* 7: 1037–1045
4 Jenne A, Hartig JS, Piganeau N, Tauer A, Samarsky DA, Green MR, Davies J, Famulok M (2001) Rapid identification and characterization of hammerhead-ribozyme inhibitors using fluorescence-based technology. *Nat Biotechnol* 19: 56–61

References

1 Gale EF, Cundliffe E, Reynolds PE, Richmond MH, Waring MJ (1981) *The molecular basis of antibiotic action*. Wiley, London
2 Kaji A, Ryoji M (1979) Tetracycline. *In*: FE Hahn (ed.): *Antibiotics*. Springer, Berlin, 304–328
3 Chopra I (1994) Tetracycline analogs whose primary target is not the bacterial ribosome. *Antimicrob Agents Chemother* 38: 637–640
4 Spahn CM, Prescott CD (1996) Throwing a spanner in the works: antibiotics and the translation apparatus. *J Molec Med* 74: 423–439
5 Hostalek Z, Vanek Z (1985) Biosynthesis of the tetracyclines. *In*: JJ Hlavka, JH Booth (eds): *The Tetracyclines*. Springer, Berlin, 137–178
6 Clive DLJ (1968) Chemistry of tetracyclines. *Quart Rev Chem Soc* 22: 435–456
7 Mitscher LA (1978) *The chemistry of tetracycline antibiotics*. Marcel Dekker, New York
8 Rogalski W (1985) Chemical modification of the tetracyclines. *In*: JJ Hlavka, JH Booth (eds): *The Tetracyclines*. Springer, Berlin, 179–316
9 Sum PE, Sum FW, Projan SJ (1998) Recent developments in tetracycline antibiotics. *Curr Pharmaceut Design* 4: 119–132
10 Dürkheimer W (1975) Tetracycline: Chemie, Biochemie und Struktur-Wirkungs-Beziehungen. *Angew Chem* 87: 751–784
11 Blackwood RK, English AR (1977) Structure-activity relationships in the tetracycline series. *In*: D Perlman (ed.): *Structure-activity relationships among the semisynthetic antibiotics*. Academic Press, New York, 397–426
12 Nikaido H, Thanassi DG (1993) Penetration of lipophilic agents with multiple protonation sites into bacterial cells: tetracyclines and fluoroquinolones as examples. *Antimicrob Agents Chemother* 37: 1393–1399
13 Chopra I (1995) Tetracycline uptake and efflux in bacteria. *In*: NH Georgopapadakou (ed.): *Drug transport in antimicrobial and anticancer chemotherapy*. Marcel Dekker, New York, 221–243
14 Salyers AA, Speer BS, Shoemaker NB (1990) New perspectives in tetracycline resistance. *Mol Microbiol* 4: 151–156
15 Johnson R, Adams J (1992) The ecology and evolution of tetracycline resistance. *Trends Ecol Evol* 7: 295–299
16 Speer BS, Shoemaker NB, Salyers AA (1992) Bacterial resistance to tetracycline: mechanisms, transfer, and clinical significance. *Clin Microbiol Rev* 5: 387–399
17 Hillen W, Berens C (1994) Mechanisms underlying expression of Tn*10* encoded tetracycline resistance. *Annu Rev Microbiol* 48: 345–369
18 Roberts MC (1996) Tetracycline resistance determinants: mechanisms of action, regulation of expression, genetic mobility, and distribution. *FEMS Microbiol Rev* 19: 1–24
19 Schnappinger D, Hillen W (1996) Tetracyclines: antibiotic action, uptake, and resistance mechanisms. *Arch Microbiol* 165: 359–369
20 Taylor DE, Chau A (1996) Tetracycline resistance mediated by ribosomal protection. *Antimicrob Agents Chemother* 40: 1–5
21 White JP, Cantor CR (1971) Role of magnesium in the binding of tetracycline to *Escherichia coli* ribosomes. *J Mol Biol* 58: 397–400
22 Smythies JR, Benington F, Morin RD (1972) On the molecular mechanism of action of the tetra-

cyclines. *Experientia* 28: 1253–1254
23 Kruger K, Grabowski PJ, Zaug AJ, Sands J, Gottschling DE, Cech TR (1982) Self-splicing RNA: autoexcision and autocyclization of the ribosomal RNA intervening sequence of *Tetrahymena*. *Cell* 31: 147–157
24 Guerrier-Takada C, Gardiner K, Marsh T, Pace N, Altman S (1983) The RNA moiety of ribonuclease P is the catalytic subunit of the enzyme. *Cell* 35: 849–857
25 Gold L, Polisky B, Uhlenbeck O, Yarus M (1995) Diversity of oligonucleotide functions. *Annu Rev Biochem* 64: 763–797
26 Cundliffe E (1990) Recognition sites for antibiotics within rRNA. *In*: WE Hill, A Dahlberg, RA Garrett, PB Moore, D Schlessinger, JR Warner (eds): *The Ribosome*. ASM Press, Washington, 479–490
27 Wallis MG, Schroeder R (1997) The binding of antibiotics to RNA. *Prog Biophys Mol Biol* 67: 141–154
28 von Ahsen U, Schroeder R (1990) Streptomycin and self-splicing. *Nature* 346: 801
29 Gesteland RF, Atkins JF (eds) (1993) *The RNA World*. Cold Spring Harbor Laboratory Press, Cold Spring Harbor
30 Gesteland RF, Cech TR, Atkins JF (eds) (1999) *The RNA World*, 2nd ed. Cold Spring Harbor Laboratory Press, Cold Spring Harbor
31 Rasmussen B, Noller HF, Daubresse G, Oliva B, Misulovin Z, Rothstein DM, Ellestad GA, Gluzman Y, Tally FP, Chopra I (1991) Molecular basis of tetracycline action: identification of analogs whose primary target is not the bacterial ribosome. *Antimicrob Agents Chemother* 35: 2306–2311
32 Lindley EV, Munske GR, Magnuson JA (1984) Kinetic analysis of tetracycline accumulation by *Streptococcus faecalis*. *J Bacteriol* 158: 334–336
33 Yamaguchi A, Ohmori H, Kaneko-Ohdera M, Nomura T, Sawai T (1991) ΔpH-dependent accumulation of tetracycline in *Escherichia coli*. *Antimicrob Agents Chemother* 35: 53–56
34 Scholtan W (1968) Die hydrophobe Bindung der Pharmaka an Humanalbumin und Ribonukleinsäure. *Arzneim Forsch-Drug Res* 18: 505–517
35 Connamacher RH, Mandel HG (1965) Binding of tetracyclines to the 30S subunit and poly-uridilic acid. *Biochem Biophys Res Commun* 20: 98–103
36 Day LE (1966) Tetracycline inhibition of cell-free protein synthesis. I. Binding of tetracycline to components of the system. *J Bacteriol* 91: 1917–1923
37 Day LE (1966) Tetracycline inhibition of cell-free protein synthesis. II. Effect of the binding of tetracycline to the components of the system. *J Bacteriol* 92: 197–203
38 Attur MG, Patel RN, Patel PD, Abramson SB, Amin AR (1999) Tetracycline up-regulates COX-2 expression and prostaglandin E2 production independent of its effect on nitric oxide. *J Immunol* 162: 3160–3167
39 Amin AR, Attur MG, Thakker GD, Patel PD, Vyas PR, Patel RN, Patel IR, Abramson SB (1996) A novel mechanism of action of tetracyclines: effects on nitric oxide synthases. *Proc Natl Acad Sci USA* 93: 14 014–14 019
40 Beekman B, Verzijl N, de Roos JA, Koopman JL, TeKoppele JM (1997) Doxycycline inhibits collagen synthesis by bovine chondrocytes cultured in alginate. *Biochem Biophys Res Commun* 237: 107–110
41 Hanemaaijer R, Sorsa T, Konttinen YT, Ding Y, Sutinen M, Visser H, van Hinsbergh VW, Helaakoski T, Kainulainen T, Ronka H et al (1997) Matrix metalloproteinase-8 is expressed in rheumatoid synovial fibroblasts and endothelial cells. Regulation by tumor necrosis factor-alpha and doxycycline. *J Biol Chem* 272: 31 504–31 509
42 Chen M, Ona VO, Li M, Ferrante RJ, Fink KB, Zhu S, Bian J, Guo L, Farrell LA, Hersch SM et al (2000) Minocycline inhibits caspase-1 and caspase-3 expression and delays mortality in a transgenic mouse model of Huntington disease. *Nat Med* 6: 797–801
43 Fey G, Reiss M, Kersten H (1973) Interaction of tetracylines with ribosomal subunits from *Escherichia coli*. A fluorometric investigation. *Biochemistry* 12: 1160–1164
44 Strel'tsov SA, Kukhanova MK, Krayevsky AA, Beljavskaja IV, Victorova LS, Gursky GV, Treboganov AD, Gottikh BP (1974) Binding of oxytetracycline to *E. coli* ribosomes. *Mol Biol Rep* 1: 391–396
45 Strel'tsov SA, Kukhanova MK, Gurskii GV, Kraevskii AA, Beliavskaia IV (1975) Oxytetracycline binding to *E. coli* ribosomes. *Mol Biol Mosk* 9: 910–921
46 Tritton TR (1977) Ribosome-tetracycline interactions. *Biochemistry* 16: 4133–4138

47 Epe B, Woolley P (1984) The binding of 6-demethylchlortetracycline to 70S, 50S and 30S ribosomal particles: a quantitative study by fluorescence anisotropy. *EMBO J* 3: 121–126
48 Bergeron J, Ammirati M, Danley D, James L, Norcia M, Retsema J, Strick CA, Su WG, Sutcliffe J, Wondrack L (1996) Glycylcyclines bind to the high-affinity tetracycline ribosomal binding site and evade Tet(M)- and Tet(O)-mediated ribosomal protection. *Antimicrob Agents Chemother* 40: 2226–2228
49 Goldman RA, Hasan T, Hall CC, Strycharz WA, Cooperman BS (1983) Photoincorporation of tetracycline into *Escherichia coli* ribosomes. Identification of the major proteins photolabeled by native tetracycline and tetracycline photoproducts and implications for the inhibitory action of tetracycline on protein synthesis. *Biochemistry* 22: 359–368
50 Buck MA, Cooperman BS (1990) Single protein omission reconstitution studies of tetracycline binding to the 30S subunit of *Escherichia coli* ribosomes. *Biochemistry* 29: 5374–5379
51 Powers T, Noller HF (1995) Hydroxyl radical footprinting of ribosomal proteins on 16S rRNA. *RNA* 1: 194–209
52 Noller HF, Hoffarth V, Zimniak L (1992) Unusual resistance of peptidyl transferase to protein extraction procedures. *Science* 256: 1416–1419
53 Khaitovich P, Mankin AS, Green R, Lancaster L, Noller HF (1999) Characterization of functionally active subribosomal particles from *Thermus aquaticus*. *Proc Natl Acad Sci USA* 96: 85–90
54 Zhang B, Cech TR (1998) Peptidyl-transferase ribozymes: trans reactions, structural characterization and ribosomal RNA-like features. *Chem Biol* 5: 539–553
55 Ehresmann C, Baudin F, Mougel M, Romby P, Ebel JP, Ehresmann B (1987) Probing the structure of RNAs in solution. *Nucl Acid Res* 15: 9109–9128
56 Stern S, Moazed D, Noller HF (1988) Structural analysis of RNA using chemical and enzymatic probing monitored by primer extension. *Methods Enzymol* 164: 481–489
57 Moazed D, Noller HF (1987) Interaction of antibiotics with functional sites in 16S ribosomal RNA. *Nature* 327: 389–394
58 Clemons WM Jr, May JL, Wimberly BT, McCutcheon JP, Capel MS, Ramakrishnan V (1999) Structure of a bacterial 30S ribosomal subunit at 5.5 Å resolution. *Nature* 400: 833–840
59 Tocilj A, Schlunzen F, Janell D, Gluhmann M, Hansen HA, Harms J, Bashan A, Bartels H, Agmon I, Franceschi F et al (1999) The small ribosomal subunit from *Thermus thermophilus* at 4.5 Å resolution: pattern fittings and the identification of a functional site. *Proc Natl Acad Sci USA* 96: 14 252–14 257
60 Steiner G, Kuechler E, Barta A (1988) Photo-affinity labelling at the peptidyl transferase centre reveals two different positions for the A- and P-sites in domain V of 23S rRNA. *EMBO J* 7: 3949–3955
61 Oehler R, Polacek N, Steiner G, Barta A (1997) Interaction of tetracycline with RNA: photoincorporation into ribosomal RNA of *Escherichia coli*. *Nucl Acid Res* 25: 1219–1224
62 Wilms C, Noah JW, Zhong D, Wollenzien P (1997) Exact determination of UV-induced crosslinks in 16S ribosomal RNA in 30S ribosomal subunits. *RNA* 3: 602–612
63 Noah JW, Dolan MA, Babin P, Wollenzien P (1999) Effects of tetracycline and spectinomycin on the tertiary structure of ribosomal RNA in the *Escherichia coli* 30 S ribosomal subunit. *J Biol Chem* 274: 16 576–16 581
64 Ross JI, Eady EA, Cove JH, Cunliffe WJ (1998) 16S rRNA mutation associated with tetracycline resistance in a gram-positive bacterium. *Antimicrob Agents Chemother* 42: 1702–1705
65 Powers T, Noller HF (1991) A functional pseudoknot in 16S ribosomal RNA. *EMBO J* 10: 2203–2214
66 Recht MI, Douthwaite S, Dahlquist KD, Puglisi JD (1999) Effect of mutations in the A site of 16 S rRNA on aminoglycoside antibiotic-ribosome interaction. *J Mol Biol* 286: 33–43
67 Fedor MJ (2000) Structure and function of the hairpin ribozyme. *J Mol Biol* 297: 269–291
68 Frank DN, Pace NR (1998) Ribonuclease P: unity and diversity in a tRNA processing ribozyme. *Annu Rev Biochem* 67: 153–180
69 Cech TR (1990) Self-splicing of group I introns. *Annu Rev Biochem* 59: 543–568
70 Saldanha R, Mohr G, Belfort M, Lambowitz AM (1993) Group I and group II introns. *FASEB J* 7: 15–24
71 Liu Y, Tidwell RR, Leibowitz MJ (1994) Inhibition of *in vitro* splicing of a group I intron of *Pneumocystis carinii*. *J Eukaryot Microbiol* 41: 31–38
72 Rogers J (1996) *Antibiotic inhibition of catalytic RNA functions*. Ph D thesis, University of Vancouver, British Columbia

73 Michel F, Westhof E (1990) Modelling of the three-dimensional architecture of group I catalytic introns based on comparative sequence analysis. *J Mol Biol* 216: 585–610
74 Lehnert V, Jaeger L, Michel F, Westhof E (1996) New loop-loop tertiary interactions in self-splicing introns of subgroup IC and ID: a complete 3D model of the *Tetrahymena thermophila* ribozyme. *Chem Biol* 3: 993–1009
75 Banerjee AR, Jaeger JA, Turner DH (1993) Thermal unfolding of a group I ribozyme: the low-temperature transition is primarily disruption of tertiary structure. *Biochemistry* 32: 153–163
76 Laggerbauer B, Murphy FL, Cech TR (1994) Two major tertiary folding transitions of the *Tetrahymena* catalytic RNA. *EMBO J* 13: 2669–2676
77 Michel F, Ferat J-L (1995) Structure and activities of group II introns. *Annu Rev Biochem* 64: 435–461
78 Wank H (1992) Über die Wirkung von Antibiotika auf autokatalytische Intron-RNA. *Diplomarbeit, Universität Wien*
79 Hertweck M (1999) Inhibition of nuclear pre-mRNA splicing and group II intron splicing by antibiotics *in vitro*. Diplomarbeit, Universität Wien
80 Birikh KR, Heaton PA, Eckstein F (1997) The structure, function and application of the hammerhead ribozyme. *Eur J Biochem* 245: 1–16
81 Symons RH (1997) Plant pathogenic RNAs and RNA catalysis. *Nucl Acid Res* 25: 2683–2689
82 Stage-Zimmermann TK, Uhlenbeck OC (1998) Hammerhead ribozyme kinetics. *RNA* 4: 875–889
83 Murray JB, Arnold JR (1996) Antibiotic interactions with the hammerhead ribozyme:tetracyclines as a new class of hammerhead inhibitor. *Biochem J* 317: 855–860
84 Hermann T, Westhof E (1998) Aminoglycoside binding to the hammerhead ribozyme: a general model for the interaction of cationic antibiotics with RNA. *J Mol Biol* 276: 903–912
85 Lott WB, Pontius BW, von Hippel PH (1998) A two-metal ion mechanism operates in the hammerhead ribozyme-mediated cleavage of an RNA substrate. *Proc Natl Acad Sci USA* 95: 542–547
86 Ferré-D'Amaré AR, Zhou K, Doudna JA (1998) Crystal structure of a hepatitis delta virus ribozyme. *Nature* 395: 567–574
87 Been MD, Wickham GS (1997) Self-cleaving ribozymes of hepatitis delta virus RNA. *Eur J Biochem* 247: 741–753
88 Rogers J, Chang AH, von Ahsen U, Schroeder R, Davies J (1996) Inhibition of the self-cleavage reaction of the human hepatitis delta virus ribozyme by antibiotics. *J Mol Biol* 259: 916–925
89 Saville BJ, Collins RA (1991) RNA-mediated ligation of self-cleavage products of a *Neurospora* mitochondrial plasmid transcript. *Proc Natl Acad Sci USA* 88: 8826–8830
90 Beattie TL, Olive JE, Collins RA (1995) A secondary-structure model for the self-cleaving region of *Neurospora* VS RNA. *Proc Natl Acad Sci USA* 92: 4686–4690
91 Olive JE, De Abreu DM, Rastogi T, Andersen AA, Mittermaier AK, Beattie TL, Collins RA (1995) Enhancement of *Neurospora* VS ribozyme cleavage by tuberactinomycin antibiotics. *EMBO J* 14: 3247–3251
92 von Ahsen U, Davies J, Schroeder R (1991) Antibiotic inhibition of group I ribozyme function. *Nature* 353: 368–370
93 von Ahsen U, Davies J, Schroeder R (1992) Non-competitive inhibition of group I intron RNA self-splicing by aminoglycoside antibiotics. *J Mol Biol* 226: 935–941
94 Stage TK, Hertel KJ, Uhlenbeck OC (1995) Inhibition of the hammerhead ribozyme by neomycin. *RNA* 1: 95–101
95 Wank H, Rogers J, Davies J, Schroeder R (1994) Peptide antibiotics of the tuberactinomycin family as inhibitors of group I intron RNA splicing. *J Mol Biol* 236: 1001–1010
96 Rogers J, Davies J (1994) The pseudodisaccharides: a novel class of group I intron splicing inhibitors. *Nucl Acid Res* 22: 4983–4988
97 Toon S, Rowland M (1979) Quantitative structure pharmocokinetic activity relationships with some tetracyclines. *J Pharm Pharmacol (Suppl)* 31: 43P
98 Chia J-S, Wu H-L, Wang H-W, Chen D-S, Chen P-J (1997) Inhibition of hepatitis delta virus genomic ribozyme self-cleavage by aminoglycosides. *J Biomed Sci* 4: 208–216
99 Waldsich C, Semrad K, Schroeder R (1998) Neomycin B inhibits splicing of the *td* intron indirectly by interfering with translation and enhances missplicing *in vivo*. *RNA* 4: 1653–1663
100 Hoch I, Berens C, Westhof E, Schroeder R (1998) Antibiotic inhibition of RNA catalysis: neomycin B binds to the catalytic core of the *td* group I intron displacing essential metal ions. *J*

Mol Biol 282: 557–569
101 Fourmy D, Yoshizawa S, Puglisi JD (1998) Paromomycin binding induces a local conformational change in the A-site of 16 S rRNA. *J Mol Biol* 277: 333–345
102 Cate JH, Yusupov MM, Yusupova GZ, Earnest TN, Noller HF (1999) X-ray crystal structures of 70S ribosome functional complexes. *Science* 285: 2095–2104
103 Ban N, Nissen P, Hansen J, Moore PB, Steitz TA (2000) The complete atomic structure of the large ribosomal subunit at 2.4 Å resolution. *Science* 289: 905–920
104 Stelzl U, Spahn CM, Nierhaus KH (2000) Selecting rRNA binding sites for the ribosomal proteins L4 and L6 from randomly fragmented rRNA: application of a method called SERF. *Proc Natl Acad Sci USA* 97: 4597–4602
105 Fredrick K, Dunny GM, Noller HF (2000) Tagging ribosomal protein S7 allows rapid identification of mutants defective in assembly and function of 30 S subunits. *J Mol Biol* 298: 379–394
106 Uhlenbeck OC (1987) A small catalytic oligoribonucleotide. *Nature* 328: 596–600
107 Hinrichs W, Kisker C, Düvel M, Müller A, Tovar K, Hillen W, Saenger W (1994) Structure of the Tet repressor-tetracycline complex and regulation of antibiotic resistance. *Science* 264: 418–420
108 Gutell RR (1994) Collection of small subunit (16S- and 16S-like) ribosomal RNA structures: 1994. *Nucl Acid Res* 22: 3502–3507
109 Noller HF (1999) On the origin of the ribosome: Coevolution of subdomains of tRNA and rRNA. *In*: RF Gesteland, TR Cech, JF Atkins (eds): *The RNA World*, 2nd ed. Cold Spring Harbor Laboratory Press, Cold Spring Harbor, 197–219
110 Stern S, Powers T, Changchien LM, Noller HF (1989) RNA-protein interactions in 30S ribosomal subunits: folding and function of 16S rRNA. *Science* 244: 783–790
111 Moazed D, Noller HF (1989) Interaction of tRNA with 23S rRNA in the ribosomal A, P, and E sites. *Cell* 57: 585–597
112 Liu Y, Rocourt M, Pan S, Liu C, Leibowitz MJ (1992) Sequence and variability of the 5.8S and 26S rRNA genes of *Pneumocystis carinii*. *Nucl Acid Res* 20: 3763–3772
113 Cech TR, Damberger SH, Gutell RR (1994) Representation of the secondary and tertiary structure of group I introns. *Nature Struct Biol* 1: 273–280
114 Damberger SH, Gutell RR (1994) A comparative database of group I intron structures. *Nucl Acid Res* 22: 3508–3510
115 Jacquier A, Michel F (1987) Multiple exon-binding sites in class II self-splicing introns. *Cell* 50: 17–29
116 Michel F, Jacquier A (1987) Long-range intron-exon and intron-intron pairings involved in self-splicing of class II catalytic introns. *Cold Spring Harbor Symp Quant Biol* 52: 201–212
117 Hertel KJ, Pardi A, Uhlenbeck OC, Koizumi M, Ohtsuka E, Uesugi S, Cedergren R, Eckstein F, Gerlach WL, Hodgson R et al (1992) Numbering system for the hammerhead. *Nucl Acid Res* 20: 3252
118 Wu H-N, Lin Y-J, Lin F-P, Makino S, Chang M-F, Lai MMC (1989) Human hepatitis delta virus RNA subfragments contain an autocleavage activity. *Proc Natl Acad Sci USA* 86: 1831–1835
119 Been MD, Perotta AT, Rosenstein SP (1992) Secondary structure of the self-cleaving RNA of Hepatitis delta virus: Applications to catalytic RNA design. *Biochemistry* 31: 11 843–11 852

Section III
Use of tetracyclines as non-antimicrobial medicinal agents

(Editor: Robert A. Greenwald)

Tetracyclines in Biology, Chemistry and Medicine
ed. by M. Nelson, W. Hillen and R.A. Greenwald

Biological, non-antibiotic properties of semi-synthetic and chemically modified tetracyclines – a structured, annotated bibliography

Robert A. Greenwald* and Lorne M. Golub

Division of Rheumatology, Long Island Jewish Medical Center, New Hyde Park, NY 11042, USA

Introduction

The first literature regarding unusual properties of tetracycline (TC) hydrochloride date back to the 1960s, with particular reference to bone growth and mild anti-inflammatory properties, e.g., inhibition of chemotaxis. However, it was not until the collagenase discovery paper by Golub et al. in 1983 [1] that intensive interest in the non-antimicrobial properties of TC-based antibiotics developed.

The initial studies on TC inhibition of collagenase were done with minocycline (Min), a semi-synthetic TC with widespread medical and dental applications. Min was in general use among dentists as an antibiotic, and so when Golub et al. planned an experiment in which they planned to abolish the normal oral flora of a diabetic rat with periodontitis, they chose Min. Since Min was readily available for human use, the results of the initial animal experiments could readily be extended to patients. Thus Min was subsequently used as a potential collagenase inhibitor for patients with adult periodontitis, rheumatoid arthritis and epidermolysis bullosa. Even though the focus of experimental studies turned toward doxycycline (Dox) and the chemically modified tetracyclines (CMTs) (*vide infra*), Min has remained the most widely used agent in human trials where excessive collagenase was implicated.

Minocycline, however, is associated with substantial toxicities: dizziness, vertigo, lupus-like syndromes and grayish discoloration of the skin, among others [2]; it also turns inner organs such as the thyroid and heart valves black. Dox was soon shown to have even better anti-MMP (matrix metalloproteinase) potency *in vitro*, and this soon became the drug of choice for most experimental studies, culminating in the commercial availability of a low-dose Dox preparation (Periostat), now marketed for adult periodontitis. The Dox literature remains the most fully developed segment of the background on non-antibiotic properties of tetracyclines.

Long-term administration of any TC, even in low doses, raises the specter of bacterial resistance and other potential side-effects. Golub and colleagues recognized that many TCs had been synthesized and discarded for lack of antimicrobial efficacy, but that such compounds might still be anti-collagenolytic. In 1987, we published the first paper describing useful medical properties of a TC derivative which had been chemically modified so as to delete its antibacterial action. Since then, over 75 additional papers and abstracts have confirmed and expanded the initial observations.

Thus, in the 19 years since the initial Golub discovery, a variety of non-antimicrobial actions have been described for these compounds. These can be broken down into the following broad categories: enzyme inhibition (MMPs, cytokine production, non-collagenolytic proteases, etc.); effects on cellular systems (bone resorption, cartilage degradation, neutrophil or macrophage functions, invasion and proliferation, collagen synthesis); animal models (rat diabetes, rat periodontal disease, rat arthritis, rabbit corneal ulceration, dog arthritis, rat aneurysm, rat osteoporosis, ischemia/reperfusion); and human diseases (adult periodontitis, rheumatoid arthritis, cancer, osteoarthritis).

This bibliography summarizes the literature in the field dealing with non-antimicrobial properties of Min, Dox and the series of CMTs. It should be noted that many papers deal with more than one TC; such papers are listed with an explanatory note in brackets at the end of the citation. The same subject headings are used in each of the three main sections of the bibliography; if there is no entry in a particular section, it means that there may be papers on that subject in one of the other sections. For example, there are no citations specifically dealing with Min inhibition of MMP-13, but there are many such citations regarding Dox and the CMTs. Within each section, papers are listed chronologically.

Up until approximately 1995, almost all CMT citations deal primarily with CMT-1 (4-dedimethylaminotetracycline, see reference [3] for structures), except as otherwise indicated. As of this writing (September, 2000), CMTs are not approved for human use in the US except under special investigational circumstances; therefore, there are no published human use citations for these agents (except in refractory cancer). Abstracts are included here only to the extent that they contain data not yet published in full-length papers. The abbreviations CMT and COL are interchangeable; "COL" was adopted by CollaGenex Pharmaceuticals, Inc. of Newtown, PA, which is the lead firm developing these compounds for medical use.

A portion of the bibliography relating to the CMTs was published previously but is updated herein [4]. Not included in this review are the following: papers dealing primarily with tetracycline hydrochloride, papers dealing exclusively with dermatological usages of TCs (including Min), the use of Dox for gene regulation in systems where specific inducers and suppressors have been constructed with Dox responsiveness as a major feature, and papers dealing with purely chemical phenomena such as metal ion binding properties of TCs, even though the latter property may indeed be quite relevant to some of the non-antimicrobial uses of these agents.

Minocycline references

A) Enzymes/biochemical

Min inhibits general/unspecified collagenolytic activity

Golub LM, Lee HM, Lehrer G, Meniroff A, McNamara TF, Ramamurthy NS (1983) Minocycline reduces gingival collagenolytic activity during diabetes: preliminary observations and a proposed new mechanism of action. *J Periodont Res* 18: 516–521 [the anti-collagenase discovery paper]

Golub LM, Ramamurthy NS, McNamara TF, Gomes B, Wolff M, Casino A, Kapoor A, Zambon J, Ciancio S, Schneir M et al (1984) Tetracyclines inhibit tissue collagenase activity. *J Periodont Res* 19: 651–655

Golub LM, Wolff M, Lee HM, McNamara TF, Ramamurthy NS, Zambon J, Ciancio S (1985) Further evidence that tetracyclines inhibit collagenase activity in human crevicular fluid and from other mammalian sources. *J Periodont Res* 20: 12–23

Zucker S, Lysik RM, Ramamurthy NS, Golub LM, Wieman J, Wilkie D (1985) Diversity of melanoma plasma membrane proteinases: inhibition of collagenolytic and cytolytic activities by minocycline. *J Natl Cancer Inst* 75: 517–525

Greenwald RA, Golub LM, Lavietes B, Ramamurthy NS, Gruber B, Laskin RS, McNamara TF (1987) Minocycline inhibits rheumatoid synovial collagenase *in vivo* and *in vitro*. *J Rheumatol* 14: 28–32

Maehara R, Hinode D, Terai H, Sato M, Nakamura R, Matsuda N, Tanaka T, Sugihara K (1988) Inhibition of bacterial and mammalian collagenolytic activities by tetracyclines. *J Jpn Assn Peridontol* 30: 182–190

Ramamurthy NS, Vernillo A, Lee HM, Golub LM, Rifkin B (1990) The effect of tetracyclines on collagenase activity in UMR 106-01 osteoblastic osteosarcoma cells. *Res Commun Chem Pathol Pharmacol* 70: 323–335

Min inhibits stromelysin

Pourtaghi N, Radvar M, Mooney J, Kinane DF (1996) The effect of subgingival antimicrobial therapy on the levels of stromelysin and tissue inhibitor of metalloproteinases in gingival crevicular fluid. *J Periodontol* 67: 866–870

Min inhibits α_1-PI degradation

Whiteman M, Halliwell B (1997) Prevention of peroxynitrate-dependent tyrosine nitration and inactivation of α_1-antiprotease by antibiotics. *Free Radical Res* 26: 49–56

Min inhibits PLA_2

Pruzanski W, Greenwald RA, Street IP, Laliberte F, Stefanski E, Vadas P (1992) Inhibition of enzymatic activity of phospholipases A_2 by minocycline and doxycycline. *Biochem Pharmacol* 44: 1165–1170

Min inhibits prostaglandins/cytokines

Elattar TM, Lin HS, Schulz R (1988) Effect of minocycline on prostaglandin formation in gingival fibroblasts. *J Periodont Res* 23: 285–286

B) Cellular systems

Min inhibits angiogenesis

Tamargo R, Bok R, Brem H (1991) Angiogenesis inhibition by minocycline. *Cancer Res* 51: 672–675

Gilberston-Beadling S, Powers EA, Stamp-Cole M, Scott PS, Wallace TL, Copeland J, Petzold G, Mitchell M, Ledbetter S, Poorman R et al (1995) The tetracycline analogs minocycline and doxycycline inhibit angiogenesis *in vitro* by a non-metalloproteinase-dependent mechanism. *Cancer Chemother Pharmacol* 36: 418–424

Min inhibits cell proliferation

Somerman MJ, Foster RA (1988) Effects of minocycline on fibroblast attachment and spreading. *J Periodont Res* 23: 154–157

Guerin C, Laterra J, Masnyk T, Golub LM, Brem H (1992) Selective endothelial growth inhibition by tetracyclines that inhibit collagenase. *Biochem Biophys Res Commun* 188: 740–745

Teicher BA, Holden S, Liu C, Ara G, Herman T (1994) Minocycline as a modulator of chemotherapy and hyperthermia *in vitro* and *in vivo*. *Cancer Lett* 82: 17–25

Teicher B, Schwartz J, Holden S, Ara G, Northey D (1994) *In vivo* modulation of several anticancer agents by β-carotene. *Cancer Chemother Pharmacol* 34: 235–241

Min inhibits cell invasion/migration

Sotomayer EA, Teicher BA, Schwartz GN, Holden SA, Menon K, Herman TS, Frei E (1992) Minocycline in combination with chemotherapy or radiation therapy *in vitro* and *in vivo*. *Cancer Chemother Pharmacol* 30: 377–384

Teicher BA, Sotomayor E, Huang Z, Ara G, Holden S, Khandekar V, Chen Y (1993) *Beta* cyclodextrin tetradecsulfate/tetrahydrocortisol ± minocycline as modulators of cancer therapies *in vitro* and *in vivo* against primary and metastatic Lewis lung carcinoma. *Cancer Chemother Pharmacol* 33: 229–238

Masumori N, Tsukamoto T, Miyao N, Kumamoto Y, Saiki I, Yoneda J (1994) Inhibitory effect of minocycline on *in vitro* invasion and experimental metastasis of mouse renal adenocarcinoma. *J Urol* 151: 1400–1404

Min affects T-cell activation

Kloppenburg M, Verweij CL, Miltenburg M, Verhoeven AJ, Daha MR (1995) The influence of tetracyclines on T cell activation. *Clin Exp Immunol* 102: 635–641

Min affects PMN function

Miyachi Y, Yoshioka A, Imamura S, Niwa Y (1986) Effect of antibiotics on the generation of reactive oxygen species. *J Invest Dermatol* 86: 449–453

Min inhibits cellular metabolism

Soory M, Tilakaratne A (2000) The effect of minocycline on the metabolism of androgens by human oral periosteal fibroblasts and its inhibition by finasteride. *Arch Oral Biol* 45: 257–265

C) Animal models/diseases

Diabetic rat

Golub LM, Lee HM, Lehrer G, Nemiroff A, McNamara TF, Kaplan R, Ramamurthy NS (1983) Minocycline reduces gingival collagenolytic activity during diabetes. *J Periodont Res* 18: 516–524

Arthritic rat

Weinberger A, Ben-Gal T, Roizman P, Abramovici A (1996) Intraarticular minocycline injection in experimental synovitis. *Clin Rheumatol* 15: 290–294

Osteoporosis

Williams S, Wakisaka A, Zeng QQ, Barnes J, Martin G, Wechter WJ, Liang CT (1996) Minocycline prevents the decrease in bone mineral density and trabecular bone in ovariectomized aged rats. *Bone* 19: 637–644

Klapisz-Wolikow M, Saffar JL (1996) Minocycline impairment of both osteoid tissue removal and osteoclastic resorption in a synchronized model of remodeling in the rat. *J Cell Physiol* 167: 359–368

Lung injury

Yamaki K, Yoshida N, Kimura T, Ohbayashi H, Takagi K (1998) Effects of cytokines and minocycline on subacute lung injuries induced by repeated injection of lipopolysaccharide. *Kansenshogaku Zasshi* 72: 75–82

Wound healing

Soory M, Virdi H (1999) Implications of minocycline, platelet-derived growth factor, and transforming growth factor beta on inflammatory repair potential in the periodontium. *J Periodontol* 70: 1136–1143

Ischemia/neurologic

Yrjanheikki J, Tikka T, Keinanen R, Goldstein G, Chan P, Koistinaho J (1999) A tetracycline derivative, minocycline, reduces inflammation and protects against focal cerebral ischemia with a wide therapeutic window. *Proc Natl Acad Sci USA* 96: 13 496–13 500

Chen M, Ona V, Li M, Ferrante R, Fink K et al (2000) Minocycline inhibits caspase-1 and caspase-3 expression and delays mortality in a transgenic mouse model of Huntington disease. *Nat Med* 6: 797–801

Mejia ROS, Ona VO, Li M, Friedlander RM (2001) Minocycline reduces traumatic brain injury mediated caspase-1 activation, tissue damage, and neurological dysfunction. *Neurosurgery* 48: 1393–1401

Cancer

Teicher BA, Holden SA, Dupuis NP, Kakeji Y, Ikebe M, Emi Y, Goff D (1995) Potentiation of cytotoxic therapies by TNP-470 and minocycline in mice bearing EMT-6 mammary carcinoma. *Breast Cancer Res Treat* 36: 227–236

Weingart JD, Sipos EP, Brem H (1995) The role of minocycline in the treatment of intracranial 9L glioma. *J Neurosurg* 82: 635–640

Sorenmo K, Barber L, Cronin K, Sammarco C, Usborne A, Goldschmidt M, Shofer F (2000) Canine hemangiosarcoma treated with standard chemotherapy and minocycline. *J Vet Intern Med* 14: 395–398

D) Human studies

Adult periodontal disease

Ciancio S (1994) Clinical experiences with tetracyclines in the treatment of periodontal disease. *Ann N Y Acad Sci* 732: 132–139

Rheumatoid arthritis

Breedveld FC, Dijkmans BAC, Mattie H (1990) Minocycline treatment for rheumatoid arthritis: an open dose finding study. *J Rheumatol* 17: 43–46

Langevitz P, Bank I, Zemer D, Book M, Pras M (1992) Treatment of resistant rheumatoid arthritis wit minocycline: an open study. *J Rheumatol* 19: 1502–1504

Kloppenburg M, Breedveld Terwiel JPH et al (1994) Minocycline in active rheumatoid arthritis. *Arthritis Rheum* 37: 629–636

Tilley BC, Alarcon GS, Heyse SP et al (for the MIRA trial group) (1995) Minocycline in rheumatoid arthritis: a double blind placebo controlled trial. *Ann Intern Med* 122: 81–89

Langevitz P, Livneh A, Bank I, Pras M (1996) Minocycline in rheumatoid arthritis. *Isr J Med Sci* 32: 327–330

O'Dell JR, Haire CE, Palmer W et al (1997) Treatment of early rheumatoid arthritis with minocycline or placebo. *Arthritis Rheum* 40: 842–848

Bluhm GB, Sharp JT, Tilley B (1997) Radiographic results from the minocycline in rheumatoid arthritis trial. *J Rheumatol* 24: 1295–1302

Lai NS, Lan JL (1998) Treatment of DMARD resistant rheumatoid arthritis with minocycline: a local experience among the Chinese. *Rheumatol Int* 17: 245–247

O'Dell JR, Paulsen J, Haire LE et al (1999) Treatment of early seropositive rheumatoid arthritis with minocycline: four year follow-up of a double-blind placebo-controlled trial. *Arthritis Rheum* 42: 1691–1695

Langevitz P, Livneh A, Bank I, Pras M (2000) Benefits and risks of minocycline in rheumatoid arthritis. *Drug Safety* 22: 405–414 [see this paper for a list of additional citations]

Miscellaneous

Lynch WS, Bergfeld WF (1978) Pyoderma gangrenosum responsive to minocycline hydrochloride. *Cutis* 21: 535–537

White JE (1989) Minocycline for dystrophic epidermolysis bullosa. *Lancet* 1: 966

Le CH, Morales A, Trentham DE (1998) Minocycline in early diffuse scleroderma. *Lancet* 352: 1755–1756

Doxycycline references

A) Enzymes/biochemical

Dox inhibits general/unspecified collagenolytic activity

Burns FR, Stack S, Gray RD (1989) Inhibition of purified collagenases from alkali-burned rabbit corneas. *Invest Ophthalmol Visual Sci* 30: 1569–1575

Yanagimura M, Koike F, Hara K (1989) Collagenase activity in gingival crevicular fluid and inhibition by tetracyclines. *J Dent Res* 68 (spec issue): 1691–1693

McCullough CAG, Birck P, Overall C et al (1990) Randomized controlled trial of doxycycline in prevention of recurrent periodontitis in high-risk patients: antimicrobial activity and collagenase inhibition. *J Clin Periodontol* 17: 616–622

Lauhio A, Nordstrom D, Sorsa T et al (1993) Long term treatment of reactive arthritis with tetracycline. *Prog Rheumatol* 5: 84–89

Hurewitz A, Wu C, Mancuso R, Zucker S (1993) Tetracycline and doxycycline inhibit pleural fluid metalloproteinases. *Chest* 103: 1113–1117 [also CMT]

Hayrinen R, Sorsa T, Pettila J et al (1994) Effect of tetracyclines on collagenase activity in patients with recurrent aphthous ulcers. *J Oral Pathol* 23: 269–272

Koivunen AL, Maisi P, Konttinen Y, Prikk K, Sandholm M (1997) Collagenolytic activity and its sensitivity to doxycycline inhibition in tracheal aspirates of horses with chronic obstructive pulmonary disease. *Acta Vet Scand* 38: 9–16

Dox inhibits MMP-1

Konttinen YT, Kangaspunta P, Lindy O et al (1994) Collagenase in Sjogren's syndrome. *Ann Rheum Dis* 53: 836–839 [Dox inhibition shows that Sjogren's collagenase is MMP-1, not MMP-8; a useful demonstration of how doxycycline inhibition curves can be used to characterize the type of collagenase]

Cakir Y, Hahn K (1999) Direct action by doxycycline against canine osteosarcoma cell proliferation and collagenase (MMP-1) activity *in vitro*. *In Vivo* 13: 327–332

Dox inhibits MMP-8

Suomalainen K, Sorsa T, Golub LM et al (1992) Specificity of the anticollagenase action of tetracyclines: relevance to their anti-inflammatory potential. *Antimicrob Agents Chemother* 36: 227–229

Sorsa T, Ingman T, Suomalainen K et al (1992) Cellular source and tetracycline inhibition of gingival crevicular fluid collagenase of patients with labile diabetes mellitus. *J Clin Periodontol* 19: 146–149

Suomalainen K, Sorsa T, Ingman T, Lindy O, Golub LM (1992) Tetracycline inhibition identifies the cellular origin of interstitial collagenases in human periodontal disease *in vivo*. *Oral Microbiol Immunol* 7: 121–123

Sorsa T, Ding Y, Salo T et al (1994) Effects of tetracyclines on neutrophil, gingival, and salivary col-

lagenases. *Ann N Y Acad Sci* 732: 112–131 [also, MMP-1, MMP-9, gingival gelatinase]
Smith GN, Brandt K, Hasty K (1994) Procollagenase is reduced to inactive fragments upon activation in the presence of doxycycline. *Ann N Y Acad Sci* 732: 436–438
Sorsa T, Ding YL, Ingman T et al (1995) Cellular source, activation and inhibition of dental plaque collagenase. *J Clin Periodontol* 22: 709–717 [Dox inhibits PMN collagenase at 20 μM]
Golub LM, Sorsa T, Lee HM, Ciancio S, Sorbi D, Ramamurthy NS, Gruber B, Salo T, Konttinen YT (1995) Doxycycline inhibits neutrophil (PMN)-type matrix metalloproteinases in human adult periodontitis gingiva. *J Clin Periodontol* 22: 100–109
Smith GN, Brandt KD, Hasty KA (1996) Activation of recombinant human neutrophil procollagenase in the presence of doxycycline results in fragmentation of the enzyme and loss of enzyme activity. *Arthritis Rheum* 39: 235–244
Smith GN, Brandt KD, Mickler E, Hasty KA (1997) Inhibition of recombinant human neutrophil collagenase by doxycycline is pH-dependent. *J Rheumatol* 24: 1769–1773
Hanemaaijer R, Sorsa T, Konttinen Y, Ding Y, Sutinen M, Visser H, van Hinsbergh V, Helakoski T, Kainulainen T, Ronka H et al (1997) Matrix metalloproteinase-8 is expressed in rheumatoid synovial fibroblasts and endothelial cells. *J Biol Chem* 272: 31 504–31 509
Shlopov BV, Smith GN, Cole AA, Hasty KA (1999) Differential patterns of response to doxycycline and transforming growth factor beta1 in the down regulation of collagenases in osteoarthritic and normal human chondrocytes. *Arthritis Rheum* 42: 719–727 [MMP-1, -8, and -13 were studied]

Dox inhibits MMP-13

Smith GN, Mickler EA, Hasty KA, Brandt KD (1996) Inhibition by doxycycline of a truncated form of recombinant human MMP-13. *Arthritis Rheum* 39: S226
Greenwald RA, Golub LM, Ramamurthy NS, Chowdhury M, Moak SA, Sorsa T (1998) *In vitro* sensitivity of three mammalian collagenases to tetracycline inhibition: relationship to bone and cartilage degradation. *Bone* 22: 33–38 [also IL-1 cartilage degradation]
Sepper R, Prikk K, Tervahartiala T, Konttinen Y, Maisi P, Lopes-Otin C, Sorsa T (1999) Collagenase-2 and -3 are inhibited by doxycycline in the chronically inflamed lung in bronchiectasis. *Ann N Y Acad Sci* 878: 683–685
Smith GN, Mickler E, Hasty K, Brandt K (1999) Specificity of inhibition of matrix metalloproteinase activity by doxycycline. *Arthritis Rheum* 42: 1140–1146

Dox inhibits gelatinase(s)

Yu LP, Smith GN, Hasty K, Brandt KD (1991) Doxycycline inhibits Type XI collagenolytic activity of extracts from human osteoarthritic cartilage and gelatinase. *J Rheumatol* 18: 1450–1452
Nip LH, Uitto V-J, Golub LM (1993) Inhibition of epithelial cell matrix metalloproteinases by tetracyclines. *J Periodont Res* 28: 379–385 [also CMT]
Hurewitz AN, Wu CL, Mancuso R et al (1993) Tetracycline and doxycycline inhibit pleural fluid metalloproteinases. *Chest* 103: 1113–1117
Burris TS, Hugh K, Orth MW, Kuettner KE, Cole AA (1995) The effects of doxycycline on gelatinase A and B in human cartilage explant cultures of metatarsals. *Trans Ortho Res Soc* 20: 335
Duivenvoorden W, Hirte H, Singh G (1997) Use of tetracycline as an inhibitor of matrix metalloproteinase activity secreted by human bone-metastasizing cancer cells. *Invas Metast* 17: 312–323 [also minocycline]

Dox inhibits MMP-2

Uitto VJ, Firth JD, Nip L, Golub LM (1994) Doxycycline and chemically modified tetracyclines inhibit Gelatinase A (MMP-2) gene expression in human skin keratinocytes. *Ann N Y Acad Sci* 732: 140–151

Dox inhibits stromelysin

Jonat C, Chung FZ, Baragi VM (1996) Transcriptional downregulation of stromelysin by tetracycline. *J Cell Bioch* 60: 341–347

Dox inhibits α_1-PI degradation

Humbert P, Faivre B, Gibey R, Agache P (1991) Use of anti-collagenase properties of doxycycline in treatment of α_1-trypsin deficiency panniculitis. *Acta Derm Venereol (Stockh)* 71: 189–194
Sorsa T, Kontinen YT, Lindo O et al (1993) Doxycycline protects serum alpha-1 antitrypsin from human neutrophil collagenase. *Agents Actions* 39: 225–229
Sorsa T, Lindy O, Kontinnen YT et al (1993) Doxycycline in the protection of serum alpha-1 antitrypsin from human neutrophil collagenase and gelatinase. *Antimicrob Agents Chemother* 37: 592–594
Lee HM, Golub LM, Chan D, Leung M, Schroeder K, Wolff M, Simon S, Crout R (1997) α_1-proteinase inhibitor in gingival crevicular fluid of humans with adult periodontitis: serpinolytic inhibition by doxycycline. *J Periodont Res* 32: 9–19

Dox inhibits plasminogen activator

Hanemaaijer R, Visser H, van den Hoogen GM, Sorsa T, van Hinsbergh VWM (1996) Inhibition of urokinase-type and tissue-type plasminogen activator mediated plasminogen activation by doxycycline. *Fibrinolysis* 10 (S2): 109–111

Dox inhibits cytokines

Shapira L, Soskolne WA, Houri Y, Barak V, Halabi A, Stabholz A (1996) Protection against endotoxic shock and lipopolysaccharide induced local inflammation by tetracycline: correlation with inhibition of cytokine secretion. *Infect Immunity* 64: 825–828 [mostly tetracycline HCl]
Kirkwood KL, Golub LM, Bradford PG (1999) Non-antimicrobial and antimicrobial tetracyclines inhibit IL-6 expression in murine osteoblasts. *Ann N Y Acad Sci* 878: 667–670
Solomon A, Roenblatt M, Li DQ, Liu ZG, Monroy D, Ji ZH, Lokeshwar BL, Pflugfelder SC (2000) Doxycycline inhibition of interleukin-1 in the corneal epithelium. *Invest Ophthalmol* 41: 2544–2557

Dox inhibits cathepsin L

Rifkin BR, Vernillo A, Golub LM, Ramamurthy NS (1994) Modulation of bone resorption by tetracyclines. *Ann N Y Acad Sci* 732: 165–180 [also gelatinase, also CMTs, also bone resorption]

Dox inhibits protein kinase C

Webster GF, Toso SM, Hegemann L (1994) Inhibition of a model of *in vitro* granuloma formation by tetracyclines and ciprofloxin: involvement of protein kinase C. *Arch Dermatol* 130: 748–752

Dox inhibits oxy radicals

Akamatsu H, Asada M, Komura J, Asada Y, Niwa Y (1992) Effect of doxycycline on the generation of reactive oxygen species. *Acta Derm Venereol* (Stock) 72: 178–179

B) Cellular systems

Dox inhibits bone resorption

Gomes BC, Golub LM, Ramamurthy NS (1984) Tetracyclines inhibit parathyroid hormone-induced bone resorption in organ culture. *Experientia* 40: 1273–1275
Grevstad HJ, Boe OE (1995) Effect of doxycycline on surgically induced osteoclast recruitment in the rat. *Eur J Oral Sci* 103: 156–159

Dox inhibits cartilage degradation

te Koppele JM, Verziil N, Beekman B (1995) Chondrocytes cultured in alginate beads: a model system to measure collagen synthesis and degradation. *Arthritis Rheum* 38 (9, suppl): S159

Cole AA, Chubinskaya S, Chlebek K, Kuettner KE, Schmid TM, Greenwald RA (1994) Doxycycline disrupts chondrocyte differentiation and inhibits cartilage matrix degradation. *Arthritis Rheum* 37: 1727–1734

Steinmeyer J, Daufeldt S, Taiwo Y (1998) Pharmacological effect of tetracyclines on proteoglycanases from interleukin-1-treated articular cartilage. *Biochem Pharmacol* 55: 93–100

Dox inhibits angiogenesis

Gilberston-Beadling S, Powers EA, Stamp-Cole M et al (1995) The tetracycline analogs minocycline and doxycycline inhibit angiogenesis *in vitro* by a non-metalloproteinase-dependent mechanism. *Cancer Chemother Pharmacol* 36: 418–424 [IC_{50} curves for several MMPs]

Dox inhibits cell attachment/proliferation

Potts RC, Hassan HA, Brown RA, MacConnachie A, Gibbs JH, Robertson AJ, Swanson Beck J (1983) *In vitro* effects of doxycycline and tetracycline on mitogen-stimulated lymphocyte growth. *Clin Exp Immunol* 53: 458–464

Tsukuda N, Gabler WL (1993) The influence of doxycycline on the attachment of fibroblasts to gelatin-coated surfaces and its cytotoxicity. *J Periodontol* 64: 1219–1224

Guerin C, Laterra J, Masmyk T, Golub LM, Brem H (1992) Selective endothelial growth inhibition by tetracyclines that inhibit collagenase. *Biochem Biophys Res Commun* 188: 740–745

Fife RS, Sledge GW, Roth B, Proctor C (1998) Effects of doxycycline on human prostate cancer cells *in vitro*. *Cancer Lett* 127: 37–41 [also gelatinase]

Fife RS, Rougraff B, Proctor C, Sledge GW (1997) Inhibition of proliferation and induction of apoptosis by doxycycline in cultured human osteosarcoma cells. *J Lab Clin Med* 130: 530–534

Dox inhibits cell invasion/migration

Fife RS, Sledge GW (1995) Effects of doxycycline on *in vitro* growth, migration, and gelatinase activity of breast carcinoma cells. *J Lab Clin Med* 125: 407–411

Teicher BA, Emi Y, Kakeji Y, Northey D (1996) TNP-470/minocycline/cytotoxic therapy: a system approach to cancer therapy. *Eur J Cancer* 32A: 2461–2466

Dox inhibits leukocyte function

Gabler W, Smith J, Tsukuda N (1992) Comparison of doxycycline and a chemically modified tetracycline inhibition of leukocyte functions. *Res Commun Chem Pathol Pharmacol* 78: 151–160

Kuzin II, Snyder TG, Ugine GD, Wu D, Lee S, Bushnell T, Insel RA, Young FM, Bottaro A (2001) Tetracyclines inhibit activated B cell function. *Int Immunol* 13: 921–931

Dox inhibits collagen synthesis

Davies S, Cole A, Schmid T (1996) Doxycycline inhibits Type X collagen synthesis in avian hypertrophic chondrocyte cultures. *J Biol Chem* 271: 25 966–25 970

C) Animal models

Arthritic rat

Greenwald RA, Moak SA, Ramamurthy NSD, Golub LM (1992) Tetracyclines suppress metalloproteinase activity in adjuvant arthritis and, in combination with flurbiprofen, ameliorate bone damage. *J Rheumatol* 19: 927–938

Ganu V, Doughty J, Spirito S, Goldberg R (1994) Elevation of urinary pyridinoline in adjuvant arthritic rats and its inhibition by doxycycline. *In*: RA Greenwald, LM Golub (ed.): *Inhibition of Matrix Metalloproteinases: Therapeutic Potential. Ann N Y Acad Sci* 732: 416–418

Ramamurthy NS, Greenwald RA, Moak S, Sciubba J, Goren A, Turner G, Rifkin B, Golub LM (1994) CMT/tenidap treatment inhibits temporomandibular joint destruction in adjuvant arthritic rats. *Ann N Y Acad Sci* 732: 427–430

Weithmann KU, Schlotte V, Jeske V, Seiffge D, Laber A, Haase B, Schleyerbach R (1997) Effects of tiaprofenic acid on urinary pyridinium crosslinks in adjuvant arthritic rats: comparison with doxycycline. *Inflamm Res* 46: 246–252

Rabbit cornea

Perry HD, Hodes LW, Seedor JA, Donnenfeld ED, McNamara TF, Golub LM (1993) Effect of doxycycline hyclate on corneal epithelial wound healing in the rabbit alkali-burn model. *Cornea* 12: 379–382

Arthritic dog

Yu LP, Smith GN, Brandt K, Myers SL, O'Connor BL, Brandt DA (1993) Reduction of the severity of canine osteoarthritis by prophylactic treatment with oral doxycycline. *Arthritis Rheum* 35: 1150–1159

Yu LP, Smith GN, Brandt KMyers SL, O'Connor BL, Brandt DA (1993) Therapeutic administration of doxycycline slows the progression of cartilage destruction in canine osteoarthritis. *Trans Ortho Res Soc* 18: 724

Yu LP, Burr DB, Brandt KD et al (1996) Effects of oral doxycycline administration on histomorphometry and dynamics of subchondral bone in a canine model of osteoarthritis. *J Rheumatol* 23: 137–142

Aneurysm/vascular

Petrinec D, Liao S, Holmes DR et al (1996) Doxycycline inhibition of aneurysmal degeneration in an elastase induced rat model of abdominal aortic aneurysm: preservation of aortic elastin associated with suppressed production of 92 kD gelatinase. *J Vasc Surg* 23: 336–346

Boyle JR, McDermott E, Crowther M, Wills AD, Thompson MM (1998) Doxycycline inhibits elastin degradation and reduces metalloproteinase activity in a model of aneurysmal disease. *J Vasc Surg* 27: 354–361

Pyo R, Lee J, Shipley M et al (2000) Targeted gene disruption of matrix metalloproteinase-9 (gelatinase B) suppresses development of experimental abdominal aortic aneurysms. *J Clin Invest* 105: 1641–1649 [doxycycline effect compared to MMP-9 knock-out]

Viellard-Baron A, Frisdal E, Eddahibi S et al (2000) Inhibition of matrix metalloproteinases by lung TIMP-1 gene transfer or doxycycline aggravates pulmonary hypertension in rats. *Circ Res* 87: 418–425 [Dox used to inhibit gelatinase, effect compared to TIMP gene transfer]

Osteoporosis

Golub LM, Ramamurthy NS, Kaneko H, Sasaki T, Rifkin B, McNamara T (1990) Tetracycline administration prevents diabetes-induced osteopenia in the rat – initial observations. *Res Commun Chem Pathol Pharmacol* 68: 27–40

Sasaki T, Ohyori N, Debari K, Ramamurthy NS, Golub LM (1999) Effects of chemically modified tetracycline, CMT-8, on bone loss and osteoclast structure and function in osteoporotic states. *Ann N Y Acad Sci* 878: 347–360

Folwarczna J, Janiec W, Firlus K, Kaczmarczyk-Sedlak I (1999) Effects of doxycycline on the development of bone damage caused by prednisolone in rats. *Polish J Pharmacol* 51: 243–251

Apoptosis

Fife RS, Sledge GW, Proctor C et al (1996) Doxycycline inhibits matrix metalloproteinase activity and enhances apoptosis in human prostate cancer cells. *J Invest Med* 44: 249A

Ischemia/reperfusion

Clark WM, Calcagno FA, Gabler WL et al (1994) Reduction of central nervous system reperfusion injury in rabbits using doxycycline treatment. *Stroke* 25: 1411–1416

Smith JR, Gabler WL (1994) Doxycycline suppression of ischemia-reperfusion induced hepatic injury. *Inflammation* 18: 193–201

Smith JR, Gabler WL (1995) Protective effects of doxycycline in mesenteric ischemia and reperfusion. *Russ Commun Mol Pathol Pharmacol* 88: 303–315

Clark et al (1994) Reduction of central nervous system reperfusion injury in rabbits using doxycycline treatment. *Stroke* 25: 1411–1416

Clark WM, Lessov N, Lauten JD, Hazel K (1997) Doxycycline treatment reduces ischemic brain damage in transient middle cerebral artery occlusion in the rat. *J Mol Neurosci* 9: 103–108

Reasoner DK, Hindman BJ, Dexter F, Subieta A, Cutkomp J, Smith T (1997) Doxycycline reduces early neurologic impairment after cerebral arterial air embolism in the rabbit. *Anesthesiology* 87: 569–576

Cancer

Fife RS, Sledge CW, Dunn J et al (1994) Suppression of growth of primary breast cancer in mice treat ed with doxycycline. *Clin Res* 42: 394A

D) Human studies

Adult periodontal disease

Golub LM, Ciancio S, Ramamurthy NS, Leung, M, McNamara TF (1990) Low dose doxycycline ther apy: effect on gingival and crevicular fluid collagenase activity in humans. *J Periodont Res* 25: 321–330

McCullough CAG, Birek P, Overall C, Aiken S, Lee W, Kulkarni R (1990) Randomized controlled trial of doxycycline in prevention of recurrent periodontitis in high-risk patients: antimicrobial activity and collagenase inhibition. *J Clin Periodontol* 17: 616–622

Bouwmsa O, Payonk G, Baron H et al (1991) Low dose doxycycline. effects on clinical parameters in adult periodontitis. *J Dent Res* 71: abstract 1119

Lee W, Aiken S, Kulkarni G, Birek P, Overall CM, Sodek J, McCullough CAG (1991) Collagenase activity in recurrent periodontitis: relationship to disease progression and doxycycline therapy. *J Periodont Res* 26: 479–483

Schroeder K, Ramamurthy NS, Seckepanek KA et al (1992) Low dose doxycycline prevents attachment loss in adult periodontitis. *J Dent Res* 72: abstract 1936

Crout RJ, Lee HM, Schroeder K, Crout H, Ramamurthy NS, Wiener M, Golub LM (1996) The "cyclic" regimen of low-dose doxycycline for adult periodontitis: a preliminary study. *J Periodontol* 67: 506–514

McCullough CA (1994) Collagonolytic enzymes in gingival crevicular fluid as diagnostic indicators of periodontitis. *Ann N Y Acad Sci* 732: 152–164

Golub LM, Lee HM, Greenwald RA, Ryan ME, Sorsa T, Salo T, Giannobile WV (1997) A matrix metalloproteinase inhibitor reduces bone-type collagen degradation fragments and bone-type collagenase in gingival crevicular fluid during adult periodontitis. *Inflamm Res* 46: 310–319 [a landmark paper the first in the medical literature in which an anticollagenase agent was given to human patients with a disorder characterized by excessive collagenase, and it was demonstrated that inhibition of collagenase as well as decreased collagen breakdown was achieved]

Ashley RA (1999) Clinical trials of a matrix metalloproteinase inhibitor in human periodontal disease. *Ann N Y Acad Sci* 878: 335–346

Caton J, Ciancio S, Blieden TM et al (2000) Treatment with subantimicrobial dose doxycycline improves the efficacy of scaling and root planing in patients with adult periodontitis. *J Periodontol* 71: 521–532

Golub LM, McNamara TF, Ryan ME, Kohut B, Bleiden T, Payonk G, Sipos T, Baron H (2001) Adjunctive treatment with subantimicrobial doses of doxycycline: effects on gingival fluid collagenase activity and attachment loss in adult periodontitis. *J Clin Periodontol* 28: 146–156

Rheumatoid arthritis

Nordstrom D, Lindy O, Lauhio A, Sorsa T, Santavirta S, Konttinen Y (1998) Anti-collagenolytic mechanism of action of doxycycline treatment in rheumatoid arthritis. *Rheumatol Int* 17: 173–180

Osteoarthritis

Yu LP, Smith GN, Brandt KD, Capello WN (1994) Preoperative oral administration of doxycycline reduces collagenase in extracts of human osteoarthritic cartilage. *Arthritis Rheum* 37 (suppl 9): S362 (abstract 1207)

Brandt KD (1993) Modification by oral doxycycline administration of articular cartilage breakdown in osteoarthritis. *J Rheumatol* (Suppl 43) 22: 149–151

Smith GN, Yu L, Brandt K, Capello W (1998) Oral administration of doxycycline reduces collagenase and gelatinase activities in extracts of human osteoarthritic cartilage. *J Rheumatol* 25: 532–535

Aneurysm

Curci JA, Mao D, Bohner DG, Allen B, Rubin BG, Reilly JM, Sicard GA, Thompson RW (2000) Preoperative treatment with doxycycline reduces aortic wall expression and activation of matrix metalloproteinases in patients with abdominal aortic aneurysms. *J Vasc Surg* 31: 325–342

CMT references

A) Enzymes/biochemical

CMTs inhibit general/unspecified collagenolytic activity

Golub LMMcNamara TF, D'Angelo G, Greenwald RA, Ramamurthy NS (1987) A non-antibacterial chemically modified tetracycline inhibits mammalian collagenase activity. *J Dent Res* 66: 1310–1314 [the CMT discovery paper]

CMTs inhibit MMP-8

Sorsa T, Ramamurthy NS, Vernillo T, Zhang X, Konttinen Y, Rifkin B, Golub LM (1998) Functional sites of chemically modified tetracyclines: inhibition of the oxidative activation of human neutrophil and chicken osteoclast pro-matrix metalloproteinases. *J Rheumatol* 25: 975–982 [also MMP-9]

CMTs inhibit MMP-13

Ramamurthy NS, Vernillo AT, Golub LM, Rifkin BR (1990) The effect of tetracyclines on collagenase activity in UMR 106-01 rat osteoblastic osteosarcoma cells. *Res Commun Chem Pathol Pharmacol* 70: 323–335

Ramamurthy NS, Vernillo AT, Greenwald RA, Lee HM, Sorsa T, Golub LMRifkin BR (1993) Reactive oxygen species activate and tetracyclines inhibit rat osteoblast collagenase. *J Bone Miner Res* 8: 1247–1253 [activation of procollagenase by HOCl]

Greenwald RA, Golub LM, Chowdhury M, Ramamurthy NS, Moak SA, Sorsa T (1998) *In vitro* sensitivity of the three mammalian collagenases to tetracycline inhibition: relationship to bone and cartilage destruction. *Bone* 22: 33–38 [Extensive comparative review for multiple CMTs and multiple collagenases.]

CMTs inhibit gelatinase(s)

Zucker S, Wieman J, Lysik R, Imhof B, Nagase H, Ramamurthy NS, Liotta L, Golub L (1989) Gelatin-degrading type IV collagenase isolated from human small cell lung cancer. *Invas Metast* 9: 167–181

Nip LH, Uitto VJ, Golub LM (1993) Inhibition of epithelial cell MMPs by tetracycline. *J Periodont Res* 28: 379–385 [CMT-1 and dox inhibit gelatinase secretion by cultured epithelial cells]

Uitto VJ, Firth JD, Nip L, Golub LM (1994) Doxycycline and chemically-modified tetracyclines inhibit gelatinase A (MMP-2) gene expression in human skin keratinocytes. *Ann N Y Acad Sci* 732: 140–151

Paemen L, Martens E, Norga K, Masure S, Roets E, Hoogmartens J, Opdenakker G (1996) The gelatinase inhibitory activity of tetracyclines and chemically modified tetracycline analogues as measured by a novel microtiter assay for inhibitors. *Biochem Pharmacol* 52: 105–111

Hendrix MJC, Seftor EA, Seftor REB, DeLarco JE, McNamara TF, Golub LM (1996) Chemically modified tetracyclines inhibit both human melanoma cell invasiveness and gelatinases A and B. *J Dent Res* 75: 152 (abstract #1075)

Maisi P, Kiili M, Raulo SM, Pirila E, Sorsa T (1999) MMP inhibition by chemically modified tetracycline-3 (CMT-3) in equine pulmonary epithelial lining fluid. *Ann N Y Acad Sci* 878: 675–677 [gelatinolytic activity]

CMTs inhibit macrophage/PMN elastase

Simon SR, Roemer EJ, Golub LM, Ramamurthy NS (1998) Serine proteinase inhibitory activity by hydrophobic tetracycline United States Patent #5, 773, 430, issued June 30 (1998)

CMTs inhibit MT-mmps

Lee HM, Golub LM, Cao C, Sorsa T, Teronen O, Laitinen M, Salo T, Zucker S (1998) CMT-3, a non-antimicrobial tetracycline (TC) inhibits MT1-MMP activity: relevance to cancer. *J Dent Res* 77: 747 (abstract #926)

CMTs inhibits PLA$_2$

Pruzanski W, Stefanski E, Vadas P, McNamara TF, Ramamurthy NS, Golub LM (1998) Chemically modified non-antimicrobial tetracyclines inhibit activity of phospholipase A$_2$. *J Rheumatol* 25: 1807–1812

CMTs inhibit/enhance cytokine production

Kirkwood KL, Golub LM, Bradford PG (1999) Non-antimicrobial and antimicrobial tetracyclines inhibit IL-6 expression in murine osteoblasts. *Ann N Y Acad Sci* 878: 667–670

Lee HM, Moak SA, Greenwald RA, Sion S, Ritchlin C, Golub LM (1999) CMT-3, a non-antimicrobial tetracycline (TC), increases the production of IL-10 in macrophages. *J Dent Res* 78: 192 (abstract #693) [IL-10 is an anti-inflammatory rather than phlogistic cytokine]

Ritchlin C, Greenwald RA, Moak SA, Haas-Smith S (1998) Pattern of metalloproteinase production in psoriatic synovial explant tissues. *Arthritis Rheum* 41: (suppl 9), S335 (abstract #1815)

Eklund KK, Sorsa T (1999) Tetracycline derivative CMT-3 inhibits cytokine production, degranulation, and proliferation in cultured mouse and human mast cell. *Ann N Y Acad Sci* 878: 689–692

Ritchlin CT, Haas-Smith SA, Schwarz EM, Greenwald RA (2000) Minocycline but not doxycycline upregulates IL-10 production in human synoviocytes, mononuclear cells, and synovial explants. *Arthritis Rheum* 43 (suppl 9): abstract 1673, S345, 2000

CMTs affect NO production

Trachtman H, Futterweit S, Greenwald RA, Moak S, Singhal P, Franki N, Amin A (1996) Chemically modified tetracyclines inhibit inducible nitric oxide synthetase expression and nitric oxide production in cultured rat mesangial cells. *Biochem Biophys Res Commun* 229: 243–248

Amin AR, Patel RN, Thakkar GD, Lowenstein CJ, Attur MG, Abramson SB (1997) Post-transcriptional regulation of inducible nitric oxide synthetase mRNA in murine macrophages by doxycycline and chemically modified tetracyclines. *FEBS Lett* 410: 259–264

Cillari E, Milano S, D'Agostini P, DiBella C, La Rosa M, Barbera C, Ferdozzo V, Cammarata G, Guinaudo S, Tolomeo M et al (1998) Modulation of nitric oxide production by tetracyclines and chemically modified tetracyclines. *Adv Dent Res* 12: 126–130 [CMTs 1, 3, and 8 but not 5 inhibited NO production in LPS stimulated cell lines]

D'Agostino P, Arcoleo F, Barbera C et al (1998) Tetracycline inhibits the nitric oxide synthetase activity induced by endotoxin in cultured murine macrophages. *Eur J Pharmacol* 346: 283–290

CMTs inhibit protein glycation

Ryan ME, Ramamurthy NS, Golub LM (1998) Tetracyclines inhibit protein glycation in experimental diabetes. *Adv Dent Res* 12: 152–158 [CMTs 1, 3, 4, 5, 7, 8 all inhibited non-enzymatic glycation of serum proteins and skin collagen without reducing hyperglycemia]

CMTs alter collagen metabolism

Yu Z, Ramamurthy NS, Leung M, Chang KM, McNamara TF, Golub LM (1993) Chemically modified tetracycline normalizes collagen metabolism in diabetic rats: a dose response study. *J Periodont Res* 28: 420–428

Sasaki T, Ramamurthy NS, Yu Z, Golub LM (1992) Tetracycline administration increases protein (presumably procollagen) synthesis and secretion in periodontal ligament fibroblasts of streptozotocin-diabetic rats. *J Periodont Res* 27: 631–639

Grossman M, Takeuchi E, Golub LM, Rifkin B, Landesberg R (1998) Regulation of Type II collagen in growth plate chondrocytes by chemically modified tetracyclines. *J Dent Res* 77: 772 (abstract #1122)

B) Cellular systems

CMTs inhibit bone resorption/osteoclasts

Chowdhury MH, Moak SA, Greenwald RA (1993) Effect of tetracycline metalloproteinase inhibitors on basal and heparin-stimulated bone resorption by chick osteoclasts. *Agents Actions* 40: 124–128 [multiple CMTs and dox]

Zaidi M, Moonga BS, Huang CLH, Alam AT, Shankar VS, Pazianos M, Eastwood JB, Datta HK, Rifkin BR (1993) Effect of tetracyclines on quantitative measures of osteoclast morphology. *Bioscience Rep* 13: 175–182 [CMT-1 and min inhibit osteoclast spreading and motility, possible effect on Ca^{2+} receptor]

Sasaki T, Ramamurthy NS, Golub LM (1994) Bone cells and matrix bind chemically modified non-antimicrobial tetracycline. *Bone* 15: 373–375

Golub LM, Ramamurthy NS, Llavaneras A, Ryan ME, Lee HS, Liu Y, Bain S, Sorsa T (1999) A chem ically modified non-antimicrobial tetracycline (CMT-8) inhibits gingival matrix metalloproteinases, periodontal breakdown, and extra-oral bone loss in ovariectomized rats. *Ann N Y Acad Sci* 878: 290–310

Bettany JT, Peet NM, Wolowacz RG, Skerry TM, Grabowski PS (2000) Tetracyclines induce apoptosis in osteoclasts. *Bone* 27: 75–80

CMTs inhibit cartilage degradation

Cole AA, Yi W, Kuettner K, Golub LM, Greenwald RA (1995) The effect of chemically modified tetracyclines on cartilage degradation in chicken tibial explants. *Trans Ortho Res Soc* 20: 337

CMTs inhibit cell proliferation

Guerin C, Laterra J, Masnyk T, Golub LM, Brem H (1992) Selective endothelial growth inhibition by tetracyclines that inhibit collagenase. *Biochem Biophys Res Commun* 188: 740–745 [CMT-1 but not CMT-5 normalizes uptake of tritium by cells]

Zhu B, Block NL, Lokeshwar BL (1999) Interaction between stromal cells and tumor cells induces chemoresistance and MMP secretion. *Ann N Y Acad Sci* 878: 642–646 [CMT-3 inhibited human and rat prostate tumor cell proliferation]

CMTs inhibit cell invasion/migration

Lokeshwar BL, Dudak SD, Selzer MG, Golub LM (1996) Chemically modified tetracyclines inhibit invasion and metastasis of carcinoma. *J Dent Res* 75: 152, (abstract #1076)

Seftor RE, Seftor EA, DeLarco J, Kleiner D, Leferson J, Stetler-Stevenson W, McNamara TF, Golub LM, Hendrix MJC (1998) Chemically modified tetracyclines inhibit human melanoma cell invasion and metastasis. *Clin Exp Metastasis* 16: 217–225

Lukkonnen A, Sorsa T, Salo T, Tervahartiala T, Koivunen E, Golub LM, Simon S, Stenman U (2000) Down-regulation of trypsinogen-2 expression by chemically modified tetracyclines: association with reduced cancer cell migration. *Int J Cancer* 86: 577–581

CMTs inhibit non-specific proteolysis

Schneider BS, Maimon J, Golub LM, Ramamurthy NS, Greenwald RA (1992) Tetracyclines inhibit intracellular muscle proteolysis *in vitro*. *Biochem Biophys Res Commun* 188: 767–772

Lee HM, Grenier D, Plamondon P, Sorsa T, Teronen O, Mayrand D, Ramamurthy NS, Golub LM (2000) Tetracycline (TC) analogs inhibit bacterial metallo- but not serine proteinases. *J Dent Res* 79: 403 (abstract #2075)

CMTs induce apoptosis

Bettany JT, Wolowacz RG, Pelt NM, Skerry TM, Grabowski PS (1998) Tetracyclines induce apoptosis in osteoclasts. ASBMR, program abstract F071, [CMT-1 and dox induced apoptosis in rabbit osteoclasts in culture]

CMTs affect leukocyte function

Gabler WL, Smith J, Tsukuda N (1992) Comparison of doxycycline and a chemically modified tetracycline on inhibition of leukocyte function. *Res Commun Chem Pathol Pharmacol* 78: 151–160 [CMT-1 less effective than Dox on PMN functions such as superoxide generation, Dox accumulated in cells to higher levels than CMT]

C) Animal models

Diabetic rat

Ryan ME, Ramamurthy NS, Sorsa T, Ding YL, Golub LM (1996) Inhibition of periodontitis as a complication of diabetes by chemically modified tetracyclines. *J Dent Res* 75: 371 (abstract #2831)

Craig RG, Yu Z, Barr R, Ramamurthy NS, Boland J, Schneir M, Golub LM (1998) A chemically modified tetracycline inhibits streptozotocin-induced diabetic depression of skin collagen synthesis and steady state type I procollagen mRNA. *Biochim Biophys Acta* 1402: 250–260

Ryan ME, Ramamurthy NS, Lambrou K, Ioannou M, Golub LM (1998) Inhibition of tooth loss and mortality by chemically modified tetracyclines in Type II diabetes. *J Dent Res* 77: 1001 (abstract #2960)

Ryan ME, Ramamurthy NS, Amin A, Husuh D, Sorsa T, Golub LM (1999) Host response in diabetes-associated periodontitis: effects of a tetracycline analogue. *J Dent Res* 78: 380 (abstract #2193) [decreased PGE_2 and NO levels in diabetic rats]

Ryan ME, Ramamurthy NS, Sorsa T, Golub LM (1999) MMP-mediated events in diabetes. *Ann N Y Acad Sci* 878: 311–334

Ryan ME, Usman A, Ramamurthy NS, Golub LM, Greenwald RA (2001) Excessive matrix metalloproteinase activity in diabetes: inhibition by tetracycline analogues with zinc reactivity. *Curr Med Chem* 8: 305–316

Experimental periodontal disease (monoinfected or LPS rat)

Chang KM, Ramamurthy NS, McNamara TF, Evans RT, Klausen B, Murray PA, Golub LM (1994) Tetracyclines inhibit *Porphyromonas gingivalis*-induced alveolar bone loss in rats by a non-antimicrobial mechanism. *J Periodont Res* 29: 242–249

Golub LM, Evans RT, McNamara TF, Lee HM, Ramamurthy NS (1994) A non-antimicrobial tetracycline inhibits gingival matrix metalloproteinases and bone loss in *Porphyromonas gingivalis*-induced periodontitis in rats. *Ann N Y Acad Sci* 732: 96–111 [also Dox and CMT inhibition of gelatinase and CGase, also CMT inhibition of elastase]

Rifkin B, Ramamurthy NS, Gwinnett A, Turner G, Vernillo A, Golub LM (1996) Inhibition of endotoxin induced periodontal bone resorption by CMTs. *J Dent Res* 75: 308 (abstract #2324)

Karimbux N, Ramamurthy NS, Golub LM, Nishimura I (1998) The expression of collagen I and XII mRNAs in *Porphyromonas gingivalis*-induced periodontitis in rats: the effect of doxycycline and chemically modified tetracyclines. *J Periodontol* 69: 34–40 [also Dox]

Llavaneras S, Golub LM, Rifkin BR, Heikkila P (1999) CMT-8/clodronate therapy synergistically inhibits alveolar bone loss in LPS-induced periodontitis. *Ann N Y Acad Sci* 878: 671–674

Rat caries

Ramamurthy NS, Schroeder KL, McNamara TF, Gwinnett AJ, Evans RT, Bosko C, Golub LM (1998) Root surface caries in rats and humans: Inhibition by a non-antimicrobial property of tetracyclines. *Adv Dent Res* 12: 43–50

Tjaderhane L, Sulkala M, Sorsa T, Teronen O, Larmas M, Salo T (1999) The effect of MMP inhibitor Metastat {COL-3} on fissure caries progression in rats. *Ann N Y Acad Sci* 878: 686–688

Arthritic rat

Greenwald RA, Moak SA, Ramamurthy NS, Golub LM (1992) Tetracyclines suppress metalloproteinase activity in adjuvant arthritis and, in combination with flurbiprofen, ameliorate bone damage. *J Rheumatol* 19: 927–938

Ramamurthy NS, Greenwald RA, Moak S, Goren A, Turner G, Rifkin B, Golub LM (1994) CMT/tenidap treatment inhibits TMJ destruction in adjuvant arthritis (AA) rats. *Ann N Y Acad Sci* 732: 427–430

Zernicke R, Wohl G, Greenwald RA, Moak S, Leng W, Golub LM (1997) Administration of systemic matrix metalloproteinase inhibitors maintains bone mechanical integrity in adjuvant arthritis. *J Rheumatol* 24: 1324–1331

Guinea pig osteoarthritis

Greenwald RA (1994) Treatment of destructive arthritic disorders with MMP inhibitors: potential role of tetracyclines. *In*: RA Greenwald, LM Golub (ed.): *Inhibition of Matrix Metalloproteinases: Therapeutic Potential. Ann N Y Acad Sci* 732: 181–198 [CMT-7]

de Bri E, Lei W, Svensson O, Chowdhury M, Moak S, Greenwald RA (1998) Effect of an inhibitor of matrix metalloproteinases on spontaneous osteoarthritis in guinea pigs. *Adv Dent Res* 12: 82–85

Aneurysm

Petrinec D, Holmes DR, Liao S, Golub LM, Thompson LM (1996) Suppression of experimental aneurysmal degeneration with chemically modified tetracycline derivatives. *Ann N Y Acad Sci* 800: 263–265

Curci JA, Petrinec D, Liao S, Golub LM, Thompson RW (1998) Pharmacological suppression of experimental abdominal aortic aneurysms: A comparison of doxycycline and four chemically modified tetracyclines. *J Vasc Surg* 28: 1082–1093

Osteoporosis/metabolic bone disease

Greenwald RA, Simonson BG, Moak SA, Rush S, Ramamurthy NS, Laskin RS, Golub LM (1988) Inhibition of epiphyseal cartilage collagenase by tetracyclines in low phosphate rickets in rats. *J Orthopaed Res* 6: 695–703 [also Dox, Min]

Sasaki T, Ramamurthy NS, Golub LM (1990) Insulin-deficient diabetes impairs osteoblast and periodontal ligament fibroblast metabolism: Response to tetracycline administration. *J Biol Buccale* 18: 215–226 [normalization of tritiated proline uptake by fibroblasts]

Kaneko H, Sasaki T, Ramamurthy NS, Golub LM (1990) Tetracycline administration normalizes the structure and acid phosphatase activity of osteoclasts in streptozotocin induced diabetic rats. *Anat Rec* 227: 427–436

Golub LM, Ramamurthy NS, Kaneko H, Sasaki T, Rifkin B, McNamara TF (1990) Tetracycline administration prevents diabetes-induced osteopenia in the rat. *Res Commun Chem Pathol Pharmacol* 68: 27–40 [CMT-1 normalized morphology of atrophic osteoblasts]

Sasaki T, Kaneko H, Ramamurthy NS, Golub LM (1991) Tetracycline administration restores osteoblast structure and function during experimental diabetes. *Anat Rec* 231: 25–34 [CMTs normalize alkaline phosphatase and ATPase]

Sasaki T, Ramamurthy NS, Golub LM (1992) Tetracycline administration increases collagen synthesis in osteoblasts of diabetic rats: a quantitative autoradiographic study. *Calcified Tissue Int* 50: 411–419 [CMTs reverse diabetic suppression of tritiated proline uptake]

Aoyagi M, Sasaki T, Ramamurthy NS, Golub LM (1996) Tetracycline/flurbiprofen combination therapy modulates bone remodeling in ovariectomized rats: preliminary observations. *Bone* 19: 629–635

Bain S, Ramamurthy NS, Impeduglia T, Scolman S, Golub LM, Rubin C (1997) Tetracycline prevents cancellous bone loss and maintains near-normal rates of bone formation in streptozotocin diabetic rats. *Bone* 21: 147–153

Ohyori N, Sasaki T, Debari K, Higashi S, Ramamurthy NS, Golub LM (1997) Chemically modified tetracycline (CMT-7,8) therapy increases bone mass in ovariectomized (OVX) rats. *J Dent Res* 76: 433 (abstract #3357)

Yokoya K, Sasaki T, Shibasaki Y, Ramamurthy NS, Golub LM (1997) Chemically modified tetracycline (CMT-7,8) therapy increases bone volume in TMJ condyles ovariectomized rats. *J Dent Res* 76: 447 (abstract #3471)

Bain SD, Ramamurthy NS, Llavaneras A, Puerner D, Strachan MJ, Scheer K, Golub LM (1999) A chemically modified doxycycline (CMT-8) alone or in combination with bisphosphonate inhibits bone loss in the ovariectomized rat: a dynamic histomorphometric study. *J Dent Res* 78: 416 (abstract #2486)

Llavaneras A, Ramamurthy NS, Wolff M, Sorsa T, Golub LM (1999) A chemically modified doxycycline (CMT-8) and bisphosphonate combination therapy inhibits periodontal breakdown in the ovariectomized (OVX) rat. *J Dent Res* 78: 431 (abstract #2605)

Wound healing

Zhang X, Kucine A, Ramamurthy NS, McClain S, Ryan ME, McNamara TF, Golub LM (1996) Chemically modified tetracycline (CMT-6) applied topically enhances diabetic wound healing. *J Dent Res* 75: 108 (abstract #723)

Ramamurthy NS, McClain SE, Pirila E, Maisi P, Sorsa T, Golub LM (1999) Wound healing in aged and normal ovariectomized rats: effects of chemically modified tetracyclines on MMP expression and collagen synthesis. *Ann N Y Acad Sci* 878: 720–723

Kucine A, McClain S, Pirila E, Maisi P, Salo T, Sorsa T, Golub LM, Ramamurthy NS (1999) Delayed wound healing and altered MMP expression in the skin of ovariectomized (OVX) rats is normalized by chemically modified tetracyclines (CMT-8). *J Dent Res* 78: 551 (abstract #3563)

Ischemia

Carney DE, Lutz C, Picone AL, Gatto LA, Ramamurthy NS, Golub LM, Simon SR, Searles B, Paskanik A, Finck C et al (1999) Matrix metalloproteinase inhibitor prevents acute lung injury after cardiopulmonary bypass. *Circulation* 100: 400–406

Cancer

Lokeshwar BL, Selzer MG, Block NL, Golub LM (1997) COL-3, a modified non-antimicrobial tetracycline, decreases prostate tumor growth and metastasis. *J Dent Res* 76: 449 (abstract #3481)

Lokeshwar BL, Houston-Clark HL, Selzer MG, Block NL, Golub LM (1998) Potential application of a chemically modified non-antimicrobial tetracycline against metastatic prostate cancer. *Adv Dent Res* 12: 149–151 [CMT-3 more cytotoxic to cancer cell lines than dox, CMT-3 induced apoptosis, CMT-3 inhibited Matrigel invasion]

Selzer MG, Zhu B, Block NL, Lokeshwar BL (1999) CMT-3, a chemically modified tetracycline, inhibits bony metastases and delays the development of paraplegia in a rat model of prostate cancer. *Ann N Y Acad Sci* 878: 678–682

Lush RM, Rudek MA, Figg WD (1999) Rev three new agents that target angiogenesis, matrix metalloproteinases, cyclin-dependent kinases Cancer Control 6: 459–465

D) Human studies

Cancer

Rudek MA, Figg WD, Dyer V, Dahut W, Turner ML, Steinberg S, LiewaHR DJ, Kohler DR, Pluda JM, Reed E (2001) Phase I clinical trial or oral COL-3, a matrix metalloproteinase inhibitor, in patients with refractory metastatic cancer. *J Clin Oncol.* 19: 584–592

Rudek MA, March CL, Bauer KS, Pluda JM, Figg WD (2000) High performance liquid chromatography with mass spectrometry detection for quantitating COL-3, a chemically modified tetracycline, in human plasma. *J Pharm Biomed Anal* 22 22: 1003–1014

Ghate JV, Turnel ML, Rudek MA, Figg WD, Dahut W, Dyer V, Pluda JM, Reed E, (2001) *Arch Dermatol* 137: 471–474

Rudek MA, Horne M, Figg WD, Dahut WM, Dyer V, Pluda JM, Reed E (2001) Reversible sideroblastic anemia associated with tetracycline analogue COL-3. *Am J Hematol* 67: 51–53

E) Pharmacokinetics and biological properties

Microbial resistance

Golub LM, Greenwald RA, Ramamurthy NS, McNamara TF, Rifkin BR (1991) Tetracyclines inhibit connective tissue breakdown: new therapeutic implications for an old family of drugs. *Crit Revs Oral Biol Med.* 2: 297–322 [review article which contains primary data]

Thomas J, Karakiozis J, Powala C, Dawson-Andoh D, Van Dyke K (1998) Establishing the non-antimicrobial activity of chemically modified tetracyclines *versus* doxycycline. *J. Dent. Res.* abstract for 78th meeting of IADR, 77: 795 [COL-3 and COL-8 showed no antimicrobial activity over a range from 0.03 to 10 μg/ml]

Distribution and cellular uptake

Donahue HJ, Ijima K, Goligorsky MS, Rubin CT, Rifkin BR (1992) Regulation of cytoplasmic calcium concentration in tetracycline-treated osteoclasts. *J Bone Min Res* 7: 1313–1318

Bax CMR, Shanker VS, Alam AS, Bax BE, Moonga BS, Huang CLH, Zaidi M, Rifkin BR (1993) Tetracycline modulates cytosolic Ca^{2+} response in the osteoclast associated with Ca^{2+} receptor activation. *Bioscience Rep* 13: 169–174

Yu Z, Leung MK, Ramamurthy NS, McNamara TF, Golub LM (1992) HPLC determination of a chemically-modified non-antimicrobial tetracycline: biological implications. *Biochem Med Metabol Biol* 47: 10–20

Sasaki T, Ramamurthy NS, Golub LM (1994) Bone cells and matrix bind chemically modified non-antimicrobial tetracycline. *Bone* 5: 373–375

Rudek MA, Figg WD, Dyer V, Dahut W, Turner ML, Steinberg S, Kohler D, Pluda J, Reed E (2001) A Phase I clinical trial of oral COL-3, a matrix metalloproteinase inhibitor, in patients with refractory metastatic cancer. *J Clin Oncol* 19: 584–592

MMP inhibition by other tetracyclines

Makinen PL, Makinen KK (1988) Near stoichiometric irreversible inactivation of bacterial collagenase by o-choranil (3,4,5,6-tetrachloro-1,2-benzoquinone). *Biochem Biophys Res Commun* 153: 74–80

Dubois B, D'Hooghe MB, De Lepeleire K, Ketelaer P, Opdenakker G, Carton H (1998) Toxicity in a double-blind placebo-controlled pilot trial with D-penicillamine, metacycline in secondary progressive multiple sclerosis Mult Scler 4: 74–78 [inhibition of MMP-9]

Review articles

Golub LM, Greenwald RA, Ramamurthy NS, McNamara TF, Rifkin BR (1991) Tetracyclines inhibit connective tissue breakdown: New therapeutic implications for an old family of drugs. *Crit Rev Oral Biol Med* 2: 297–322

Ryan M, Greenwald RA, Golub LM (1996) Potential of tetracyclines to modify cartilage breakdown in osteoarthritis. *Curr Opin Rheumatol* 8: 238–247

Rifkin BR, Golub LM, Ramamurthy NS, Vernillo T (1966) Inhibition of bone resorption by tetracyclines. *In*: Z Davidovitch, LA Norton (eds): *Biological Mechanisms of Tooth Movement and Carniofacial Adaptation*. Harvard Soc for the Advancement of Orthodontics, Boston, pp 357–363

Greenwald RA, Golub LM (eds) (1998) *Non-antibiotic Properties of Tetracyclines*, proceedings of the Garden City conference. Adv Dent Res 12 [published proceedings of a 2-day symposium with numerous CMT papers, including a comprehensive review]

Golub LM, Lee HM, Ryan ME, Giannobile WV, Payne J, Sorsa T (1998) Tetracyclines inhibit connective tissue breakdown by multiple non-antimicrobial mechanisms. *Adv Dent Res* 12: 12–26 [comprehensive review addressing structure-function relationships of CMTs, role as MMPIs, direct and indirect mechanisms on expression and activation of MMPs, including cytokines, iNOS, arachidonic acid metabolism, calmodulin, protein kinase C, etc.]

References

1 Golub LM, Lee HM, Lehrer G, Nemiroff A, McNamara TF, Kaplan R, Ramamurthy NS (1983) Minocycline reduces gingival collagenolytic activity during diabetes. *J Periodont Res* 18: 516–521

2 Schlienger RG, Bircher AJ, Meier CR (2000) Minocycline-induced lupus. *Dermatology* 200: 223–231

3 Greenwald RA, Golub LM (eds) (1998) Non-antibiotic Properties of Tetracyclines, proceedings of the Garden City conference. *Adv Dent Res 12*

4. Greenwald RA, Golub LM (2001) An annotated bibliography of the non-antibiotic properties of chemically modified tetracyclines. *Curr Med Chem* 8: 237–242

Abbreviations used: CMT, chemically modified tetracycline(s); TC, tetracycline; Min, minocycline; Dox, doxycycline; MMP, matrix metalloproteinase; MMPI, matrix metalloproteinase inhibitor; NO, nitric oxide; IL, interleukin; LPS, lipopolysaccharide; PMN, polymorphonuclear leukocyte.

Tetracyclines in Biology, Chemistry and Medicine
ed. by M. Nelson, W. Hillen and R.A. Greenwald

Tetracycline consumption in prehistory

George J. Armelagos*, Kristi Kolbacher, Kristy Collins, Jennifer Cook and Maria Krafeld-Daugherty

*Emory University, Atlanta, USA

Debra Martin was visiting the Calcified Tissue Laboratory at Henry Ford Hospital in Detroit, Michigan, learning to make thin sections of undecalcified[1] archaeological bone. The objective of her visit was to make slides of bone thin enough to allow the light from the microscope to be transmitted through them. Although modern equipment speeds up the process, Martin, at the time a graduate student at the University of Massachusetts, was using a manual method for grinding the bone to the desired 100-micron thickness. The femoral cross-section is removed from an area just below the lesser trochanter (a few inches below its neck).

In order to ensure that the thickness of the section is uniform, it is frequently examined with a standard light-transmitting microscope. While preparing to examine a thin-section under the light-transmitting microscope, Martin (who is now a professor of anthropology at Hampshire College in Amherst, Massachusetts) found that another researcher was using it. The only available microscope was one that transmitted ultraviolet (UV) light. The Calcified Tissue Laboratory employed an UV microscope to detect tetracycline in bone by the unique yellow-greenish fluorescence emitted when bone is exposed to ultraviolet light at the 490-nanometer wavelength. A tetracycline "label" provides a way to measure bone formation. Since tetracycline chelates (binds) calcium, the antibiotic will leave its mark or label in any bone that is being mineralized. In such a study, tetracycline is administered at 2-week intervals and, with a biopsy, the amount of bone formed during that interval can be measured. To Dr. Martin's surprise, the ancient bones showed marked fluorescence indicative of tetracycline.

The bones under study which are the focus of this report were obtained from a population that inhabited the shore of the Nile River in Sudanese Nubia from 350 CE to 550 CE. The people lived during the X-Group period and represented a localized cultural development following the fall of the Meroitic Kingdom [1] prior to the subsequent reunification of Nubia under Christianity [2, 3].

[1] The traditional method for preparing bone is to use chemicals to remove the minerals in the bone (decalcification). Once this is accomplished a microtome can be used to cut a thin section. Since we are interested in measuring the degree of mineralization, it was necessary to use a method that allows us to make a thin section of mineralized bone.

The serendipitous discovery of the tetracycline stimulated an intense research project. The identification of tetracycline as the fluorophor (the substance that emitted the light) had to be established histologically. The telltale light yellow-green to bright yellow-green fluorescence in the ancient bone occurred at the diagnostic wavelength [4, 5]. The pattern of fluorescence in the Nubian bone was found to be identical to that of tetracycline-labeled bone found in modern clinical applications [4, 6, 7].

Evidence that tetracycline was incorporated into Nubian bone prior to death is revealed by histological features that can only occur during life (*in vivo*). Tetracycline labeling has been found in the cement lines of Nubian bone that we have analyzed. These cement lines are reversal lines (the point where osteoclasts stop resorbing bone and where osteoblasts begin the process of bone deposition) [8]. In addition, unlabeled young osteons formed within their fully fluorescing predecessors are also evidence for an *in vivo* process [4–10]. For example, there is also histological evidence of osteons that show interrupted mineralization in which only part of the osteon is labeled (when calcium-bound tetracycline was available) while the other portion is not (calcium-bound tetracycline was not available). Some of the Nubian and Egyptian bones display what has been described as "concentrically labeled osteons" [4–10]. Megan Cook and coworkers [9], analyzing remains from the Dakhleh Oasis, Egypt [11] from the Roman Period (CE 400–500), found osteons with differential labeling in which there were three rings of labeled bone within a single osteon. This pattern of labeling has also been discovered in a Christian population (CE 550–1450) from the island of Kulubnarti in the Batn el-Hagar region, south of Wadi Halfa [5]. Cook and colleagues also found a 19-year-old individual with labeling of alternating layers of the lamellar bone (evidence of rapidly growing bone) and evidence of tetracycline labeling in the tooth enamel [9].

Subsequent extraction of tetracycline from the Nubian bone and chemical analysis supported our contention that the fluorophors were in fact tetracycline. James Boothe, one of the scientists who developed the initial commercial applications of tetracycline for American Cyanamid, was able to extract the antibiotic from the Nubian bone. Most impressively, Boothe showed that extracted tetracycline retained its antibiotic potential to kill bacteria after 14 centuries[2]. Recently, Mark Nelson, Paratek, has extracted the tetracycline and is in the process of characterizing it.

Our initial reports of ancient tetracycline were met with skepticism by some researchers. In our original publication [4], we were well aware of the possibility that diagenesis (*post mortem* deterioration) may result in the production of fluorophors that would appear similar to the tetracyclines. After our publication, Piepenbrink and coworkers [12, 13] argued that most of the tetracycline labeling that we reported in the Nubian bone was in fact due to *post mortem* changes that occurred when the bone was invaded by bacteria and fungus.

[2] Joe Hlavaka a pioneer in tetracycline chemistry confirms the recovery of tetracycline from the Nubian bone. He was working with James Boothe at the time and observed the experiment.

Despite the evidence that we presented for an *in vivo* incorporation of the tetracycline label, Piepenbrink and co-workers continued to suggest that the fluorescence in Nubian bone was merely the result of taphonomic processes. They argued that the "tetracycline-like" fluorescence was introduced by microbes invading buried bone [12, 13]. Furthermore they claimed that soil organisms such as *Strachybotrys* (a fungus) can invade and destroy bone and leave a tetracycline-like fluorescence. Piepenbrink and his coworkers also maintained that they were able to add fluorescent labels by directly applying tetracycline to previously unlabelled bone, resulting in tetracycline *in vitro* labeling which they suggested supports their argument for a *post mortem* incorporation in the Nubian material.

Even though we believed that we had adequately dealt with the possibility for *post mortem* diagenesis in our original publication [4], we reevaluated the possibility of diagenesis in light of the persistent argument by Piepenbrink and his colleagues. Chemical and histological analyses have confirmed our original hypothesis. *In vivo* incorporation of tetracycline with subsequent deposition of a permanent, discrete fluorescent label in the bone is now firmly established. The pattern of fluorescence was identical to tetracycline labeling that is incorporated during life *(in vivo)* and very different from *post mortem* mold infestation produced in the test tube (*in vitro*).

The taphonomic changes described by Piepenbrink result in collagen degradation and bone destruction described as tunneling and cuffing. Tunneling is a destructive process in which microorganisms bore out sharply bordered canals that are easily recognized in thin sections [14, 15]. Cuffing occurs during the process of tunneling when the invading microorganisms re-deposit excreted minerals. Cuffing may fluctuate with differences in moisture that exist in the burial environment. This redeposited mineral is frequently leached out, leaving a sharply defined tunnel.

The excellent preservation of X-Group Nubian bone is clearly evident. Compact and trabecular bone, as well as the dentition, have been examined and reveal no macroscopic evidence of *post mortem* degradation. Thin sections of compact X-Group bone confirm structural integrity at the histological level [16]. Thin sections were specifically stained with haematoxylin to determine the state of preservation of the organic portion of bone [17] and there is little evidence of deterioration at the cellular level [10]. We have recently examined thin sections under polarized light. Polarization microscopy is used to check the structural preservation of bone collagen (Collagen I) on the submolecular level. Viewed under a light microscope to which a polarizer has been added, the osteons from modern or well-preserved bone show the characteristic birefringence, seen as alternating bright and dark fields or "Maltese crosses."[3] Birefringence is due to the alternating orientation of the collagen fibers; in mature lamellar bone, the collagen fibers are highly

[3] *Birefringence* is the term used to describe an optical property in which the axis of light passing through a matter, in this case the collagen fibers, is rotated.

ordered in their arrangement. Those within each lamella of a Haversian system are parallel in their orientation, but the direction of the fibers in successive lamellae changes (some disagreement persists as to the precise arrangement of the fibers). The deterioration of collagen results in a loss of birefringence. Checking the bone samples from Nubia under a polarizing microscope has confirmed their excellent state of preservation, showing no tunneling and no loss of birefringence.

Extraction of collagen also shows biochemical preservation comparable to autopsy or biopsy material [18]. The percentage of carbon (16.92%) and nitrogen (4.14%), as well as the carbon to nitrogen ratio (C/N = 3.39) is similar to that found in modern collagen. *Post mortem* infestation by fungal or bacterial invasion is a destructive process and the Nubian material shows no evidence of such deterioration.

Finally, we have evaluated Piepenbrink and colleagues' claims of duplicating labeling by placing archaeological bone in contact with a solution of tetracycline. However, *in vitro* exposure results in labeling on surfaces only [10]. Only those areas that are in direct contact with the fluorophor are labeled and do not result in the "hot spots" as seen in the *in vivo* incorporation of the label [19]. We specifically stained under-mineralized osteons with fuchsin and these osteons were not tetracycline-labeled. Piepenbrink and colleagues [12, 13] claim that they produced "concentrically labeled osteons" but we have found their evidence unconvincing. Piepenbrink and co-workers have produced surface labeling, i.e., merely the Haversian canal surface and contents of the canal retained a label. This pattern is unlike the interrupted osteonal labeling found in the Nubian material. Interrupted osteons represent periods of cessation of growth or cessation of exposure to tetracycline and cannot occur after death. The Nubian osteons are found with labeled cement lines and unlabeled lamellae and canals. Reaffirming our position, bone which is inaccessible to an invading microorganism (no tunnels are present) is labeled and surfaces most accessible to a *post mortem* contaminant are not. Furthermore, the cement line is hyper-mineralized bone, effectively eliminating the argument that only under-mineralized bone is labeled – as would be the case with *in vitro* or *post mortem* labeling. Keith and Armelagos [10] firmly rejected Piepenbrink's suggestion that the occurrence of tetracycline in prehistoric Nubians is the result of diagenesis.

Given the evidence of the significant levels of tetracycline that Nubians consumed during their lifetime, it takes a little detective work to show how it was ingested. We know that streptomycetes, mold-like bacteria commonly found in soils, produce tetracycline. The streptomycetes are actinomycetes [20], the bacteria that give some newly turned soil its "musty" odor. In some soils, 60–70% of microorganisms are streptomycetes [20]. They are slow-growing bacteria that are at a selective disadvantage in the usually wet, acidic soil in which most bacteria flourish. In a hot (28 °C to 38 °C), dry (15% or less humidity) and alkaline or neutral environment (optimal pH is 7.00 to 8.00) [20], streptomycetes have a distinct reproductive advantage. They survive desiccation and there are reports of 10-year-old spores found in dry soil that were easily cultured.

Selman Waksman, who received the 1952 Nobel Prize in Medicine and Physiology for his work with antibiotics, claims that there is little evidence of tetracycline in the soil and therefore questions that streptomycetes produce the antibiotic in their natural environment. While Waksman's surprising observation has been supported by recent research, Alexander [21] notes that antibiotic-resistant bacteria are frequently found in soil. However, when bacteria that produce antibiotics are placed in soil, there is little evidence of antibiotic production [22]. Recent studies using more sensitive probes detected evidence of antibiotic production [23].

The question remains as to how the Nubians consumed the tetracycline. Initially, we thought that during famine or drought, they might have been forced to eat moldy grain. Storing grain in mud bins produces an ideal environment for the streptomycetes. The warm, dry and alkaline environment that made streptomycetes so successful in the soil is replicated in the mud storage bin and their ability to grow on solid substrates would allow them to grow on the grain. However, given the frequency and extent of tetracycline labeling, we had to consider other possibilities. Since many of the Nubians received continuous doses of tetracycline, it had to be a food or medicine that was commonly consumed. Given the minimal production of tetracycline by streptomycetes in the soil, the process that produces physiologically significant levels in the bone must be determined.

To complete our story, we had to determine how the contaminated grain could have been processed to yield high enough quantities to account for the levels of tetracycline found in Nubian bone. We thought that ethnographic or historical sources could provide insights into how grain was converted into a food that was also producing tetracycline. In their search of ancient and historical texts, Everett Bassett and Margaret Keith and other team members soon realized that the texts (often translations of hieroglyphic inscriptions) described the grain processing linking bread baking and beer brewing. In Egyptian art, baking and brewing are in "constant association" [24]. Baked bread is an essential part of the brewing process. From the time of ancient Egyptians to contemporary villagers who still live along the Nile, the same recipe is being used in brewing beer – the term "booza" is used to describe the beer in much of Africa.

Our focus on ancient Egyptian sources was necessary since the written records of the ancient Nubians remain to be deciphered. In addition, while we have found references to Sudanese Nubian beer making [25, 26], we have also relied on traditional Egyptian villagers and their recipes for brewing. Many of the Egyptians are Coptic Christians and continue to brew beer. In Sudanese Nubia, the Muslims have religious prohibitions that preclude the use of alcohol.

The beer produced in Nubia is unlike the recreational drink that many of us have consumed. Barry Kemp [27] emphasizes this point:

> Ancient beer…was rather different from its modern counterpart. It was probably an opaque liquid looking like a gruel or soup, not necessarily very alcoholic but highly nutritious. Its prominence in Egyptian diet reflects its food value as much as the mildly pleasurable sensation that went with drinking it.

One of the keys to successful brewing is having a grain that provides sufficient sugar for fermentation. Malting is the process that converts the starches in grains to sugar. Malting converts starch to maltose sugar and diastase. In modern recipes, grain is germinated, heated and dried to halt the process. The malt is boiled, strained and incubated with yeast in the final stage. In the traditional method, the bread dough is processed and then set out in air to capture airborne yeast. It is also possible that airborne streptomycetes spores are potentially captured at the same time. Other traditional recipes actually add booza that was held back from previous batches for that purpose (the yeast would be in the liquid). The bread is baked enough to form a crust, but it is removed from the oven in time to leave the center uncooked, allowing the yeast to grow in the "raw" dough. The streptomycetes in the bread would be producing tetracycline during the process. The partially baked bread is then broken up and added to a broth of malted grain to make the beer.

Cummings [26] describes Sudanese Nubian recipes for brewing. In an account derived from Culwick [25], Cummings describes a 6-day process for processing grain into beer. *Sorghum* is malted by soaking the grain for 24 h, and then the water is drained from the grain and is spread out and covered with a cloth (the Ancient Egyptians referred to this as "spreading out the bed of Osiris."). The sprouting grain grows under the cloth for 4 days and a thick layer of mold covers the plants and the cloth. On the fifth day, the sprouted grain is spread out to dry. After drying, it can be used or stored.

The malted grain can be coarsely ground for brewing or finely ground for use as ordinary flour. Coarsely ground malted grain is used to produce beer by putting it in water for 4 days, creating a soft dough that is allowed to grow a green mold. The dough is then formed into a loaf, covered with suet and placed on a baking sheet. The suet-covered dough is scorched and then put in old beer pots for an additional 30 h in order to allow more molds to form. After this extensive processing, it is ready for brewing.

Daniel Popowich and Brennan Posner, undergraduates at Emory University, modified Egyptian and Sudanese Nubian recipes to brew a beer from bread and grain. Their objective was to investigate the potential for producing tetracycline at the various stages of the brewing process. Popowich and Posner added streptomycetes during two phases of the brewing process. In the first experiment a small colony of streptomycetes was added to the just-baked bread. In the second experiment, they added the streptomycetes to the mix of the grain and bread. The second experiment was the most successful and produced significant amounts of tetracyclines.

Beer can be brewed from any grain. In Egypt, barley, wheat and emmer have been found in baking and brewing areas of archaeological sites. Beer and bread have had an important part in Egyptian culinary culture and religion. James O. Mills, a former colleague at the University of Florida, has written about the importance of beer and bread in Egypt [28]. Beer and bread were standards for wages and rations and used as a measure for labor exchange [27]. Accounts that some ancient Egyptian officials received payments of 500 loaves of bread and 50 jugs of beer a day show that they were essentially an earned line of credit. Given its importance in ancient Egyptian culture, it's not surprising that beer has superhuman powers.

It is difficult to determine the dosage of tetracycline ingested by the ancient Nubians. Frost et al. [29, 30] note that as little as 25 mg of tetracycline per kilogram of a person's weight is capable of labeling bone. Even one or two grams of tetracycline consumed by a human in a single day will produce gross and microscopic fluorescence in bone [31]. The critical factor for labeling bone is not the dosage of tetracycline, but the rate of mineralization and the duration of the dosage [29, 30]. From 3% to 11% of ingested tetracycline in clinical studies is incorporated into the inorganic phase of bone [32].

Given that tetracycline can impair bone development [33, 34] and inhibit spermatogenesis [35], Oklund [32] analyzed the amount of tetracycline sequestered in bone. He was concerned that tetracycline in bone ("yellow bone") might represent a problem when used in transplants[4]. He concluded that it did not represent a problem.

While the dosage of tetracycline is impossible to determine, labeling within the Nubian skeletons provides clues to the amount consumed. Initially, we were sensitive to issues of preservation and decided to measure only the osteons that were fully labeled. In a combined X-Group population (NAX and 24-I-3), 58,423 osteons were sampled and 4865 (8.01%) were labeled with tetracycline (Tab. 1). Of the 78 individuals studied, 74 (94.9%) had labeled osteons and (56.4%) exhibited levels of tetracycline labeling that we arbitrarily designated as "therapeutic" – greater than 4.99% (Tab. 2)[5]. A comparison by sex showed 22 males to have a mean of 9.7% of their osteons labeled, while 38 females have an average labeling of 7.1% (differences are not significant, p = 153). Sub-adults (those under age 16) averaged 7.7% labeled osteons. The difference in labeled osteons by sex and age is not significant (Fig. 1). It is worth noting that in sub-adults labeled osteons decline with age. This may be

[4] Oklund investigated the amount of tetracycline in the bones of a young man who had consumed as much as 500 mg of tetracycline daily for 18 years (Armelagos calculated that over the course of these 18 years, he consumed 6.8 pounds of tetracycline). The concentration of tetracycline in his bone averaged 239 micrograms/g of wet bone weight. In a 70 kilogram human (150 lbs.), the skeleton comprises 15% (10.5 kgs or 23 lbs) of the body weigh, the sequestered tetracycline would only be 3 g and the release would only represent 1 mg per day. Only a small amount of this released tetracycline will be re-sequestered in bone.

[5] We are well aware of the problem in defining what we mean by therapeutic since we do not have a standard for determining the physiological effects of the various levels of tetracycline labeling. We have arbitrarily used the 5% as a therapeutic level.

Table 1. The frequency of tetracycline label in combined X-group sample (NAX and 24-I-3)

Total	N	Minimum	Maximum	Sum	Mean	S.D.
Osteon total	78	0.0	1343	58,423	749	368.9
Labeled osteons	78	0.0	246	4,865	62.4	64
% Labeled osteons	78	0.0	34.2		8.02	
% Labeled bone	78	0.63	25.1		6.98	4.82

Table 2. Frequency of tetracycline labeled osteons and labeled bone by percent grouping in combined X-group sample (NAX and 24-I-3)

Total Percent grouped	Labeled osteons	Percent	Labeled bone	Percent
None	4	5.1	0	0.0
0.1–1	5	6.4	4	5.1
0.1–4.9	25	32.1	31	39.7
05–9.9	18	23.1	28	35.9
10–14.9	14	17.9	10	12.8
15–19.9	7	9.0	3	3.8
20–24.9	3	3.8	1	1.3
25–29.9	1	1.3	1	1.3
30–34.9	1	1.3	0	0.0
Total	78	100	78	100

an artifact of osteon formation since there is a high production of osteons in the infants and children and these newly formed osteons are labeled. Interestingly, Jennifer Cook [11] has shown that two infants (7 months and 8 months) had 1.9% and 14.70% of their osteons labeled (a 5-month old infant had no labeling). All three infants had total bone labeling of 4.1%, 3.6% and 20.2%. These infants with labeled osteons and labeled bone indicate tetracycline was passed in mother's milk.

We soon realized that our approach in measuring just the intact osteon overlooked tetracycline that was present in osteon fragments and lamellar bone. Kristi Kohlbacher, an undergraduate from SUNY, Binghamton, examined the prevalence of all labeled bone, including intact and fragmentary osteons and lamellar bone, in the NAX population. Jennifer Cook did the same for the 24-I-3 population[6]. Using projected images of bone on a grid, a "hit and miss" method was used to determine the percentage of bone that was labeled. It was determined that on the average 6.98% of all bone was labeled with tetracycline

[6] Both Collins and Kohlbacher were SURE (Summer Undergraduate Research Experience) funded by Howard Hughes Initiative.

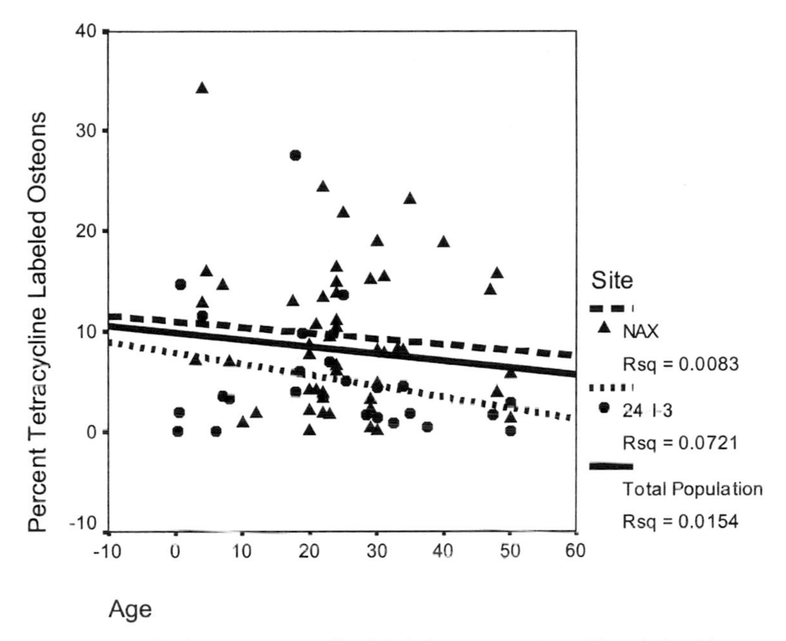

Figure 1. The Relationship between tetracycline labeled osteons and age. The relationships were not statistically significant: $p = 0.205$ for 24-I-3 and $p = 0.530$ for NAX.

and 55.2% of the individuals had more than 5% of their total bone labeled (Tab. 2). There does not appear to be a relationship between total bone labeled and age (Fig. 2).

The use of labeled intact osteons and the total labeled bone offers related but distinct measures of tetracycline ingestion. The intact osteons are restricted to the most recent bone formation events and the total labeled bone represents a lifetime of bone modeling and remodeling (some labeled bone has been resorbed and has disappeared). Finally, there is an obvious relationship between the two measures. The labeled osteons predict variation that exists in the total labeled bone (Fig. 3).

At Kulubnarti, 63% of 110 individuals randomly drawn from the population showed evidence of labeling. Among the individuals with tetracycline labeling, 2% to 20% (with an overall mean of 3.6%) of the osteons were labeled [5]. At Kulubnarti, tetracycline was found in individuals in all groups from age two to over 50 years of age. Hummert and Van Gerven found that there was no detectable pattern with age. The number of labeled osteons did not increase with age, as one would expect if the tetracycline were accumulating as an individual grows older. It would appear that as labeled osteons are resorbed, new-labeled osteons are formed (if the tetracycline is ingested when the osteons are mineralized). There appeared to be no significant differences in frequency between the sexes.

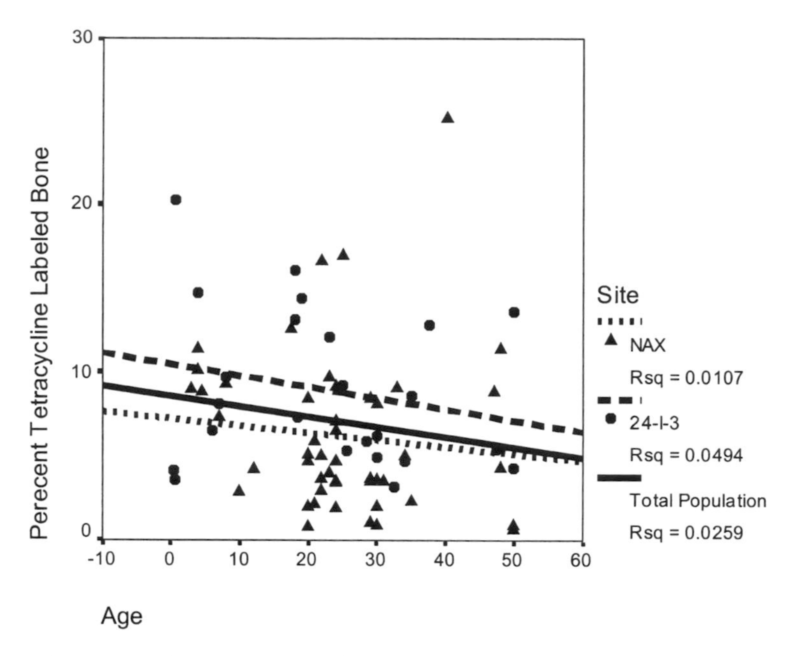

Figure 2. The relationship between tetracycline labeled bone and age. Age is not a predictor of tetracycline labeled osteon: p = 0.297 for 24-I-3 and p = 0.475 for NAX.

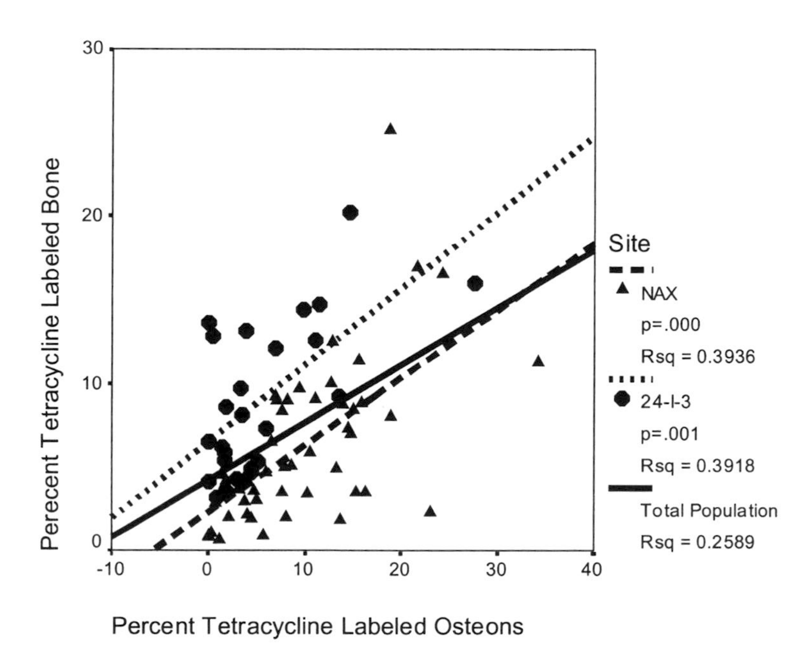

Figure 3. The relationship between total tetracycline labeled bone and tetracycline labeled osteons. The relationships are significant, p = 0.001.

Of the 25 individuals Megan Cook and coworker [9] examined from Dakhleh Oasis, all exhibited tetracycline labeling. These individuals displayed concentric labeling indicating intermittent use. Cook and her colleagues compared the pattern of labeling in Egyptians with a modern clinical sample. Clinically tetracycline is prescribed in two patterns. For acne it is given continuously; for infection it is given intermittently. The Dakhleh Oasis patterns are consistent with doses given intermittently. The distance between labeling ranges from 13.5 to 20.67 micrometers. Frost and co-authors [29] estimate that the normal appositional rate is 1 micrometer per day in modern populations. This would indicate that there was a two- to 3-week interval between the doses of tetracycline. In comparison to the intermittent dosage at Dakhleh, the Nubian sample from the X-Group period followed a pattern more consistent with the modern pattern of individuals receiving a continuous dosage.

Anne Grauer and George Armelagos found the only evidence of tetracycline in bone from an archaeological site outside of Africa [36]. They discovered traces of tetracycline labeling in osteons from bone recovered from the Early Roman (198–63 BCE) to Early Byzantine (365–400 CE) periods at the site of Hesbon. Thin sections were made from nine juvenile and one adult femora. The adult specimen displayed definite fluorescence in a well-mineralized osteon. Of the nine juvenile specimens, four of them displayed fluorescence, two showed no sign of fluorescence and two are too poorly preserved to be analyzed.

The dose of tetracycline in the four populations is quite different. The X-Group Nubians received physiological doses of this antibiotic. The rate of infection was quite low and the tetracycline was likely to be responsible for inhibiting infection. Only 6% of the X-Group people showed evidence of a periosteal reaction that indicates infection. In most of these cases the periosteal reaction was localized and there was little evidence for systemic infection that undoubtedly contributed to their low rate of infectious bone lesions.

According to Hummert and Van Gerven [5], the pattern of labeling at Kulubnarti suggests sporadic use and low levels of ingestion. They report that around 43% of the Kulubnarti population shows evidence of a mild periosteal reaction, suggestive of low-grade infection. There was no significant association between the occurrence of tetracycline labeling and periosteal infection, suggesting that the amount of the antibiotic they were receiving had little therapeutic effect. The Dakhleh populations not only received a lower dosage than is evident in the Wadi Halfa population but also its use was seasonal. Skeletal evidence shows that infection is absent in over 100 adults sampled and high in the 11 sub-adults. They suggest the low infection rates among adults support the hypothesis of antibiotic surveillance, but an alternative hypothesis is that the low infection is due to an absence of the pathogen in the dry environment.

The differences in tetracycline use at Kulubnarti and Wadi Halfa are due to differences in ecological productivity between the two regions. The living conditions at Kulubnarti were harsher than those found in Wadi Halfa and this could have decreased the potential for storage. Hummert and Van Gerven also suggest that at Kulubnarti the grain tax of the Church would have further reduced storage.

Cook and coworkers [9] suggest that the Dakhleh people had a system of storage similar to their Nubian neighbors to the south. They argue that contamination was most likely to occur when the grain was at the bottom of the bins. This, they note, corresponds to the time when it would have been most needed – when nutritional and disease stress would have been most likely.

Ethnohistorical evidence showing that the Egyptians believed that beer had medicinal properties supports the contention that beer was considered more than food. From accounts found in the work of von Dienes and Grapow [37] cited by Darby and coworkers [24], there are 12 pages of beer remedies for various illnesses. Egyptians used beer:

To treat the gums by rinsing the mouth
To strengthen the gums and treat the gums
As enemata
As vaginal douches
As dressings
For anal fumigation to treat disease of the anus

We have no evidence that the microbes which infected the Nubians became resistant to tetracyclines. The skeletal evidence of infection as revealed by periosteal reaction remains low throughout the Nubian occupation of this region. If the microbes that attacked the Nubians became resistant, we would have expected that the periosteal reactions would have become more intense with time.

Tetracyclines appear to restrain the breakdown of collagen by inhibiting the action of collagenase. There is now a concerted effort to produce chemically modified tetracyclines (CMT) [38–42] that have the collagenase inhibitory effects without the antibiotic qualities. They are very effective in blocking the full effects of matrix metalloproteinases (MMPs) [43–51]. The MMPs are involved in a number of the pathological processes that affect connective tissue disease and that result in osteoarthritis [43, 45], rheumatoid arthritis [52, 53], periodontal disease [50, 54], osteoporosis [38] and even cardiovascular disease [55]. The new thrust in research on the non-antibiotic properties of the tetracyclines provides one of the most interesting aspects of this amazing molecule.

The final piece of our story examines the impact of tetracyclines on the health of Nubian populations. The ancient Nubians who received the antibiotic benefited from their tetracycline ingestion; it protected many of them from age-related bone loss. To demonstrate the effect of tetracycline, we examined its impact on the age-related changes as measured by the mineralization of the femoral heads of the X-Group population [56]. A 1.5 cm cube of trabecular bone was cut from the femur head 0.5 cm dorsally or ventrally from the midsagittal plane of the femur head. We determined the ash and bone density to measure the effect of age and tetracycline use on change in bone mineralization. We used two volume determinations (bone organ and bone tissue) in the study. Whole bone volume refers to an entire specific bone (i.e., femur), or a specific portion of bone (i.e., femur head) and includes both cortical and trabecular bone and interstitial spaces. The bone organ volume refers to the major constituents of bone (trabecular and/or cortical tissues with interstitial space) of a specific skeletal site (trabecular bone of the femur head or cortical and trabecular bone of the rib). Bone tissue, in contrast, refers to the protein and mineral portions of the bone excluding interstitial space.

The method for quantifying bone organ and bone tissue is described by Armelagos and colleagues [56] who modified the methods of Arnold and coworkers [57, 58]. The bone organ volume was determined by paraffin-coating the trabecular cube and weighing it in the air and under water. The difference between weight in the air and under water converted to volume in cubic centimeters is the volume of bone organ. The samples were then ashed in a muffle furnace for 48 h at 580 °C and allowed to cool in a desiccator for 24 h and weighed to determine the ash weight.

The ash weight data (ash weight in grams/bone organ volume in cubic centimeters) for the NAX females (Tab. 4) and males (Tab. 5) by age groups shows the expected decrease with age in both sexes. Using the 17–25 age group as the reference point, females show a decrease of 25.4% by their 42–55th year and males experienced a 16.5% decrease. Regression analysis (Fig. 4) shows that the decrease in ash density of femur head trabecular bone density with age is significant in females ($R^2 = .1862$, $p = .014$) and not significant in males ($R^2 = .0967$, $p = .094$).

Table 3. Ash weight data summary for NAX females

Age group	N	Minimum	Maximum	Mean	Std. Dev.
17–25	18	.1975	.4184	.3000	.0617
26–31	8	.2393	.2970	.2393	.0488
32–41	8	.2626	.4173	.2626	.0742
42–55	8	.2239	.3508	.2239	.0712

Table 4. Ash weight data summary for NAX males

Age group	N	Minimum	Maximum	Mean	Std. Dev.
17–25	10	.2569	.4378	.3317	.0751
26–31	9	.1407	.3594	.2763	.0769
32–41	9	.2186	.3010	.2764	.0261
42–55	6	.2367	.3601	.2771	.0488

Table 5. Percent decrease in ash weight for NAX females

Age group	N	Mean
17–25	18	.3000
26–31	8	–20.23
32–41	8	–12.46
42–55	8	–25.36

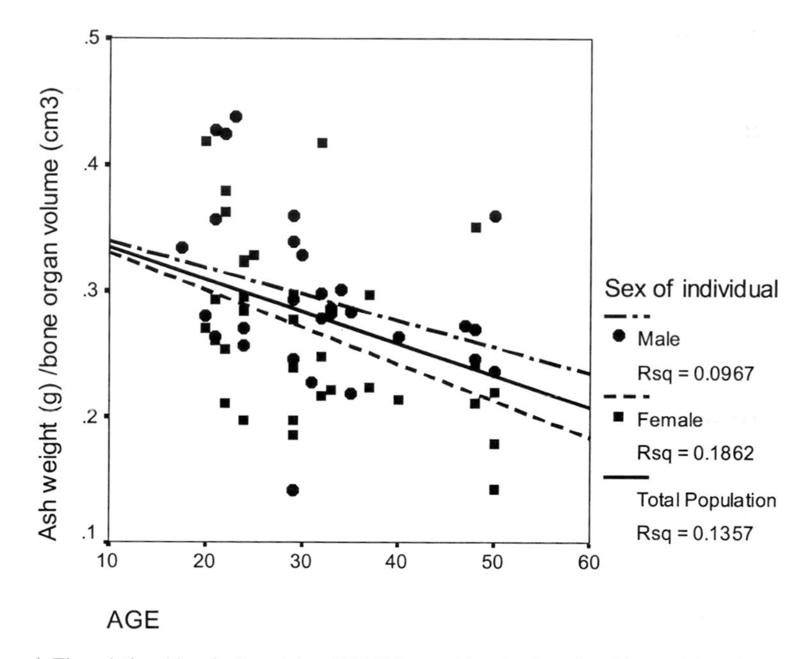

Figure 4. The relationship of ash weight of NAX femoral head trabecula with age. The relationship is significant in females (p = 0.014) but not in males.

The relationship between the percent of tetracycline and total bone and ash weight density shows an interesting association. There is a positive relationship between ash density and the percent of tetracycline in the bone of females (Fig. 5). This association is especially noteworthy since the decrease of ash weight with age in females is significant.

To ensure that the interaction of age, total tetracycline label and ash weight was not confounding these findings, we performed a multiple regression analysis. The relationship in females between total bone tetracycline and ash weight is significant (t = 3.335, p = .013) and the interaction of total tetracycline and age is also significant (t = –2.775, p = .027). The relationship of total tetracycline and ash weight in males is not significant (t = –.747, p = .489) and the interaction of tetracycline and age is not significant (t = 358, p = 735). This

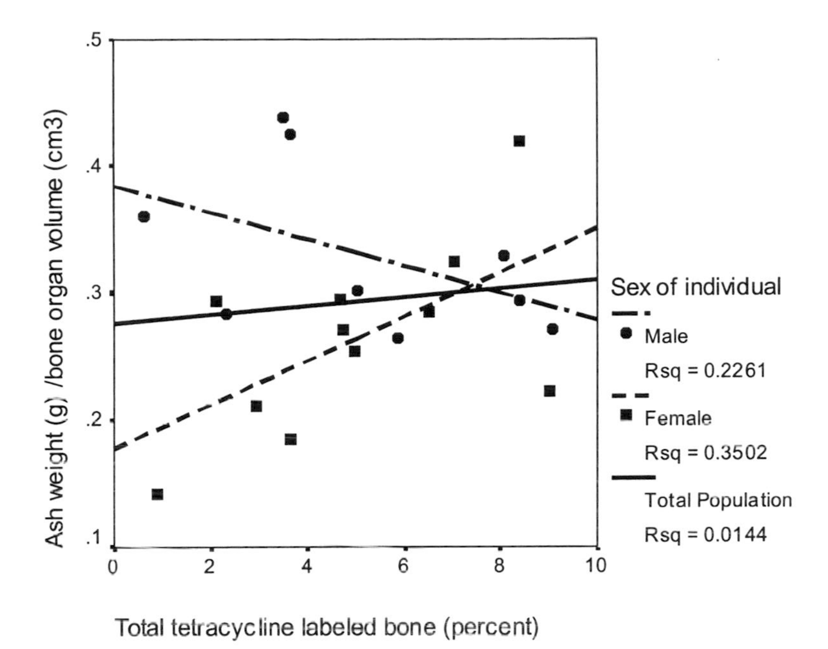

Figure 5. Percent tetracycline labeled bone and ash weight. The relationship approaches significance in females ($p = 0.055$) but is not significant in males.

analysis shows that tetracycline protects females against the loss of bone mineral with age.

The serendipitous discovery of tetracycline in prehistory is one of the most interesting and unexpected discoveries in palaeopathology. The fact that tetracycline can be shown to have a therapeutic effect in the reduction of infectious lesions in Nubians is noteworthy. The fact that the tetracycline affected bone density represents an unexpected finding and demonstrates the potential that ancient remains have for understanding modern medical problems. The motto of the Palaeopathology Association *"Morturi Vientes Docent"* (translated, "The dead are our teachers") is an apt axiom. The skeletal and mummified remains have much to teach us and we are ready to learn.

Table 6. Percent decrease in ash weight for NAX males

Age group	N	Mean
17–25	10	.3317
26–31	9	–16.70
32–41	9	–16.67
42–55	6	–16.45

References

1 Strouhal E (1977) The physical anthropology of the Meroitic area. Third International Meroitic Conference, Toronto, Canada, 1977
2 Adams WY (1967) Continuity and change in Nubian culture history. Sudan Notes and Records 48: 1–32
3 Adams WY (1977) *Nubia: Corridor to Africa*. Princeton: Princeton University Press
4 Bassett EJ, Keith MS, Armelagos GJ, Martin DL, Villanueva A (1980) Tetracycline-labeled human bone from ancient Sudanese Nubia (A.D. 350). *Science* 209: 1532–34
5 Hummert JR, VanGerven DP (1982) Tetracycline-labeled human bone from a medieval population in Nubia's Batn el Hajar (550–1450 A. D.). *Hum Biol* 54: 355–371
6 Keith M, Armelagos GJ (1981) Naturally occurring dietary antibiotics and human health. *In*: L Romannuci-Ross, D Moerman, RR Tancreddi (eds): *The Anthropology of Medicine*, 1st ed. Preager Press, New York
7 Keith M, Armelagos GJ (1983) Naturally occurring dietary antibiotics and human health. *In*: L Romannuci-Ross, D Moerman, RR Trancreddi (eds): *The Anthropology of Medicine*. Preager Press, New York
8 Ortner DJ, Putschar WGJ (1981) *Identification of Pathological Conditions in Human Skeletal Remains*, Vol 28. Smithsonian Institution Press, Washington D.C.
9 Cook M, Molto EL, Anderson C (1989) Fluorochrome labelling in Roman Period skeletons from Dakhleh Oaisis, Egypt. *Amer J Phys Anthropol* 80: 137–143
10 Keith M, Armelagos GJ (1988) An example of *in vivo* tetracycline labelling: reply to Piepenbrink. *J Archaeol Sci* 15: 595–601
11 Cook JA (1998) Tetracycline labeling in an ancient Nubian X-Group (24-I-3) population. Senior Honors, Emory University
12 Piepenbrink H (1986) Two examples of biogenous dead bone decomposition and their consequences for taphonomic interpretation. *J Archaeol Sci* 13: 417–430
13 Piepenbrink H, Herrmann B, Hoffmann P (1981) Tetracyclintypische Fluoreszenzen an bodengelagerten Skeletteilen. *Z Rechtsmedizin* 91: 71–74
14 Hackett CJ (1981) Microscopical focal destruction (tunnels) in exhumed human bone. *Med Sci Law* 21: 243–265
15 Marchiafava V, Bonnucci E, Ascenzi A (1974) Fungal osteoclasia: a model of dead bone resorption. *Calcified Tiss Res* 14: 195–210
16 Martin D (1983) *Paleophysiological Aspects of Bone Remodeling in Meroitic, X-Group and Christian Populations from Sudanese Nubia*. Doctoral Dissertation, University of Massachusetts, Amherst
17 Ascenzi A (1963) Microscopy and prehistoric bone. *In*: DR Brothwell, E Higgs (eds): *Science in Archaeology*. Thames and Hudson, London, 526–538
18 Baker BJ (1992) *Collagen Composition in human Skeletal Remains from the NAX Cemetery (A.D. 350–550) in Lower Nubia*. Doctoral Dissertation, University of Massachusetts, Amherst
19 Urist MR, Ibsen KH (1963) Chemical reactivity of mineralized tissues with oxytetracycline. *Arch Pathol* 75: 484–496
20 Waksman SA (1967) *The Actinomycetes: A Summary of Current Knowledge*. The Ronald Press Company, New York
21 Alexander M (1977) *Introduction to Soil Microbiology*. John Wiley and Sons, New York
22 Williams ST (1982) Are antibiotics produced in soil? *Pedobiologia* 23: 427–435
23 Howie WJ, Suslow TV (1991) Role of antibiotic biosynthesis in the inhibition of *Pythium ultimum* in the cotton spermosphere and rhizosphere by *Pseudomonas fluorescens*. *Mol Plant Microbe Interact* 4: 393–399
24 Darby WJ, Ghalioungui P, Grivetti L (1977) *Food: The Gift of Osiris*. 2 vols. Academic Press, London
25 Culwick GM (1951) Diet in the Gezira Irrigated Area, Sudan. Khartoum: Sudan Survey Department, No. 304
26 Cummings LS (1983) *Coprolites from Medieval Christian Nubia: An Interpretation of Diet and Nutritional Stress*. Doctoral Dissertation, University of Colorado
27 Kemp BJ (1989) *Ancient Egypt: Anatomy of a Civilization*. Routledge, London
28 Mills J (1992) Beyond nutrition: Antibiotics produced through grain storage practices, their recognition and implication for the Egyptian Predynastic. *In*: R Friedman, B Adams (eds): *The*

Followers of Horus: Studies Dedicated to Michael Allen Hoffman, (1944–1990). Egyptian Studies Association Publication No. 2, Oxbow Monograph 20, 28–35

29 Frost HM, Villanueva A, Roth H, Stanisavlevic S (1961) Experimental multiband tetracycline measurements of lamellar osteoblastic activity. *Henry Ford Hospital Med Bull* 9: 312–329

30 Frost HM, Villanueva A, Roth H, Stanisavlevic S (1961) Tetracycline bone labelling. *New Drugs* 1: 206–211

31 Milch R, Koll DP, Tobie JE (1958) Fluorescence of tetracycline antibiotics in bone. *J Bone Joint Surg* 40A: 897–909

32 Oklund SA, Prolo DJ, Guittierrez RV (1981) The significance of yellow bone: Evidence for tetracycline in adult human bone. *JAMA* 246: 761–763

33 Cohlan SQ, Bevelander G, Tiamsic T (1963) Growth inhibition of prematures receiving tetracycline: A clinical and laboratory investigation of tetracycline-induced bone fluorescence. *Amer J Dis Child* 20: 275–290

34 Demers P, Fraser D, Goldbloom RB, Haworth JC, LaRochelle J, McLean R, Murray TK (1968) Effects of tetracycline on skeletal growth and dentition. *Can Med Assn J* 99: 849–854

35 Timmermans L (1974) Influence of antibiotics on spermatogenesis. *J Urol* 112: 348–49

36 Grauer A, Armelagos GJ (1998) Skeletal Biology of Hesban: a biocultural interpretation. *In*: SD Waterhouse (ed.): *The Necropolis of Hesban: A Typology of Tombs. Hesban 10.* Andrews University Press, Berrin Springs, 107–131

37 Von Deines H, Grapow H (1959) *Wörterbuch der Aegyptischen Drogennamen*, Vol 6. Akadenie-Verlag, Berlin

38 Golub LM, Lee HM, Ryan ME, Giannobile WV, Payne J, Sorsa T (1998) Tetracyclines inhibit connective tissue breakdown by multiple non-antimicrobial mechanisms. *Adv Dent Res* 12: 12–26

39 Golub LM, McNamara TF, D'Angelo G, Greenwald RA, Ramamurthy NS (1987) A non-antibacterial chemically-modified tetracycline inhibits mammalian collagenase activity. *J Dent Res* 66: 1310–4

40 Greenwald RA, Golub LM, Lavietes B, Ramamurthy NS, Gruber B, Laskin RS, McNamara TF (1987) Tetracyclines inhibit human synovial collagenase *in vivo* and *in vitro*. *J Rheumatol* 14: 28–32

41 Ramamurthy N, Leung BM, Moak S, Greenwald R, Golub L (1993) CMT/NSAID combination increases bone CMT uptake and inhibit bone resorption. *Ann N Y Acad Sci* 696: 420–421

42 Yu Z, Ramamurthy NS, Leung M, Chang KM, McNamara TF, Golub LM (1993) Chemically-modified tetracycline normalizes collagen metabolism in diabetic rats: a dose-response study. *J Periodont Res* 28: 420–8

43 de Bri E, Lei W, Svensson O, Chowdhury M, Moak SA, Greenwald RA (1998) Effect of an inhibitor of matrix metalloproteinases on spontaneous osteoarthritis in guinea pigs. *Adv Dent Res* 12: 82–5

44 Golub L, Greenwald R, Ramamurthy N, Zucker S, Ramsammy L, McNamara T (1992) Tetracyclines (TCs) inhibit matrix metalloproteinases (MMPs): *in vivo* effects in arthritic and diabetic rats and new *in vitro* studies. *Matrix Suppl* 1: 315–6

45 Greenwald RA (1994) Treatment of destructive arthritic disorders with MMP inhibitors. Potential role of tetracyclines. *Ann N Y Acad Sci* 732: 181–198

46 Golub LM, Lee HM, Greenwald RA, Ryan ME, Sorsa T, Salo T, Giannobile WV (1997) A matrix metalloproteinase inhibitor reduces bone-type collagen degradation fragments and specific collagenases in gingival crevicular fluid during adult periodontitis. *Inflamm Res* 46: 310–9

47 Greenwald RA, Golub LM, Ramamurthy NS, Chowdhury M, Moak SA, Sorsa T (1998) *In vitro* sensitivity of the three mammalian collagenases to tetracycline inhibition: relationship to bone and cartilage degradation. *Bone* 22: 33–8

48 Myers SA, Wolowacz RG (1998) Tetracycline-based MMP inhibitors can prevent fibroblast-mediated collagen gel contraction *in vitro*. *Adv Dent Res* 12: 86–93

49 Ramamurthy NS, Schroeder KL, McNamara TF, Gwinnett AJ, Evans RT, Bosko C, Golub LM (1998) Root-surface caries in rats and humans: inhibition by a non-antimicrobial property of tetracyclines. *Adv Dent Res* 12: 43–50

50 Ryan ME, Ramamurthy S, Golub LM (1996) Matrix metalloproteinases and their inhibition in periodontal treatment. *Curr Opin Periodontol* 3: 85–96

51 Suomalainen K, Halinen S, Ingman T, Lindy O, Saari H, Konttinen YT, Golub LM, Sorsa T (1992) Tetracycline inhibition identifies the cellular sources of collagenase in gingival crevicular fluid in different forms of periodontal diseases. *Drug Exp Clin Res* 18: 99–104

52 Lauhio A, Salo T, Tjaderhane L, Lahdevirta J, Golub LM, Sorsa T (1995) Tetracyclines in treatment of rheumatoid arthritis. *Lancet* 346: 645–646
53 Nordstrom D, Lindy O, Lauhio A, Sorsa T, Santavirta S, Konttinen YT (1998) Anti-collagenolytic mechanism of action of doxycycline treatment in rheumatoid arthritis. *Rheumatol Int* 17: 175–180
54 Golub LM, Ramamurthy N, McNamara TF, Gomes B, Wolff M, Casino A, Kapoor A, Zambon J, Ciancio S, Schneir M (1984) Tetracyclines inhibit tissue collagenase activity. A new mechanism in the treatment of periodontal disease. *J Periodont Res* 19: 651–5
55 Dollery CM, McEwan JR, Henney AM (1995) Matrix metalloproteinases and cardiovascular disease. *Circ Res* 77: 863–868
56 Armelagos GJ, Mielke JH, Owen KH, VanGerven DP, Dewey JR, Mahler PE (1972) Bone growth and development in prehistoric populations from Sudanese Nubia. *J Hum Evol* 1: 89–119
57 Arnold JS (1960) Quantification of mineralization of bone as an organ and tissue in osteoporosis. *Clinl Orthopaed* 49: 167–175
58 Arnold JS, Bartley MH, Tont SA, Jenkins DP (1965) Skeletal changes in aging and disease. *Clinl Orthopaed* 49: 17–38

Tetracyclines in Biology, Chemistry and Medicine
ed. by M. Nelson, W. Hillen and R.A. Greenwald

Tetracycline treatment of periodontal disease: antimicrobial and non-antimicrobial mechanisms

Maria Emanuel Ryan* and Danielle M. Baker

* *Department of Oral Biology and Pathology, School of Dental Medicine, State University of New York at Stony Brook, Stony Brook, New York 11794-8702, USA*

Introduction

Before the evolution of the role of tetracyclines in the management of periodontal diseases can be understood, a short historical review of the history of periodontitis as a disease and the therapeutic strategies to manage it is necessary. This review will help us to understand how the tetracyclines have come to play a major role as an adjunctive chemotherapy to the mechanical/surgical modalities of therapy used at present to manage chronic progressive periodontitis. The proven efficacy of this group of drugs in the management of periodontal diseases has been found to be related not only to their antibacterial actions, but also to a number of additional properties that have recently been identified. These include the inhibition of matrix metalloproteinases, anti-inflammatory properties, the inhibition of bone resorption, and their ability to promote the attachment of fibroblasts to root surfaces and increase the production of collagen and bone. Consequently, tetracyclines have not only been used as an adjunct to non-surgical therapy such as scaling and root planing, but they have also been used as an adjunct to bone grafting in periodontal defects, and as root conditioning agents to enhance attachment of regenerated periodontal tissues. The potential for tetracyclines to play a prominent role in the newly emerging discipline of periodontal medicine will also be addressed.

From the earliest times, the human race has been plagued by a variety of dental problems and has attempted to alleviate these, often painful if left untreated, problems. The first dental healers were physicians, with Hippocrates describing a malady of loose teeth and bleeding gums [1]. In Rome, Celsus (30 B.C.) recommended vinegar as a remedy for gingival disorders [2], perhaps the first documented attempt at managing the disease with chemotherapeutics. Islamic physicians also noted this condition and Abucalsis, in the 11th century, designed a set of "scrapers" for the removal of calcific deposits from the teeth [3]. Physicians were replaced in the Middle Ages by the barber-surgeons of Europe who specialized in dental care [4]. The surgeons continued to make progress as dental practitioners through the 18th century when the treatise "Le chirurgien dentiste", written by Pierre Fauchard in 1728, established dentistry as a profession based on rational and scientific principles [5]. Fauchard in "The Surgeon Dentist" described periodontal disease as "scurvy of the gums". John Hunter, in 1778, gave the first comprehensive description of periodontal disease, a condition which initiates at the gingival border and which may involve the deeper periodontal tissues [6]. It was not until 1840 that the first dental college in the world, the Baltimore College of Dental Surgery, was established [7].

Beyond the direct treatment and management of dental disease, one of the great contributions that dentistry has made to human welfare is the discovery of anesthesia by the dentist, Dr. Horace Wells, in 1844 [8]. Dr. Wells proved his theory that nitrous oxide gas was an anesthetic to be used during surgical procedures by asking his colleague Dr. John Riggs to extract one of his own molars after inhalation of nitrous oxide, thus demonstrating the use of this gas for elimination of pain during dental surgery. Periodontal disease had been regarded as incurable, and teeth attacked by it doomed to extraction, until Dr. Riggs later went on to introduce his techniques for the treatment of periodontal disease at the 1881 International Medical Congress in London, which led to the term "Riggs' Disease" [9]. Based on the work of Riggs, Dr. Younger and later Dr. Hutchinson advocated surgical treatment of periodontal inflammation by pocket obliteration, which was undertaken by numerous dentists in the early part of the twentieth century. At the same time, C.M. Carr developed scalers for the removal of subgingival deposits wherever found and for the planing of root surfaces, by which most of modern periodontics actually got its start. Several dentists, encouraged by initial treatment successes, began around 1910 to limit their practices to periodontal treatment or, as it was then known, "pyorrhea" treatment. At the same time a British physician, Dr. William Hunter, introduced the concept that chronic infection is the cause of certain systemic diseases, with remarks on etiology directed largely to mouth infections, with special emphasis on periodontal infections [10]. Occasionally medical, rather than surgical, treatment was attempted, including attempts to develop suitable vaccines for "pyorrhea". In 1914, Drs. Stillman and McCall of New York, Spaulding of Michigan and Hayden of Ohio formed the American Academy of Periodontology [11]. This group renamed the disease "periodontoclasia" and the term "periodontology" was accepted as the generic name for the branch of dentistry dealing with diseases of the supporting tissues of the teeth. In 1915, recognizing that bacterial deposits around the teeth were involved in this degenerative process, Dr. Thomas Hartzell, before the First District of New York Dental Society, urged comprehensive methods of treatment that included deep scaling as well as surgery [12]. The first book in the field of periodontology, "A Textbook of Clinical Periodontia", written by Drs. Paul Stillman and John Oppie McCall, was published in 1922.

A major step forward was taken in 1948 when the National Institute of Dental Research, now the National Institute of Dental and Craniofacial Research, was established by the United States Public Health Service, which was ultimately incorporated into the National Institutes of Health (NIH). Through this act Congress recognized the importance of dental health [13]. This provided a means by which the epidemiology as well as the pathogenesis of periodontal disease could be investigated. It was also recognized by the 1960s that systemic disorders often play an important role in the etiology of certain periodontal diseases. For this reason it became clear that periodontal disease in some of its phases constituted a medico-dental problem. It was recognized that the mouth was part of the body and could not be divorced from it, although the vital connection seemed at times to be rather tenuous. It was stated in a report presented to the World Health Organization in 1961 by Drs. A.L. Russell of the U.S. Public Health Service and J. Kostlan of the Czechoslovakian Institute of Dental Research that periodontitis is a "spectacular disease affecting the majority of the world's population [10]." They emphasized the importance of periodontal disease as a world health threat in view of its deleterious systemic effects and its world-wide distribution.

Theories about the pathogenesis of periodontitis have evolved from considering it to be purely a plaque-associated disease to the more recent hypotheses, which place considerable emphasis on the host's response to the bacteria [14]. The first-ever Surgeon General's Report on "Oral Health in America", published in 2000, has once again recognized the importance of dental health in the overall general health and well-being of a patient [15]. Recent research findings pointing to possible associations between chronic oral infections, such as periodontitis, and diabetes, heart and lung diseases, stroke, and low-birth-weight, premature births are addressed in this report which assesses these emerging associations and explores factors that may underlie these oral-systemic disease connections. Along with these findings and the emergence of the discipline of periodontal medicine there have been many developments in therapeutic approaches to the management of periodontitis.

No class of drugs has made more of an impact on periodontal therapy than the tetracyclines. These compounds have been used in a medical chemotherapeutic approach as adjuncts to accepted mechanical procedures to treat periodontal disease. They have been used in conjunction with scaling and root planing, the gold standard of non-surgical therapy, as well as with surgical procedures, both resective and regenerative. The tetracyclines have been used locally and systemically as antimicrobial agents, and, more recently, systemically as host-modulatory agents. The tetracyclines have been used not only to address chronic adult periodontitis but also for the management of specific, often more aggressive, types of periodontitis. Most recently, the tetracyclines have been advocated for the management of patients with systemic diseases, such as diabetes. The use of tetracyclines has lead not only to improvements in the periodontal health of compromised diabetic patients, but also to improvements in long-term markers of glycemic control, such as glycated hemoglobin [16]. Clinically, the purpose of using tetracyclines as adjunctive agents has been to kill the pathogens in the tissues and pockets, modulate the host response and increase the predictability of a variety of non-surgical periodontal therapies, including scaling and root planing. This strategy may result in the need for less extensive surgical procedures or ultimately decrease the need for periodontal surgery. As an adjunct to mechanical therapies, the goal has been to enhance reattachment or even to establish new attachment of the supporting apparatus and to stimulate osseous formation. The literature is replete with references to the use of tetracyclines in the management of periodontal disease, so the objective of this review is to give an overview of the use of these pleiotropic compounds for the treatment of periodontitis.

Antimicrobial therapy with tetracyclines

Since 1965 there have been at least 3 eras in modern periodontics [17], two of which have been dominated by the bacteria which initiate the disease process. The decade from 1965–1975 is often referred to as the "Non-Specific Plaque Theory Era" anchored by the Loe et al. [18] discovery that plaque causes inflammation of the gingival tissues known as gingivitis. During this era it was believed that untreated gingivitis would inevitably progress very slowly to become the more destructive and irreversible form of the disease known as periodontitis. However, it has been shown that some patients can have chronic gingivitis that never progresses to periodontitis. Furthermore periodontitis develops most aggressively in susceptible individuals and, rather than having a slow steady course, may be episodic in nature, and also appears to be site-specific. During the "Non-Specific Plaque Era" the conclusion was that all plaque was bad and that excessive accumulations led to disease. Consequently, therapies were directed at reducing the quantity of plaque on the tooth surfaces. Treatment failures were perceived to be due to poor oral hygiene compliance by the patient.

The decade of 1975–1985, known as the "Bacterial Specificity Era", resulted from the identification of specific microbes which could be related to specific forms of periodontal disease. There was also recognition of the differences between sub-gingival and supra-gingival plaque and a realization that the quality and not just the quantity of plaque was of critical importance. One consequence of this was that there was a move from previously accepted classifications of periodontal disease to an understanding of new types of disease based on the microbial pathogens identified in certain individuals. Since this time, the importance of the ecosystem in which microbial species reside, known as the biofilm, has also become apparent, and the predominance of certain species within the biofilm has come to the forefront of periodontal microbiology. The periodontal pocket can contain over 300 species of microorganisms, which are not all present in every pocket [19]. Also there was a realization that approximately 50% of the bacteria which are located in the gingival pocket are non-cultivable and have not yet been identified, although through DNA analyses further identification of microbes has become possible. There is still much to be explored with regard to the periodontal microbiota. Despite this, the antimicrobial approach to periodontal therapy has been utilized for many years, recognizing that the prevalence and severity of these diseases can be reduced by mechanical plaque removal or by the use of a variety of systemic or topically applied antimicrobial agents, aimed at inhibiting pathogenic bacteria.

Mechanical removal of plaque and calculus, both surgical and non-surgical, is time-consuming, operator and patient-dependent, and difficult to master [20]. Although mechanical and surgical interventions continue to be the most widely used methods of controlling disease progression, instrumentation inevitably leaves behind significant numbers of microorganisms, including putative pathogens. Recolonization of these pathogens can occur within 60 days of scaling and root planing. The need for the use of chemotherapeutic agents as adjuncts to mechanical and surgical debridement is compelling. In addition to innovative anti-infective treatments with the tetracyclines to be discussed here, considerable research and the introduction of host-modulatory therapeutic uses of tetracyclines as adjuncts to block the progression of periodontitis will be discussed in a later section.

It has become apparent in certain types of periodontitis, such as juvenile periodontitis or prepubertal periodontitis, that patients have chemotactic defects affecting neutrophil function, a defect which may be genetic in nature [21, 22]. However, adding to the complexity of this disorder, the major pathogen associated with these forms of periodontitis is *Actinobacillus actinomycetemcomitans*, which secretes a leukotoxin, which is known to have deleterious effects on neutrophil function. These bacteria, unlike most associated with periodontal disease, are also known to invade the soft tissues, indicating that the mechanical removal of plaque may be insufficient to arrest the disease and that systemic use of antimicrobial agents would be judicious. These findings led to the appreciation that some antibiotics may serve as adjuncts in the treatment of certain types of periodontitis.

A few specific microorganisms have been associated with progressive forms of periodontal disease. These include *Actinobacillus actinomycetemcomitans*, *Porphyromonas gingivalis*, *Prevotella intermedia*, *Bacteroides forsythus*, *Eikinella corrodens*, *Fusobacterium nucleatum*, *Peptostreptococcus micros*, *Selenomonas* species, *Campylobacter rectus*, and *Treponema* species [23]. While most of these organisms are sensitive to commonly used antibiotics such as the tetracyclines, including tetracycline and minocycline, many strains of *Actinobacillus actinomycetemcomitans*, *Porphyromonas gingivalis*, *Prevotella intermedia*, *Eikinella corrodens* and *Fusobacterium nucleatum* are resistant to chlortetracycline, oxytetracycline, and doxycycline [24, 25].

Most of the subgingival microorganisms are susceptible to tetracycline at a minimal inhibitory concentration (MIC) of >1–2 μg/ml [26, 27]. However, it should be emphasized that most studies on microbial susceptibility are conducted *in vitro* and do not take into account the existence of these microbes in the biofilm *in vivo*. For minocycline, a MIC of 1 μg/ml was found to inhibit 85% of bacterial strains [28]. Many of the Gram-positive oral streptococci show marked resistance to tetracycline (MIC >64 µg/ml) and some Gram-positive *actinomycetes* exhibit intermediate (MIC 4–8 µg/ml) tetracycline resistance [29–31]. There are some species that demonstrate resistance to tetracycline (MIC >16 µg/ml), for example: *Eikenella corrodens, Prevotella oralis, Selenomonas sputigena* and some strains of *Campylobacter* and *Veillonella* [27]. Similarly, a number of black pigmented bacteroides exhibit resistance to minocycline at MICs greater than the level attainable in the gingival crevicular fluid (GCF) after systemic dosing with the drug [28]. At concentrations of doxycycline ranging from 5–25 µg/ml, bacterial resistance has been reported in 2–6.6% of the species in the subgingival plaque obtained from healthy periodontal sites, which increases to 18.5% when the plaque is obtained from patients with advanced periodontal disease [32].

Systemic tetracyclines in the management of periodontitis

For the most part, systemic antimicrobial therapy has been reserved for refractory cases of periodontitis, which demonstrate progressive periodontal destruction. Systemic antibiotics may be recommended as adjuncts to conventional mechanical therapy, but strong evidence for their use as a monotherapy has not been developed. A consensus has developed that systemic antimicrobial therapy should be reserved for situations that cannot be managed with mechanical therapy alone, such as severe or acute infections, early-onset periodontal diseases and refractory cases. For these special situations, randomized double-blinded clinical trials, as well as longitudinal assessments of patients, indicate that systemic antimicrobials may be useful in slowing disease progression [33].

Refractory forms of periodontal disease, demonstrating a progressive destruction of the periodontal attachment apparatus despite conventional therapy, can be managed with the following recommended dosing regimens for antibiotics in the tetracycline family: 250 mg qid of tetracycline for 2–3 weeks, or 100 mg bid for day one followed by 100 mg qd of doxycycline for 2–3 weeks [34], with 2 weeks of therapy as the usual regimen for advanced periodontitis not responding to standard treatment. Microbiological testing in a non-responsive patient reporting high levels of *Actinobacillus actinomycetemcomitans* requires 3 weeks of dosing to ensure eradication of this species of bacteria [35–38]. In addition, there are individuals who are infected with *Actinobacillus actinomycetemcomitans* who have juvenile periodontitis where the recommendation is for tetracycline hydrochloride, 250 mg qid for 12–21 days as adjunct to either non-surgical or surgical therapy. Minocycline is prescribed at 50 mg tid for 2–3 weeks in juvenile periodontitis, rapidly progressing periodontitis, or severe adult periodontitis. The absorption of doxycycline is not affected by food as much as are other commercially available tetracyclines and compliance with a once a day, 100 mg regimen is better than for bid, tid or qid regimens. Treatment of refractory forms of periodontitis, especially those which have been treated with multiple courses of antibiotics, often requires culture and testing to determine which microbes are present and what is their antimicrobial susceptibility. Other periodontal diseases in which antibiotics may be of use include AIDS-periodontitis or gingivitis, periodontitis in compromised patients such as those with diabetes, Papillon-Lefevre syndrome, as well as periodontitis in neutropenic patients [39, 40].

Major adverse reactions to tetracycline antimicrobial therapy which need to be considered include anaphylactic reactions, gastrointestinal disturbances, photosensitivity, tooth pigmentation and delayed fontanelle closure in children [41]. Tetracyclines should also be avoided during pregnancy and in children under the age of 8. Photosensitivity in patients taking doxycycline has been shown to be dose-dependent. The gastrointestinal side-effects of antibiotic doses of the tetracyclines include nausea, heartburn, epigastric pain, vomiting and diarrhea. *Entero colitis* rarely occurs. *Candida albicans* superinfection can occur in debilitated patients. Antimicrobials have been reported to interfere with hormonal birth control. With doxycycline usage, side-effects of the lower bowel, particularly diarrhea, are infrequent. Minocycline can cause vestibular disturbances characterized by vertigo, reversible dizziness, ataxis, and tinnitus associated with weakness. With regard to antibiotic resistance induced by the use of tetracyclines, many members of the normal oral flora are naturally resistant to tetracycline [24, 42, 43]. After treatment with tetracycline, however, tetracycline resistance of several bacteria can be found. Tetracycline resistance is uncommon with *Actinobacillus actinomycetemcomitans* and *Porphyromonas gingivalis*. However, it is common in *Prevotella intermedia* strains. Because antibiotic resistance can be induced after periodontal therapy with antibiotics, indiscriminate or widespread use of systemic antibiotics is contraindicated for the treatment of common forms of periodontal disease [27]. Antibiotics should be reserved for refractory periodontitis or unique forms of the disease such as localized juvenile periodontitis where other modes of therapy are not effective.

Controlled local delivery of tetracyclines

In order to have a therapeutic effect on the microflora, antimicrobial agents must reach adequate concentrations to kill or inhibit the growth of target organisms. The drug of choice must reach the site where the organisms exist and stay there long enough to get the job done and it must cause no harm. Mouth rinses do not reach the depths of periodontal pockets, whereas irrigation can deliver drugs to the base of the pocket. However, the duration of exposure is short because the GCF in the pocket is replaced about every 90 s and topically applied agents are rapidly washed out. Early research suggested that doxycycline administered systemically [44, 45] was highly concentrated in the GCF at levels 5–10 times greater than those found in serum. Furthermore, tetracyclines show substantivity because they bind to the tooth structure and are slowly released as still-active agents. However, even this supposed hyperconcentration of the drug in the GCF still resulted in a level to which many organisms were not susceptible. More recent work has challenged earlier findings of hyperconcentration of tetracyclines in the GCF. In the 2 h after the administration of a single dose of tetracycline (250 mg), minocycline (100 mg) or doxycycline (100 mg), the concentration of these tetracyclines was found to be highest in the plasma, intermediate in the GCF (doxycycline achieving the highest levels) and lowest in the saliva [46]. Further experimentation may be required to resolve this issue as there was a great deal of variability in the average GCF concentrations (0–8 µg/ml) in this study and steady-state levels of the drug were never achieved. To address these issues, controlled local delivery of

Table 1. Comparison of periodontal antimicrobial delivery systems

	Mouthrinse	Local irrigation	Systemic delivery	Controlled delivery
Reach the site	Poor	Good	Good	Excellent
Adequate concentration	Poor	Good	Fair	Excellent
Adequate duration	Poor	Poor	Fair	Good

antimicrobials was developed. In Table 1, a comparison between periodontal antimicrobial delivery systems is made.

Dental research has provided us with a better understanding of the microbial etiology and the nature of periodontitis. Periodontitis, initiated by bacteria, frequently either appears in localized areas in the patient's mouth or is confined to localized areas by treatment. These infected localized areas lend themselves well to treatment with a controlled local delivery system using an antimicrobial agent [47]. Antimicrobial agents may be applied directly to the pocket, thereby eliminating many of the adverse side-effects associated with systemic delivery of antibiotics. Both non-resorbable and resorbable intrapocket delivery systems have been used. There is evidence that local delivery of sustained-release antimicrobials may lead to improvements in periodontal health, although a few side-effects, such as transient discomfort, erythema, recession, transient resistance and allergy have been reported. Oral candidiasis has been reported in a small number of cases with local tetracycline delivery.

Systems have been developed for the release of all three commercially available tetracyclines at high doses and at a regular rate over a 10–14-day period. The first such Food and Drug Administration (FDA)-approved system, Actisite™, was developed by Dr. Max Goodson in 1983 [48]. Actisite™ consisted of a non-resorbable polymer fiber of ethyl vinyl acetate, 25% saturated with tetracycline hydrochloride. Use of this product results in substantially higher doses of tetracycline in the pocket than could be achieved by systemic dosing, 1590 μg/ml in the GCF and 43 μg/ml in tissue *versus* 2–8 μg/ml for systemic administration. 30 μg/ml will eliminate most pathogenic bacteria associated with periodontal diseases. The area being treated is saturated with doses of the therapeutic agent that can be sustained for prolonged periods of time. Despite these high doses of drug that are achieved locally, serum levels of the drug did not exceed 0.1 μg/ml. The primary endpoint of most periodontal trials is a change in the levels of clinical attachment and pocket depth (Fig. 1) from the baseline to a designated time point after therapy. The use of a singly applied tetracycline fiber as an adjunct to scaling and root planing (SRP) proved to be more effective than scaling and root planing alone at reducing bleeding on probing (BOP), pocket depths (PD) and achieving attachment level (AL) gain as early as 60 days after placement, with continued significant improvements at 6 months. At 6 months after a single application of Actisite™, the average results for SRP and Fiber *versus* SRP only respectively were: 1.81 mm *versus* 1.08 mm for PD; 1.56 mm *versus* 1.08 mm for AL and 63% *versus* 50% for BOP reductions [49]. In recent studies it was concluded that SRP combined with full mouth Actisite™ therapy *versus* SRP alone resulted in increased bone density (+2.43 CADIA *versus* –2.13 CADIA) and alveolar bone height (+0.24 mm *versus* –0.29 mm) at 6 months after therapy [50]. However Actisite™ was difficult to use, requiring considerable operator skill, and because it was not resorbed, a second visit had to be scheduled to remove it. In attempts to improve upon ease of placement of the local antimicrobial agent into the pocket and to obviate the need for a second visit to remove the product, other delivery systems were explored.

The second FDA-approved, locally delivered tetracycline developed, Atridox™, is a 10% formulation of doxycycline in a bioabsorbable, flowable

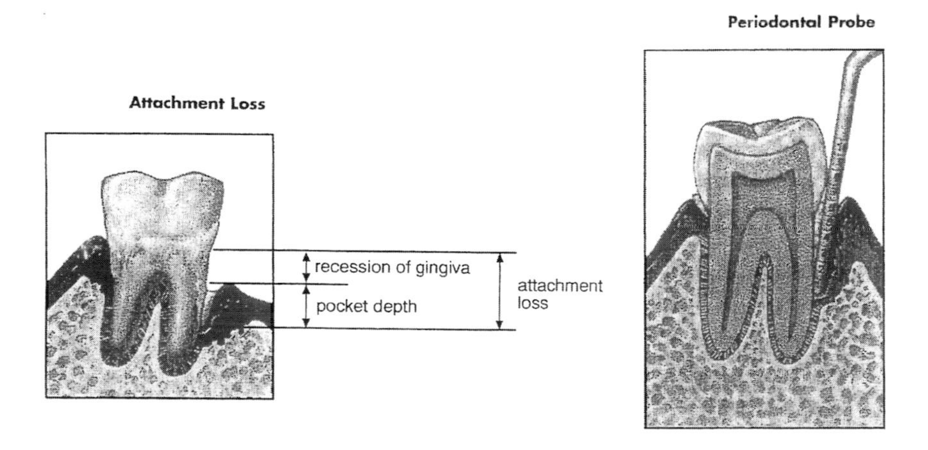

Figure 1. Clinical parameters for diagnosis of periodontal disease. The periodontal probe is the primary tool used for diagnosing periodontal disease. The periodontal probe is marked off in multiples of 1 mm increments. It is used to assess signs of periodontal disease including: pocket depth (PD), attachment loss (AL), and inflammation based upon bleeding on probing (BOP).

poly DL-lactide and N-methyl-2-pyrrolidone mixture delivery system. In subjects with chronic adult periodontitis, the application of this doxycycline gel twice, at baseline and 4 months later, resulted in reductions in probing depths (1.3 mm) and gains in clinical attachment (0.8 mm) equivalent to SRP alone at 9 months after baseline [51]. An important finding of these studies was that smoking status did not seem to affect the outcome of clinical parameters such as probing depth reductions and clinical AL gains for the Atridox™ treatment group, whereas smokers and even former smokers did not respond as well to mechanical therapy alone [52]. Despite these results, it is unlikely that this agent will ever be used as a monotherapy for the management of periodontal disease. Removal of the offending plaque and calculus deposits by SRP has proven to be effective. Additional studies to support improved outcomes by using this therapeutic agent as an adjunct to SRP have been initiated. Despite the fact that this vehicle is biodegradable, a periodontal adhesive dressing is recommended to retain the system, necessitating a second visit to remove the dressing 7–10 days after placement.

With regard to minocycline there is a non-FDA cleared ointment product of 2% (w/w) minocycline hydrochloride known as Dentamycin or Perio Cline marketed in a number of countries. In a 4-center randomized double-blind trial conducted in Belgium, the minocycline ointment was applied once every 2 weeks for 4 applications due to insufficient sustained-release properties. Probing depth reductions were significantly greater in the SRP and minocycline group *versus* SRP alone, whereas there was only a trend towards improvements in clinical AL and bleeding indices in the minocycline treatment group [53]. In a long-term, 15-month study, after placement of the gel subgingivally at baseline, 2 weeks, 1, 3, 6, 9 and 12 months, results showed a statistically significant improvement for all clinical and microbiological parameters for adjunctive minocycline ointment [54]. A new minocycline microsphere system (Arestin™) is currently under consideration at the FDA. The data from these studies with this product indicate that the use of minocycline microspheres at baseline, 3 and 6 months results in a 0.25 mm improvement above average probing depth reductions. If the data are stratified in accordance with the severity of baseline probing depths, there are 20% improvements in mild sites, 40% in moderately diseased sites and 100% in severely diseased sites compared with SRP alone. There were no differences in the microflora between the SRP alone group and those administered the minocycline microspheres [55].

The tetracyclines are clearly effective as antibiotics against many periodontal pathogens, as are many other antibiotics. However, despite the fact that other antimicrobials, such as metronidazole, have proven to be more effective against the putative periodontal pathogens, the tetracyclines have dominated as chemotherapeutic adjuncts for the management of periodontitis. The efficacy of tetracyclines administered either systemically or locally may be dictated by far more than their antimicrobial properties, as will be discussed below.

Non-antimicrobial, host-modulatory therapy with tetracyclines

Periodontal disease does not appear to act like a classical infection, but more like an opportunistic infection. There is no way to eliminate bacteria from the oral cavity, so bacteria are always present in the periodontal milieu. When certain more virulent species exist in an environment that allows for them to be present in greater proportion, there is the opportunity for periodontal destruction to occur. However, while it is apparent that plaque is essential for the development of the disease, the severity and pattern of the disease are not explained solely by the amount of plaque present. In 1985 research began to focus very closely on the bacterial-host interactions, leading to the "Host-Bacteria Inter-Relationship Era" [17]. During this era it has been recognized that although there are bacterial pathogens that initiate the periodontal inflammation, the host response to these pathogens is equally if not more important in mediating connective tissue breakdown, including bone loss. It has become clear that it is the host-derived enzymes known as the matrix metalloproteinases (MMPs), and changes in osteoclast activity driven by cytokines and prostanoids, that cause the majority of the tissue destruction in the periodontium [56]. This shift in paradigms with a concentration on the host response has led to the development of host-modulatory therapies to improve upon therapeutic outcomes, slow the progression of the disease, allow for more predictable management of patients and possibly even work as preventive agents against the development of periodontitis.

There are a number of environmental and acquired risk factors that increase a patient's susceptibility to periodontitis. The risk factors that can affect onset, rate of progression and severity of periodontal disease as well as response to therapy include: heredity, smoking, hormonal variations such as those seen in pregnancy where there are increased levels of estradiol and progesterone which may change the environment and permit the virulent organisms to become more destructive, or in menopause where we find the reductions in estrogen levels leads to osteoporosis. Other risk factors include systemic diseases such as diabetes, immunocompromised state, stress, nutrition, medications such as calcium channel blockers, faulty dentistry and a multitude of others including a previous history of periodontal disease [57–59]. Some of these risk factors can be modified to reduce a patient's susceptibility. Risk management may include smoking cessation, improved control of diabetes, nutritional supplementation and stress management. The field of "perioceutics" is emerging to aid in the management of these susceptible patients who develop periodontal disease. Host-modulatory therapy, which can be used to bring down excessive levels of enzymes, cytokines and prostanoids as well as modulate osteoclast function, is the key to addressing many of these risk factors which have adverse effects on the host response. Tetracyclines work so well as host-modulatory agents because of their pleiotropic effects on multiple components of the pathological host response (Fig. 2).

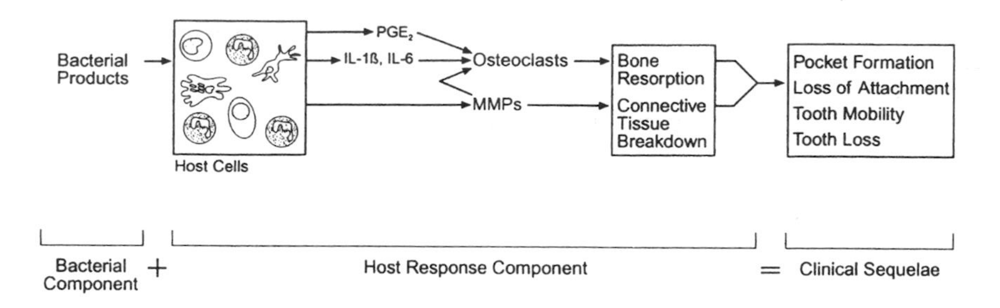

Figure 2. Periodontal braekdown cascade. Simplified schematic depicting etiologic factors and cascade of events contributing to periodontitis.

The role of MMPs in both physiological and pathological processes has been thoroughly discussed in the periodontal [60] and other literature. The role of inhibitors is particularly important because it is an imbalance between the activated MMPs and their endogenous inhibitors that leads to pathological breakdown of the extracellular matrix (ECM) in diseases such as periodontitis, arthritis, cancer invasion, etc. This rationale has led to the development of a number of synthetic matrix metalloproteinase inhibitors (MMPIs), not only as "tools" in the study of the mechanisms involved in MMP-associated pathology, but also as potential therapeutic agents. Compensating for the deficit in the naturally occurring inhibitors or tissue inhibitors of MMPs (TIMPs) to block or retard the proteolytic destruction of connective tissues is of therapeutic significance. Conceptually, this can be accomplished with the use of drugs that can: (i) inhibit the synthesis and/or release of these enzymes; (ii) block the activation of precursor (latent) forms of these MMPs (pro-MMPs); (iii) inhibit the activity of mature MMPs; (iv) stimulate the synthesis of endogenous TIMPs; or (v) protect the host's endogenous inhibitors from proteolytic inactivation [61, 62]. The tetracyclines, which may modulate many of these matrix-protective mechanisms, have been found to be effective inhibitors of MMP-mediated connective tissue destruction in a variety of pathological processes. In addition to the possible effects of tetracyclines on expression, activation and catalytic activity of MMPs, these compounds may act on other processes involved in the overall pathophysiology of multiple disease states, including regulation of release of inflammatory cytokines, glycosylation of connective tissue proteins and even the upregulation of the expression of matrix constituents which are produced at a deficient rate during diabetes and other diseases [63–65]. These multiple modes of action listed in Table 2, not all of which have yet been sufficiently defined, may account for the positive results obtained with use of tetracycline antibiotics (TCs) as therapeutic agents in models of periodontitis (and in other collagenolytic diseases), as this disease process involves a complex, multifactorial pathogenesis. Although a number of MMPIs have been developed over the past decade, few have been found to be safe and effective, particularly after oral administration. Doxycycline is currently the only FDA-approved MMPI being used clinically and, as discussed

Table 2. Tetracyclines inhibit connective tissue breakdown: pleiotropic mechanisms

- Mediated by extracellular mechanisms
  - Direct inhibition of active MMPs – dependent on Ca^{2+} and Zn^{2+} binding properties of tetracyclines
  - Inhibition of oxidative activation of pro-MMPs – independent of cation binding properties of tetracycliness
  - Tetracyclines disrupt activation by promoting excessive proteolysis of pro-MMPs into enzymatically-inactive fragments – dependent(?) on cation binding of tetracyclines
  - Inhibition of MMPs protects α_1-proteinase inhibitor, thus indirectly reducing serine proteinase (such as neutrophil elastase) activity

- Mediated by cellular regulation
  - Tetracyclines reduce levels of cytokines associated with inflammation and accelerated bone loss (e.g., TNF, IL-1), inducible nitric oxide synthase, phopholipase A_2, prostaglandin synthase.
  - Effects on protein kinase C, calmodulin

- Mediated by pro-anabolic effects
  - Tetracyclines increase collagen production
  - Tetracyclines stimulate osteoblast activity & new bone formation

below, it is the only proteinase inhibitor that has been clinically tested in humans for efficacy in periodontal therapy [62, 66–69].

The initial demonstration that TCs can inhibit host-derived MMPs, and do so by a mechanism independent of the antimicrobial properties of the drugs, was made in germ-free rats with experimentally induced diabetes, a model of *in vivo* excess collagenase activity [70]. These results were confirmed in early studies using TCs in humans, in additional animal models, and in organ culture bone resorption experiments [71–73]. In 1987, Golub et al. [74] described a new use for the first chemically modified tetracycline (4-dedimethylamino tetracycline or CMT-1), which is devoid of antibacterial activity due to the removal of the dimethylamino group from the carbon-4 position of the "A" ring of the drug molecule, but which retains its anticollagenase activity. Many different chemically modified TCs (known as CMTs) have since been identified, most of which were found to retain their anti-collagenase but to have lost their antimicrobial properties [71, 75, 76]. The one CMT found to have lost its anti-collagenase property was CMT-5, or the pyrazole analogue, in which the carbon-11 carbonyl oxygen and carbon-12 hydroxyl groups were replaced by nitrogen atoms, which eliminated this important Zn^{2+} or Ca^{2+} binding site on the TC molecule.

The first mechanism proposed for the anticollagenase properties of CMTs was their ability to inhibit already active MMPs (collagenase and gelatinase) in the ECM, a mechanism found to be associated with the Zn^{2+} or Ca^{2+} binding properties of the TC molecule. This proposed mechanism has been supported by the following observations: (1) adding excess Ca^{2+} (μM concentrations) or excess Zn^{2+} (μM concentrations) eliminated the ability of the TC analogues to inhibit collagenase activity [70, 77]; (2) structural evaluation of collagenase has revealed that the enzyme contains a secondary Zn^{2+}, outside the active site of the enzyme (in addition to an active site Zn^{2+}), which in addition to a secondary Ca^{2+}, helps maintain the conformation and catalytic activity of the enzyme [78] and (3) TCs such as doxycycline block the MMPs *in vitro* apparently by non-competitive inhibition (Stetler-Stevenson et al., 1990s personal communication and [79, 80]). These findings suggest that the TCs (except for CMT-5) may bind to the secondary Zn^{2+} (and to a lesser extent, Ca^{2+}) in collagenase, thus altering the conformation of the enzyme molecule and blocking its catalytic activity in the ECM. This is supported by a more recent study which demonstrated that CMT-5, unlike the other CMTs and doxycycline, did not show any shifts in absorption maxima or any increase in its absorption peak height by addition of zinc cations as analyzed using an *in vitro* spectrophotometric technique [81]. A maximum "dilution" of doxycycline's ability to inhibit collagenase activity was seen at zinc concentrations of 1–5 μM, in agreement with previous reports. Of interest is that the greater zinc reactivity of CMTs 3 and 8, compared to doxycycline (and the lack of zinc reactivity of CMT-5), may explain, at least in part, their greater potency as inhibitors of MMPs. The zinc may be interfering with the ability of these tetracycline analogues to bind to the zinc at the primary site (needed for enzyme activity), or at the secondary site (needed for structural stability).

Additional inhibitory mechanisms of these drugs include their ability to prevent the conversion of pro-MMPs in the ECM into active MMPs. Two different mechanisms of doxycycline inhibition of recombinant human pro-MMP-8 have been detected [82, 83]. When doxycycline is added to an incubation mixture after activation of pro-MMP-8 (activation is achieved either by limited proteolysis with trypsin or by the use of the organomercurial agent, APMA), 30 μM of the drug is required to inhibit the collagenase activity by 50% (IC_{50}). In contrast, when doxycycline is added during activation, the IC_{50} drops to 5–12 μM. Based on Western blot analysis of the different molecular species of MMP-8, the authors proposed that, during activation, doxycycline binds to the pro-MMP (complexing with Ca^{2+}) thus altering the enzyme's conformation and resulting in excessive degradation of the proteinase to small enzymatically inactive fragments. However, this second mechanism may only apply to recombinant MMPs since Sorsa et al. [80] were unable to detect any fragmentation of either native pro-MMP-8 or pro-MMP-9 during their activation by APMA in the presence of doxycycline and, based on Michaelis-Menten kinetics, found that the drug, as stated above, acted as a non-competitive inhibitor of collagenase *in vitro*. At the National Cancer Institute (NIH), Stetler-Stevenson et al. recently examined the interaction kinetics of gelatinase A with either doxycycline or CMT-3 and found that the former tetracycline analogue functioned as an uncompetitive inhibitor, whereas the latter exhibited a mixed mechanism of inhibition and was more potent (Personal communication with Dr. Stetler-Stevenson). In addition, structural features in the hemopexin-like domain of the MMPs may modify the response of the MMPs to doxycycline and differences within the catalytic domain of the MMPs may also contribute to their susceptibility to tetracycline inhibition (e.g., of the collagenases, MMP-13 and MMP-8 appear to have a wider catalytic cleft and are more sensitive to tetracycline inhibition than MMP-1 [84]).

Scavenging of PMN-generated reactive oxygen metabolites (e.g., HOCl) by TCs may prevent the oxidative conversion of pro-MMPs in the ECM into active MMPs, and this property of TCs appears not to depend on the metal-ion binding properties of these drugs. As evidence, CMT-5, which unlike other TC compounds lacks the important Ca^{2+} and Zn^{2+}-binding site on the TC molecule, can prevent pro-MMP activation like other TC analogues [85]. The drugs also appear to down-regulate expression of MMPs, at least by keratinocytes, endothelial cells and osteoblasts in culture [86, 87].

TCs appear to inhibit ECM breakdown by indirect mechanisms as well. In this regard, the serum protein α_1-antitrypsin (also called α_1-proteinase inhibitor or α_1-PI), is the host's major defense against another family of tissue-destructive proteinases, the serine proteinases (particularly neutrophil elastase). MMPs are now known to degrade and inactivate α_1 antitrypsin, so that TC-inhibition of the MMPs could protect elastase-susceptible substrates (e.g., elastic fibers, fibronectin, proteoglycans and TIMPs) from proteolytic attack as well [88–90]. Another potential indirect mechanism by which the TCs may inhibit ECM breakdown could be through inhibition of activation of pro-TNF-α thereby leading to a decrease in the formation of the powerful cytokine, tumor necrosis factor (TNF-α) [91, 92]. Gearing et al. [93] have demonstrated that processing of the TNF-α precursor can be mediated by MMP enzymes. Recently three different MMP inhibitors including BB2275 (British Biotech, Inc.) [93], GI129471 (Glaxo, Inc.) [94] and doxycycline [92], were found to inhibit MMP-mediated activation of pro-TNF-α, *in vivo*, in endotoxin-challenged rats. In addition, there is recent evidence to support inhibition of the production and/or activation of another important cytokine, interleukin-1β (IL-1β) or osteoclast activating factor, as well [88]. In cell culture, as well as animal model systems, the tetracyclines have been found to reduce elevated prostanoid levels [95] as well as inhibit osteoclast activity [96]. The tetracyclines can clearly inhibit connective tissue breakdown by multiple non-antimicrobial mechanisms (Tab. 2) [88, 97]. These findings have led to the development of a series of novel, chemically modified tetracyclines known as Inhibitors of Multiple Proteases and Cytokines or IMPACs®.

Tetracyclines have evolved in the periodontal literature as not only antimicrobials but also MMPIs. In a review on future periodontal therapy, Page [98] addressed 3 components central to periodontal treatment: 1) scaling and root planing, with or without surgery; 2) antibiotic treatment (both 1 and 2 reduce the bacterial load) and 3) use of drugs aimed at modulating host response mechanisms in order to suppress or inhibit soft-tissue destruction and alveolar bone resorption. A major advantage of tetracyclines as inhibitors of periodontal breakdown is the fact that these drugs have been used safely for many years and, at antimicrobially effective doses, they target 2 of the 3 therapeutic strategies just listed, although microbial resistance does develop after prolonged use of commercially available formulations [34]. For this and other reasons, including adverse events such as gastrointestinal upset and the overgrowth of yeast, systemic antimicrobial agents have not been advocated for the routine treatment of adult periodontitis [99]. A significant advantage of the low-dose doxycycline regimens (Periostat® introduced by Golub et al. [75] and approved by the U.S. FDA for human use in 1998) and the CMTs is that both regimens lack antimicrobial efficacy and can be administered for long periods of time (up to 27 months) without the emergence of antibiotic-resistant microorganisms [100–102]. These and other ongoing studies on the use of this subantimicrobial low dose of doxycycline for the treatment of periodontitis have been designed to address the concerns of many in the field regarding the issue of resistance. Certain CMTs have advantages over commercially available tetracyclines because they are absorbed more rapidly, can reach higher levels in the blood, have longer serum half-lives, and are more potent inhibitors of MMPs. A number of comprehensive reviews have recently been written on the use of tetracyclines and their analogues in periodontal treatment [60–62, 71, 76, 103–105].

The only MMPIs which have been tested for the treatment of periodontitis are members of the tetracycline family of compounds. Because the CMTs are not yet approved for human use, all of the following references to clinical trials involve the use of commercially available tetracyclines and their semi-synthetic analogues, minocycline and, most importantly, doxycycline. In an early study using these different tetracyclines, Golub et al. [72] reported that the semi-synthetic compounds were more effective than tetracycline HCl in reducing excessive collagenase activity in the GCF of adult periodontitis patients. Because doxycycline was found to be a more effective inhibitor of collagenase than either minocycline or tetracycline [61, 106], recent clinical trials have focused on this compound. In an effort to eliminate the side-effects of long-term TC therapy, especially the emergence of TC-resistant organisms, low-dose doxycycline (LDD) capsules were prepared and tested [107]. Each capsule contained 20 mg of doxycycline, compared to the commercially available 50 and 100 mg, antimicrobially effective, capsules. In multiple clinical studies conducted using sub-antimicrobial dose doxycycline there has not been a difference in the composition or resistance level of the oral flora [101, 102] and more recent studies demonstrate no appreciable differences in either fecal or vaginal microflora samples [102]. In addition, these studies have also demonstrated no overgrowth of opportunistic pathogens such as *Candida* in the oral cavity, gastrointestinal or genitourinary systems.

Early studies indicated that LDD reduced the peak blood level of the drug by 92% compared to the regular dose doxycycline regimen based on a bioassay [85, 108]. Subsequent studies using HPLC analysis of total doxycycline in the serum indicate that blood levels (Cmax) range from 0.5–0.7 μg/ml rather than 0.29 μg/ml previously reported by bioassay. The discrepancy in these findings may be due to the fact that much of the doxycycline measured by HPLC in the serum is bound to serum proteins and would not be available and active when measured by bioassay techniques.

With regard to MMP inhibition, Golub et al. [75] reported that a 2-week regimen of LDD reduced collagenase in GCF and in the adjacent gingival tissues surgically excised for therapeutic purposes. Subsequent studies using this LDD therapy adjunctive to routine scaling and prophylaxis indicated that, after 1 month of treatment, there were continued reductions in the excessive levels of collagenase in the GCF, but after cessation of LDD administration there was a rapid rebound of collagenase activity to placebo levels, suggesting that a 1-month treatment regimen with this host-modulatory agent was insufficient to produce a long-term benefit [109]. In contrast, during the same study, a 3-month regimen produced a prolonged drug effect without rebounding to baseline levels during the no-treatment phase of the study. The mean levels of GCF collagenase were significantly reduced (47.3% from baseline levels) in the LDD-treated group *versus* the placebo group which received scaling and prophylaxis alone (29.1% from baseline levels). Accompanying these reductions in collagenase levels were gains in the relative AL in the LDD-treated group [109, 110], as seen in Figure 3. Continuous drug therapy over a period of several months appears to be necessary for maintaining collagenase levels near normal over prolonged periods of time. However, it is reasonable to speculate that these MMPs will eventually reappear in the more susceptible patients, and those individuals having the most risk factors and the greatest microbial challenge will require more frequent host-modulatory therapy than other patients.

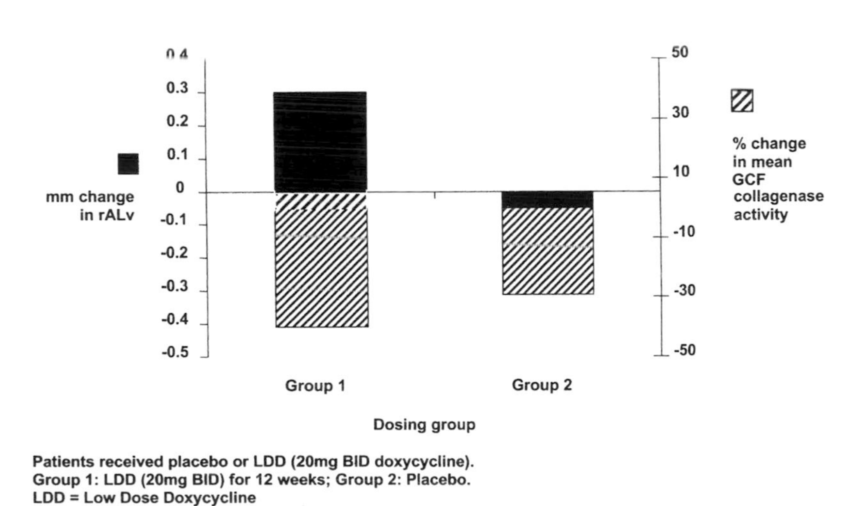

Figure 3. The effect of low-dose doxycycline on gingival crevicular fluid collagenase activity and relative attachment level (rALv). The patients in Group 1 received 20 mg of doxycycline twice daily and the patients in Group 2 received a placebo.

A series of double-blind, placebo-controlled studies of 3, 6 and 9 months' duration all showed clinical efficacy based on the reduction of PD and inhibition of gingival attachment loss as well as biochemical efficacy, based on the inhibition of collagenase activity, and protection of serum α_1-antitrypsin from collagenase attack, in the periodontal pocket [61, 69, 89]. Golub et al. [111] showed that a 2-month regimen of LDD significantly decreased both the level of bone-type collagen breakdown products (ICTP; a pyridinoline-containing crosslinked peptide of Type I collagen) and MMP-13 enzyme levels (bone-type collagenase) in adult periodontitis subjects, providing biochemical evidence of reduction of bone resorption to support computer-assisted subtraction radiography data [67, 112], the latter providing evidence of a reduction in the loss of alveolar bone height after 12 months of therapy with LDD.

There is clear evidence that modulation of the host response, whether it be by effects on MMPs, cytokines, prostanoids or osteoclast function, can play a role in slowing periodontal disease progression. In the case of enzyme suppression, LDD has been found to be useful for preventing disease progression as can be seen in Figure 4, which shows the percentage of tooth sites which lost 3 mm or more of attachment during the 12-month course of a Phase III clinical trial [66, 67]. Adjunctive treatment with LDD in conjunction with dental scaling and prophylaxis was shown to reduce disease progression by 85%, 73% and 36% in sites with severe, mild-to-moderate, and essentially no disease, respectively. This demonstrates that host-modulatory therapy can aid in the maintenance of normal sites, preventing them from developing disease, as well as in the prevention of further progression of disease in already diseased sites. Only with the systemic administration of a drug would it be possible to see effects on the normal sites which are usually ignored due to inadequate diagnostics that would otherwise alert the practitioner to incipient disease activity at these "normal" sites.

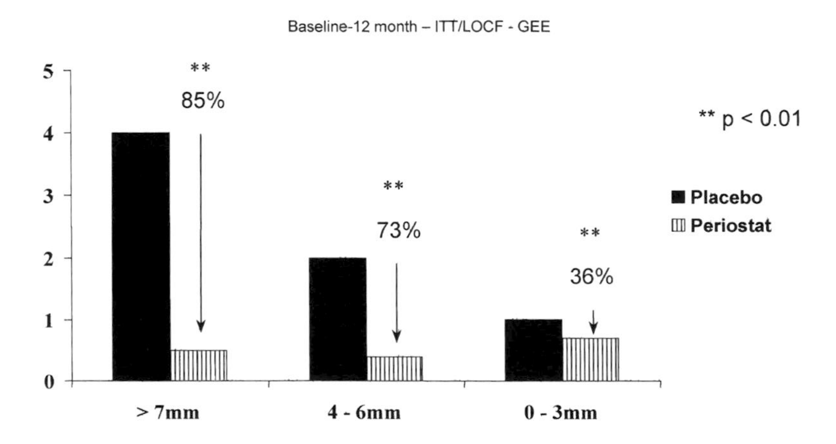

Figure 4. Effect of low-dose doxycycline therapy on periodontal disease progression as measured by the percentage of tooth sites losing 3 mm or more of clinical attachment over 12 months.

Table 3. Need for intervention (SRP) during the 12-month Periostat® phase III scaling trial

	Placebo	Periostat®	p-value
# of patients	9	9	-
# of tooth sites	52	14	-
CAL change (mm) post SRP	0.78	2.16	0.005

Upon analysis for attachment loss of 3 mm or more over the first 6 months of this 12-month study, a subset of susceptible patients in each treatment arm became evident. As seen in Table 3, there are far fewer sites in the group treated with the host-modulatory agent LDD (14 sites) that experienced this rapid loss of attachment compared to the placebo group (52 sites), corresponding to a 73% reduction in the incidence of rapid progression of periodontitis. In addition, when recovery therapy was performed (in this case SRP with anesthesia) the average attachment gain was significantly greater in the adjunctive LDD group (2.16 mm) than in the adjunctive placebo group (0.78 mm) as measured 6 months later, 12 months following baseline [109]. The adjunctive use of an MMPI clearly made these most susceptible patients, with active disease sites, more responsive to the more definitive therapy (SRP), demonstrating almost a reversal in these rapidly progressive sites in the LDD-treated group of patients. It is likely that patients who are susceptible to rapidly progressive periodontitis have dysfunctional host responses that would benefit by host-modulatory therapy. The same may be true of tooth sites that are refractory to traditional therapies.

A 9-month randomized, double-blind, placebo-controlled trial conducted at 5 dental centers demonstrated clinical efficacy and safety of LDD *versus* placebo adjunctive to the gold standard of periodontal therapy, SRP. Once again, the benefits of host-modulatory therapy in addition to mechanical therapy were seen with statistically significant reductions in probing depths, BOP and gains in clinical AL, as well as the prevention of disease progression [66, 68]. A clear effect of LDD therapy can be seen with regard to the need for recovery therapy and tooth extraction as seen in Table 4. There are reductions

Table 4. Tooth loss and need for recovery therapy during the 9 month Periostat® phase III SRP trial

	Placebo	Periostat®
Tooth extraction		
Number of teeth	23	5
Number of patients	14	4
Recovery therapy		
Number of tooth sites	59	24
Number of patients	21	11

in the numbers of individuals and tooth sites requiring recovery therapy due to rapid attachment loss of 2 mm or more along with less need for extraction in the LDD-treated group. In a discontinuation study, where LDD administration was discontinued after 9 months of continuous therapy, the incremental improvements demonstrated in the LDD group were maintained for at least 3 months post-treatment. There was no rebound effect in either the PD reductions or clinical AL gains; in fact there appeared to be slight continued improvements in both of these clinical parameters [66, 68]. The clinical relevance of such findings confirm the utility of an MMPI in the management of adult periodontitis.

More recent Phase IV clinical studies have revealed success using LDD in very susceptible individuals. One such patient population is one with a specific variation in the genes that regulate IL-1 [112]. Currently there is a PST Genetic Susceptibility Test for Periodontal Disease to determine whether or not a patient has this susceptible genotype. PST-positive patients have an increased inflammatory response in the presence of bacteria, producing 2–4 times more IL-1 with microbial challenge [113]. Therefore these patients are at greater risk for developing severe periodontal disease [112, 114] and subsequently at greater risk for tooth loss [115]. A 5-month preliminary investigation by Ryan and co-workers [116] was designed to evaluate the impact of treatment on IL-1 and MMP levels of PST-positive patients who presented with elevated levels of these biochemical markers in their GCF. These patients were initially treated with SRP, resulting in no change in the levels of these biochemical markers after 1 month. They were then placed on a subantimicrobial dose of doxycycline and these biochemical markers where monitored at 2 and 4 months. A significant decrease (50–61%) in the IL-1β and MMP-9 levels was noted after treatment with Periostat®. Correspondingly, gains in clinical attachment and reduced probing depths were also observed. The conclusions of the study were that a subantimicrobial dose of doxycycline may provide PST-positive patients with a therapeutic strategy that specifically addresses their exaggerated host response. Another recent study was conducted in susceptible patients with severe generalized periodontitis using host modulation as an adjunct to a mechanical therapy known as "repeat sub-gingival debridement" (Novak et al. Personal communication). Greater than 80% of the patients who participated in this 9-month double-blind, placebo-controlled study were smokers. Subantimicrobial dose doxycycline (SDD) as an adjunct to mechanical therapy *versus* mechanical therapy alone, respectively, resulted in significant improvements in probing depth reductions as early as 1 month after therapy (2.52 mm *versus* 1.25 mm) which were maintained during the 5.25 months of therapy (2.85 mm *versus* 1.48 mm) and even after 3 months of drug therapy cessation (3.02 mm *versus* 1.41 mm), demonstrating that there was no rebound effect. Due to all of these beneficial effects of host-modulatory therapy in susceptible patients, multicenter studies are anticipated using SDD in diabetic and osteoporotic patients as well as institutionalized patients.

Regenerative procedures

The ultimate periodontal therapeutic would be one that not only slows down loss of attachment but one that actually promotes regeneration of lost cementum, bone and connective tissue attachment. The tetracyclines have been used for many years as adjuncts to regenerative procedures requiring advanced skills and careful case selection. Tetracyclines have been incorporated into surgically placed resorbable and non-resorbable membranes used to inhibit the downward growth of the epithelium and to allow for regeneration to occur in vertical defects. Tetracyclines have been painted on root surfaces at the time of surgery as well as incorporated into allogeneic, freeze-dried bone as well as other graft materials such as tricalcium phosphate and hydroxyapatite. Tetracyclines are routinely systemically administered to patients undergoing regenerative surgical procedures for varying periods of time. Regenerative procedures are not as predictable as we would like them to be, however, the use of tetracyclines may help to improve on regenerative outcomes.

Rifkin et al. [96] recently reviewed the inhibition of bone resorption by tetracyclines and its therapeutic implications. They found, on the basis of several cell culture and *in vivo* studies, that tetracycline and CMTs can inhibit bone resorption by inhibiting osteoblast and osteoclast-derived MMPs, and that other mechanisms may also play a role, including the alteration by tetracycline of the osteoclast response to extracellular Ca^{2+} concentrations and reduced secretion of acid cathepsins. Tetracyclines have also been found to enhance osteoblast activity and to increase collagen and bone formation when these processes are suppressed during disease [64, 65] (Golub et al., Paper presented at the 8th International Conference on Periodontal Research, San Antonio, 1990). Although the mechanisms remain speculative, an intriguing therapeutic outcome may be bone regeneration, not just prevention of alveolar bone loss.

Most recently, Uitto et al. [86] and Nip et al. [117] demonstrated that tetracyclines and CMTs can inhibit MMPs generated by epithelial cells (including those with characteristics similar to junctional epithelium), not only by blocking the extracellular activity of these enzymes but also by inhibiting the intracellular (i.e., genetic) expression or synthesis. (Note that these effects were not toxic side-effects.) It is tempting to speculate that if the same effect were seen *in vivo*, it could provide another therapeutic mechanism for prevention of pocket formation by blocking gelatinases (i.e., type IV collagenases) produced by junctional epithelium, preventing the degradation of the basement membrane and preventing epithelial migration along the root surface. The avidity of tetracyclines for bone and tooth surfaces, and their subsequent prolonged release, may enhance the efficacy of these drugs as inhibitors of both epithelial migration along the root and alveolar bone loss.

The tetracyclines may facilitate periodontal ligament fibroblast migration and attachment to the root surface as well as fibronectin binding. Root surface modification has been used to improve upon reattachment of tissues and regeneration. 100 mg/ml of tetracycline has been shown to eliminate the smear layer, opening the dentinal tubules. The theory is that collagen fibers can penetrate into the open dentinal tubules and favor reattachment. Tetracyclines may also aid in the inhibition of the downward migration of the epithelium by inhibiting MMP-13 being secreted by epithelial cells. Tetracyclines have also been shown to upregulate osteoblast activity, collagen synthesis (protein and mRNA) and bone formation in animal models of diabetic osteopenia [64, 65, 118]. The same tetracycline-induced increase in collagen synthesis, as well as improvements in bone density, have been seen in ovariectomized animal models of osteoporosis [119–122] and more recently in osteoporotic women treated with SDD [123]. In LJP, the tetracyclines have been found to stimulate bone formation. A number of reports in patients with LJP have shown that thorough SRP and tetracycline therapy not only killed the pathogens associated with the disease, but also resulted in increased bone formation and reattachment resulting in decreased probing depths. Moskow and Tannenbaum [124] noted favorable osseous changes and repair of "hopeless" teeth with 2–3 wall defects in patients with adult periodontitis after administration of 1 g/day of tetracycline after 21 days of therapy.

Conclusions

The executive summary of the Surgeon General's Report on Oral Health reveals that most adults show signs of periodontal or gingival disease, with severe periodontitis affecting about 14% of adults aged 45 to 54; 23% of 54–74 year old adults have severe periodontal disease. Adding the population of Third World countries to this statistic could increase the prevalence of disease to above 33%. At all ages men are more likely than women to have severe periodontitis and people at the lowest socioeconomic levels have more severe periodontal disease. About 30% of adults age 65 and older are edentulous with higher figures for those living in poverty. It is also known that 17–28% of the treated periodontal patients either develop refractory or recurrent periodontitis. Clearly there is a need for improved diagnosis and cost-effective means of managing this chronic progressive disease which has been linked to other systemic diseases.

Periodontal pathogens and destructive host responses are involved in the initiation and progression of periodontitis. Therefore, the successful long-term management of this disease may require a treatment strategy that integrates therapies that address both etiological components. There is much evidence for the role of MMPs, cytokines and other mediators in the destructive processes of periodontal disease, distinguishing them as a viable target for a chemotherapeutic approach. The introduction of novel, adjunctive therapies to enhance the efficacy of existing mechanical procedures would contribute favorably to an integrated approach for the long-term, clinical management of periodontitis, particularly in susceptible patients where a prolonged and excessive host response to the presence of bacteria promotes the activity of MMPs.

Preclinical and clinical studies have demonstrated that inhibition of these MMPs, cytokines and other mediators through a host-modulatory approach results in favorable changes in biochemical markers of the disease such as ICTP, as well as clinical markers of therapeutic efficacy. Clearly these strategies to reduce the bacterial load with SRP, with or without systemic or locally applied antimicrobials, and to modulate the host response by inhibiting MMPs and inflammatory mediators are complementary. It is clear that standard therapy, such as the removal of supra and subgingival plaque and calculus deposits by SRP is more predictable for mild-moderate periodontitis than for moderate-severe periodontitis. The use of a host-modulatory agent, or a combination of host-modulatory agents, capable of inhibiting MMPs, cytokines and other mediators of the disease, can assist with the conventional treatment for periodontitis. When used adjunctively, chemotherapeutics can enhance and make clinical therapeutic responses more predictable in the more susceptible patient. With regard to a number of the tetracycline derivatives, these agents appear to be pleiotropic, affecting many of these pathways. This is why the tetracyclines have been found to be so effective as host-modulatory agents.

Finally, as the era of periodontal medicine becomes apparent, the use of local antimicrobial and systemic host-modulatory approaches needs to be considered. Host modulators used to manage periodontal disease, such as MMP, cytokine and prostanoid inhibitors, may have additional beneficial effects on systemic diseases which have been associated with periodontal disease, such as cardiovascular disease, diabetes and premature births. In the case of cardiovascular disease, preliminary studies have indicated that individuals with periodontal disease are almost twice as likely to suffer from a fatal heart attack and nearly three times more likely to suffer from a stroke [125]. MMPs and cytokines have been found to play a major role in weakening the plaques formed with cardiovascular disease leading to rupture and eventual thrombosis and infarction [126]. In fact, Golub et al. [127] suggested that TCs could reduce the incidence of acute myocardial infarction [128] by blocking collagenase and stabilizing the collagen cap on the athero-scleromatous arterial plaques. In diabetes, the same MMPs and cytokines involved in the development of periodontitis as the sixth long-term complication of diabetes [129] have also been associated with other well-known complications of diabetes such as nephropathy, angiopathy, retinopathy and wound healing problems [63]. Finally, periodontal disease has been found to put pregnant women at an increased risk of giving birth to pre-term, low-weight babies, as endotoxins drive up the levels of prostanoids involved in contraction [130]. In addition, MMPs have been found to play a role in the cervical effacement and dilation that precede labor and are markers of risk for pre-term delivery [131] since they are involved in pre-term premature rupture of the membranes playing a role in pre-term parturition [132]. Clearly an inhibitor of MMPs, cytokines and prostanoids might not only help in the management of periodontitis which has been associated with these and other systemic diseases by having an indirect effect on these processes, but may also directly aid in the treatment and prevention of cardiovascular disease, diabetic complications and premature labors.

The Surgeon General's Report recognizes "the mouth as a mirror of health or disease, as a sentinel or early warning system, as an accessible model for the study of other tissues and organs, and as a potential source of pathology affecting other systems and organs" [15]. The findings discussed in this chapter with regard to the use of tetracyclines as therapeutics to better manage chronic periodontal disease may have applications to other chronic progressive tissue-destructive diseases such as arthritis. The effects of tetracyclines on periodontitis in diabetics may have implications for other long-term complications of diabetes, such as cardiovascular disease, retinopathy, nephropathy, neuropathy and wound healing. The history of tetracyclines and periodontal disease may prove to have an impact on general health, making a contribution to human welfare.

References

1 Adams FR (1891) *The Genuine Works of Hippocrates*. William Wood, New York
2 Elliot JS (1972) *Outlines of Greek and Roman Medicine*. Milford House, Boston
3 Leatherman GH (1961) Dentistry and its future. *J Amer Coll Dent* 28: 33–51
4 Young S (1890) *The Annals of the Barber-Surgeons of London*. Blades, East and Blades, London
5 Deeley WS (1977) Modern dentistry began in 18th Century France, thanks to the leadership of Pierre Fauchard. Current knowledge is the result. *Dent Stud* 55: 56–86
6 Qvist G (1981) *John Hunter*. William Heinemann Medical Books, London
7 Allen DL (1998) Professional dental education…the beginnings. *J Hist Dent* 46: 40–45
8 Jacobsohn PH (1995) Horace Wells: discoverer of anesthesia. *Anesth Prog* 42: 73–75
9 Bremner MDK (1939) *The Story of Dentistry From the Dawn of Civilization*. Dental Items of Interest, New York
10 Russell SL, Kostlan J (1961) Spectacular report from world health organization. *N Y J Dent* 31: 98–125
11 The American Academy of Periodontology (1966) Principles of ethics. *J Periodontol* 37: 514–516
12 Ring ME (1985) *Dentistry: An Illustrated History*. Harry N. Abrams, New York and The C.V. Mosby Company, St. Louis
13 Prinz H (1945) *Dental Chronology: A Record of the More Important Historic Events in the Evolution of Dentistry*. Lea and Febiger, Philadelphia
14 Carrenza FA, Odont Glickman I, Newman MG (1996) *Clinical Periodontology*, 8th Edition. W.B. Saunders Company, Philedelphia
15 US Department of Health, Human Services (USD-HHS) (2000) Oral Health in America: A Report of the Surgeon General (Executive Summary)
16 Grossi SG, Skrepcinski FB, DeCaro T, Robertson DC, Ho AW, Dunford RG, Genco RJ (1997) Treatment of periodontal disease in diabetics reduces glycated hemoglobin. *J Periodontol* 68: 713–719
17 Maynard JG (1993) Eras in periodontics. *In*: *Periodontal Disease Management: A Conference for the Dental Team*. The American Academy of Periodontology, Boston, MA, 3–10
18 Loe H, Anerud A, Boysen H, Smith M (1978) The natural history of periodontal disease in man. *J Periodont Res* 13: 550–562
19 Forgas LB, Nilius AM (1991) Assessing periodontal disease activity. The role of bacteriological, immunological and DNA assays. *J Dent Hyg* 65: 188–193
20 Greenstein G (1992) Periodontal response to mechanical non-surgical therapy: A review. *J Periodontol* 63: 118–130

21 Biasi D, Bambara LM, Carletto A, Caramaschi P, Andrioli G, Urbani G, Bellavitc P (1999) Neutrophil migration, oxidative metabolism and adhesion in early onset periodontitis. *J Clin Periodontol* 26: 563–568
22 Ozmeric N, Bol B, Balos K, Berker E, Bulut S (1998) The correlation of gingival crevicular fluid interleukin-8 levels and periodontal status in localized juvenile periodontitis. *J Periodontol* 69: 1299–1304
23 Zambon J (1996) Periodontal Diseases: microbial factors. *Ann Periodontol* 1: 879–925
24 Baker PJ, Evans RT, Slots J, Genco RJ (1985) Antibiotic susceptibility of anaerobic bacteria from the human oral cavity. *J Dent Res* 64: 1233–1244
25 Baker PJ, Evan RT, Slots J, Genco RJ (1985) Susceptibility of human oral anaerobic bacteria to antibiotics suitable for topical use. *J Clin Periodontol* 12: 201–208
26 Kleinfelder JW, Muller RF, Lange DE (1999) Antibiotic susceptibility of putative periodontal pathogens in advanced periodontitis patients. *J Clin Periodontol* 26: 347–351
27 Slots J, Ram TE (1990) Antibiotics in periodontal therapy: advantages and disadvantages. *J Clin Periodontol* 17 (pt2): 479–493
28 O'Connor BC, Newman HN, Wilson M (1990) Susceptibility and resistance of plaque bacteria to minocycline. *J Periodontol* 61: 228–233
29 Lacroix JM, Walker GB (1996) Development of resistence determinant Tet Q in the microbiota associated with adult periodontitis. *Oral Microbiol Immunol* 11: 282–288
30 Sutter VL, Jones MJ, Ghoneim ATM (1983) Antimicrobial susceptibilities of bacteria associated with periodontal disease. *Antimicrob Agents Chemother* 23: 483–486
31 Williams BL, Osterberg SKA, Jorgensen J (1979) Subgingival microflora of periodontal patients on tetracycline therapy. *J Clin Periodontol* 6: 210–221
32 Fiehn N-E, Westergaard J (1990) Doxycycline-resistant bacteria in periodontally diseased individuals after systemic doxycycline therapy and in healthy individuals. *Oral Microbiol Immunol* 5: 219–222
33 Haffajee AD, Sucransky SS, Dzink JL, Taubman MA, Ebersole JL (1988) Clinical, microbiological, and immunological features of subjects with refractory periodontal disease. *J Clin Periodontol* 15: 390–398
34 Kornman K, Karl E (1982) The effect of long-term low-dose tetracycline therapy on the subgingival microflora in refractory adult periodontitis. *J Periodontol* 53: 604–610
35 Kornman KS, Robertson PB (1985) Clinical and microbiological evalulation of therapy for juvenile periodontitis. *J Periodontol* 56: 443–446
36 Lindhe J (1982) Treatment of localized juvenile periodontitis. *In*: RJ Genco, S Mergenhagen (eds). *Host Parasite Interactions in Periodontal Diseases.* American Society of Microbiology, Washington DC, 382–394
37 Slots J, Rosling BG (1983) Suppression of the periodontopathic microflora in localized juvenile periodontitis by systemic tetracycline. *J Clin Periodontol* 10: 465–486
38 Zambon JJ, Christersson LA, Genco RJ (1986) Diagnosis and treatment of localized juvenile periodontitis. *J Amer Dent Assn* 113: 295–299
39 Greenspan JS (1994) Periodontal complications of HIV infection. *Compendium* Suppl 18: S694–698; quiz S714–717
40 McCulloch CA, Birek P, Overall C, Aitken S, Lee W, Kulkarni G (1990) Randomized controlled trial of doxycycline in prevention of recurrent periodontitis in high-risk patients: antimicrobial activitiy and collagenase inhibition. *J Clin Periodontol* 17: 616–622
41 Genco RJ (1981) Antibiotics in the treatment of periodontal diseases. *J Periodontol* 52: 345
42 Baker PJ, Slots J, Genco RJ, Evans RT (1983) Minimal inhibitory concentration of various antimicrobial agents for human oral anaerobic bacteria. *Antimcrob Agents Chemother* 24: 420–424
43 Mashimo PA, Yamamoto Y, Slots J, Evans RT, Genco RJ (1981) *In vitro* evaluation of antibiotics in the treatment of periodontal disease. *Pharmacol Ther* 6: 45–56
44 Gordon J, Walker C, Murphy J, Goodson Socransky S (1981) Tetracycline levels achievable in gingival crevice fluid and *in vitro* effect on subgingival organisms. I. Concentrations in crevicular fluid after repeated doses. *J Periodontol* 52: 609–612
45 Pascale D, Gordon J, Lamster I, Mann P, Seiger M, Arndt W (1986) Concentration of doxycycline in human gingival fluid. *J Clin Periodontol* 13: 841–844
46 Sakellari D, Goodson J, Socransky S, Kolokotronis A, Konstantinidis A (1997) Concentration of 3 tetracyclines in plasma, GCF and saliva. *J Dent Res* 76: 176
47 Killoy WJ, Bobb CM (1992) Controlled local delivery of tetracycline in the treatment of peri-

odontitis. *Compend Cont Educ Dent* 12: 1150–1160
48 Goodson JM, Holborrow D, Dunn R, Hogan P, Dunham S (1983) Monolithic tetracycline containing fibers for the controlled delivery to periodontal pockets. *J Periodontol* 54: 575–579
49 Goodson JM, Cugini MA, Kent RL, Armitage CG, Scobb CM, Fine D, Fritz ME, Greene Imoberdorf MJ, Killoy WJ et al (1991b) Multicenter evaluation of tetracycline fiber therapy (II). Clinical response. *J Periodont Res* 26: 371–379
50 Formousis I, Tonetti MS, Mombelli A, Lehmann B, Lang NP, Bragger U (1998) Evaluation of tetracycline fiber therapy with digital image analysis. *J Clin Periodontol* 25: 737–745
51 Garrett S, Johnson L, Drisko CH, Adams DF, Bandt C, Beiswanger B, Bogle G, Donly K, Hallmon WW, Hancock EB et al (1990) Two multi-center studies evaluating locally delivered doxycycline hyclate, placebo control, oral hygiene, and scaling and root planing in the treatment of periodontitis. *J Periodontol* 70: 490–503
52 Ryder MI, Pons B, Adams D, Beiswanger B, Blanco V, Bogle G, Donly K, Hallmon W, Hancock EB, Hanes P et al (1999) Effects of smoking on local delivery of controlled-release doxycycline as compared to scaling and root planing. *J Clin Periodontol* 26: 683–691
53 van Steenberghe D, Bercy P, Kohl J, De Boever J, Adriaens P, Vanderfaeillie A, Adriaenssen C, Rompen E, De Vree H, McCarthy EF et al (1993) Subgingival minocycline hydrochloride ointment in moderate to severe chronic adult periodontitis: a randomised, double-blind, vehicle-controlled, multi-centre study. *J Periodontol* 64: 637–644
54 van Steenberghe D, Rosling B, Soder PO, Landry RG, van der Velden U, Timmerman MF, McCarthy EF, Vandenhoven G, Wouters C, Wilson M et al (1999) A 15-month evaluation of the effects of repeated subgingival minocycline in chronic adult periodontitis. *J Periodontol* 70: 657–667
55 Paquette DW (2000) Locally delivered antimicrobials: A medical model to complement the mechanical model. *In*: *Medicine and the Treatment of Periodontal Disease: A New Era*. Academic Dissertation held at the 86th Annual Meeting of the American Academy of Periodontology in Conjuction with the Japanese Society of Periodontology, Honolulu, Hawaii
56 Offenbacher S (1996) Periodontal diseases: pathogenesis. *Ann Periodontol* 1: 821–878
57 Genco RJ (1992) Host responses in periodontal diseases: current concepts. *J Periodontol* 63: 338–355
58 Grossi SG, Zambon JJ, Ho AW, Koch G, Dunford RG, Machtei EE et al (1994) Assessment of risk for periodontal disease. I. Risk indicators for attachment loss. *J Periodontol* 65: 260–267
59 Salvi GE, Lawrence HP, Offenbacher S, Beck JD (1997) Influence of risk factors on the pathogenesis of periodontitis. *Periodontol 2000* 14: 173–201
60 Ryan ME, Golub L (2000) Modulation of matrix metalloproteinase activity in periodontitis as a treatment strategy. *Periodontol 2000* 24: 226–238
61 Golub L, Evans R, McNamara T, Lee H, Ramamurthy N (1994) A non-antimicrobial tetracycline inhibits gingival matrix metalloproteinases and bone loss in *Porphyromonas gingivalis*-induced periodontitis in rats. *Ann N Y Acad Sci* 732: 96–111
62 Ryan ME, Ramamurthy N, Golub L (1996) Matrix metalloproteinases and their inhibition in periodontal treatment. *Curr Opin Periodont* 3: 85–96
63 Ryan ME (1998) *Host Response in Diabetes-Associated Periodontitis: Effects of Tetracycline Analogues*. Academic Dissertation, State University of New York at Stony Brook, New York
64 Sasaki T, Ramamurthy N, Golub L (1992) Tetracycline administration increases collagen synthesis in osteoblasts of streptozotocin-induced diabetic rats: a quantitative radiographic study. *Calcified Tissue Int* 50: 411–419
65 Schneir M, Ramamurthy N, Golub L (1990) Minocycline treatment of diabetic rats increases skin collagen production and mass; possible causative mechanisms. *Matrix* 10: 112–123
66 Caton J (1999) Evaluation of Periostat for patient management. *Compendium* 20: 451–462
67 Caton J, Blieden T, Adams D, Crout R, Hefti A, Killoy W, Nagy R, O'Neal R, Quinones C, Taggart E et al (1997) Subantimicrobial doxycycline therapy for periodontitis. *J Dent Res* 76: 177
68 Caton J, Ciancio S, Blieden T, Bradshaw M, Crout R, Hefti A, Massaro J, Polson A, Thomas J, Walker C (2000) Treatment with subantimicrobial dose doxycycline improves the efficacy of scaling and root planing in patients with adult periodontitis. *J Periodontol* 71: 521–532
69 Crout R, Lee H, Schroeder K, Crout H, Ramamurthy N, Wiener M, Golub L (1996) The cyclic regimen of low dose doxycycline for adult periodontitis: A preliminary study. *J Periodontol* 67: 506–514
70 Golub LM, Lee HM, Lehrer G, Nemiroff A, McNamara TF, Kaplan R, Ramamurthy NS (1983)

Minocycline reduces gingival collagenolytic activity during diabetes: preliminary observations and a proposed new mechanism of action. *J Periodont Res* 18: 516–526

71 Golub L, Ramamurthy N, McNamara T, Greenwald R, Rifkin B (1991) Tetracyclines inhibit connective tissue breakdown: New therapeutic implications for an old family of drugs. *Crit Rev Oral Biol Med* 2: 297–322

72 Golub L, Wolff M, Lee H, McNamara T, Ramamurthy N, Zambon J, Ciancio S (1985) Further evidence that tetracyclines inhibit collagenase activity in human crevicular fluid and from other mammalian sources. *J Periodont Res* 20: 12–23

73 Gomes B, Golub L, Ramamurthy N (1984) Tetracyclines inhibit parathyroid hormone induced bone resorption in organ culture. *Experientia* 40: 1273–1275

74 Golub L, McNamara T, D'Angelo G, Greenwald R, Ramamurthy N (1987) A non-antibacterial chemically modified tetracycline inhibits mammalian collagenase activity. *J Dent Res* 66: 1310–1314

75 Golub L, Ciancio S, Ramamurthy N, Leung M, McNamara T (1990) Low dose doxycycline therapy: effect on gingival and crevicular fluid collagenase activity in humans. *J Periodont Res* 25: 321–330

76 Rifkin B, Vernillo A, Golub L (1993) Blocking periodontal disease progression by inhibiting tissue-destructive enzymes: A potential therapeutic role for tetracyclines and their chemically-modified analogs. *J Periodontol* 64: 819–827

77 Yu L, Smith G, Hasty K, Brandt K (1991) Doxycycline inhibits Type XI collagenolytic activity of extracts from human osteoarthritic cartilage and of gelatinase. *J Rheumatol* 18: 1450–1452

78 Lovejoy B, Cleasby A, Hassell A, Longely K, Luther M, Weigl D, McGeehan G, McElroy A, Drewry D, Lambert M et al (1994) Structure of the catalytic domain of fibroblast collagenase complexed with an inhibitor. *Science* 263: 375–377

79 Seftor R, Seftor E, DeLarco J, Kleiner D, Leferson J, Stetler-Stevenson W, McNamara T, Golub L, Hendrix M (1998) Chemically modified tetracyclines inhibit human melanoma cell invasion and metastasis. *Clin Exp Metastasis* 16: 217–225

80 Sorsa T, Ding Y, Salo T, Lauhio A, Teronen O, Ingman T, Ohtani H, Andoh N, Takeha S, Konttinen Y (1994) Effects of tetracyclines on neutrophil, gingival and salivary collagenases: a functional and western blot assessment with special references to their cellular sources in periodontal diseases. *Ann N Y Acad Sci* 732: 112–131

81 Ryan ME, Usman A, Ramamurthy NS, Golub LM, Greenwald RA (2001) Excessive matrix metalloproteinase activity in diabetes: Inhibition by tetracycline analogues with zinc reactivity. *Curr Med Chem* 8: 305–316

82 Brandt K (1994) Insights into the natural history of osteoarthritis provided by the cruciate-deficient dog. *Ann N Y Acad Sci* 732: 199–205

83 Smith G, Brandt K, Hasty K (1994) Procollagenase is reduced to inactive fragments upon activation in the presence of doxycycline. *Ann N Y Acad Sci* 732: 436–438

84 Greenwald R, Golub L, Ramamurthy N, Chowdhury M, Moak S, Sorsa T (1998) *In vitro* sensitivity of three mammalian collagenases to tetracycline inhibition: Relationship to bone and cartilage degradation. *Bone* 22: 33–38

85 Golub L, Wolff M, Roberts S, Lee HM, Leung M, Payonk G (1994) Treating periodontal diseases by blocking tissue-destructive enzymes. *J Am Dent Assoc* 125: 163–169

86 Uitto VJ, Firth J, Nip L, Golub L (1994) Doxycycline and chemically modified tetracyclines inhibit gelatinase A (MMP-2) gene expression in human skin keratinocytes. *Ann N Y Acad Sci* 732: 140–151

87 Vernillo A, Rifkin B (1998) Effects of tetracyclines on bone metabolism. *Adv Dent Res* 12: 56–62

88 Golub L, Lee H, Ryan M, Giannobile W, Payne J, Sorsa T (1998) Tetracyclines inhibit connective tissue breakdown by multiple non-antimicrobial mechanisms. *Adv Dent Res* 12: 12–26

89 Lee H, Golub L, Chan D, Leung M, Schroeder K, Wolff M, Simon S, Crout R (1997) α1-Proteinase inhibitor in gingival crevicular fluid of humans with adult periodontitis: serpinolytic inhibition by doxycycline. *J Periodont Res* 32: 9–19

90 Mallya S, Hall J, Lee HM, Roemer E, Simon S, Golub L (1994) Interaction of matrix metalloproteinases with serine proteinase inhibitors: New potential roles for matrix metalloproteinase inhibitors. *Ann N Y Acad Sci* 732: 303–314

91 Shapira L, Soskolne W, Houri Y, Barak V, Halabi A, Stabholz A (1996) Protection against endotoxic shock and lipopolysaccharide-induced local inflammation by tetracycline: correlation with inhibition of cytokine secretion. *Infect Immunity* 64: 825–828

92 Thompson R, Harvin H, Liao S, Callery M (1996) *Doxycycline suppresses tumor necrosis factor-a (TNF-a) levels in lipopolysaccharide-challenged rats*. Academic dissertation held at Surg Infect Soc Meeting
93 Gearing A, Beckett P, Christodoulou M, Churchill M, Clements J, Davidson A, Drummond A, Galloway W, Gilbert R, Gordon J et al (1994) Processing of tumor necrosis factor-a precursor by metalloproteinases. *Nature* 370: 555–557
94 McGeehan G, Becherer J, Bast RJ, Boyer C, Champion B, Connolly K, Conway J, Furdon P, Karp S, Kidao S et al (1994) Regulation of tumour necrosis factor-processing by a metalloproteinase inhibitor. *Nature* 370: 558–561
95 Ryan ME, Ramamurthy NS, Amin A, Husuh D, Sorsa T, Golub LM (1999) Host response in diabetes-associated periodontis: Effects of a tetracycline analogue. *J Dent Res* 78: abstract#2193
96 Rifkin B, Vernillo A, Golub L, Ramamurthy N (1994) Modulation of bone resorption by tetracyclines. *Ann N Y Acad Sci* 732: 165–180
97 Ryan ME, Ashley R (1998) How do tetracyclines work? *Adv Dent Res* 12: 149–151
98 Page R (1998) The pathobiology of periodontal disease may affect systemic diseases: Inversion of a paradigm. *Ann Periodontol* 3: 108–120
99 Drisko C (1996) Non-surgical pocket therapy: Pharmacotherapeutics. *Ann Periodontol* 1: 491–566
100 Thomas J, Metheny RJ, Karakiozis JM, Wetzel JM, Crout R (1998) Long-term subantimicrobial doxycycline (Periostat) as adjunctive management in adult periodontitis: effects of subgingival bacterial population dynamics. *Adv Dent Res* 12: 32–39
101 Thomas J, Walker C, Bradshaw M (2000) Long-term use of subantimicrobial dose doxycycline does not lead to changes in antimicrobial susceptibility. *J Periodontol* 71: 1472–1483
102 Walker C, Thomas J, Nango', Lennon J, Wetzel J, Powala C (2000) Long-term treatment with subantimicrobial dose doxycycline exerts no antibacterial effect on the subgingival microflora associated with adult periodontitis. *J Periodontol* 71: 1465–1471
103 Golub L, Suomalainen K, Sorsa T (1992) Host modulation with tetracyclines and their chemically modified analogues. *Curr Opin Dent* 2: 80–90
104 Ingman T (1994) *Neutral Proteinases in Periodontal Diseases: Functional and Immunohistochemical Assessment*. Academic Dissertation, University of Helsinki, Finland
105 Vernillo A, Ramamurthy N, Golub L, Rifkin B (1994) The nonantimicrobial properties of tetracyclines for the treatment of periodontal disease. *Curr Opin Periodont* 1: 111–118
106 Burns F, Stack M, Gray R, Paterson C (1989) Inhibition of purified collagenase from alkaline-burned rabbit corneas. *Invest Ophthalmol Visual Sci* 30: 1569–1575
107 Golub L, Sorsa T, Lee HM, Ciancio S, Sorbi D, Ramamurthy N (1995) Doxycycline inhibits neutrophil (PMN)-type matrix metalloproteinases in human adult periodontitis gingiva. *J Clin Periodontol* 21: 1–9
108 McNamara T, Golub L, Yu Z, Ramamurthy N (1990) Reduced doxycycline blood levels in humans fail to promote resistant organisms. *In*: *International Conference on Periodontal Disease: Pathogens and Host Immune Responses*. Osaka, Japan,100
109 Ashley R, The SDD Clinical Research Team (1999) Clinical trials of a matrix metalloproteinase inhibitor in human periodontal disease. *Ann N Y Acad Sci* 878: 335–346
110 Golub LM, McNamara TF, Ryan ME, Kohut B, Blieden T, Payonk G, Sipos T, Baron HJ (2001) Adjunctive treatment with subantimicrobial doses of doxycycline: effects on gingival fluid collagenase activity and attachment loss in adult periodontisis. *J Clin Periodontol* 28: 146–156
111 Golub L, Lee H, Greenwald R, Ryan M, Sorsa T, Salo T, Giannobile W (1997) A matrix metalloproteinase inhibitor reduces bone-type collagen degradation fragments and bone-type collagenase in gingival crevicular fluid during adult periodontitis. *Inflamm Res* 4: 310–319
112 Ciancio S, Ashley R (1998) Safety and efficacy of sub-antimicrobial-dose doxycycline therapy in patients with adult periodontitis. *Adv Dent Res* 12: 27–31
113 McDevitt MJ, Wang HY, Knobelman C, Newman MG, di Giovine FS, Timms J, Duff GW, Kornman KS (2000) Interleukin-1 genetic association with periodontitis in clinical practice. *J Periodontol* 71: 156–163
114 Gore EA, Sander JJ, Pandey JP, Palesch Y, Galbraith GM (1998) Interleukin-1beta+3953 allele 2: association with disease status in adult periodontitis. *J Clin Periodontol* 25: 781–785
115 McGuire MK, Nunn ME (1999) Prognosis *versus* actual outcome. IV. The effectiveness of clinical parameters and IL-1 genotype in accurately predicting prognoses and tooth survival. *J Periodontol* 70: 49–56

116 Ryan ME, Lee HM, Bookbinder MK, Sorsa T, Golub LM (2000) Treatment of genetically susceptible patients with a subantimicrobial dose of doxycycline. *J Dent Res* 79: abstract #3719
117 Nip L, Uitto VJ, Golub L (1993) Inhibition of epithelial cell matrix metalloproteinases by tetracyclines. *J Periodont Res* 28: 379–385
118 Craig RG, Yu Z, XuL, Bana R, Ramamurthy N, Boland J, Schneir M, Golub LM (1998) A chemically-modified tetracycline inhibits Streptozotocin-induced diabetic depression of skin collagen synthesis and steady-state type I procollagen mRNA. *Biochim Biophys Acta* 1402: 250–260
119 Bain SD, Ramamurthy NS, Llavaneras A, Puerner D, Strachan MJ, Scheer K, Golub LM (1999) A chemically-modified doxycycline (CMT-8) alone or combined with bisphosphonate inhibits bone loss in the ovariectomized rat: a dynamic histomorphometric study. *J Dent Res* 78: abstract #2486
120 Golub LM, Ramamurthy NS, Llavaneros A, Ryan ME, Lee HM, Liu Y, Bain S, Sorsa T (1999) A chemically modified tetracycline (CMT-8) inhibits gingival matrix metalloproteinases, periodontal breakdown, and extra-oral bone loss in ovariectomized rats. *Ann N Y Acad Sci* 878: 290–310
121 Williams S, Wakisaka A, Zeng QQ, Barnes J, Martin G, Wechter WJ, Liang CT (1996) Minocycline prevents the decrease in bone mineral density and trabecular bone in ovariectomized aged rats. *Bone* 19: 637–644
122 Williams S, Wakisaka A, Zeng QQ, Barnes J, Seyedin S, Martin G, Wechter WJ, Liang CT (1998) Effect of minocycline on osteoporosis. *Adv Dent Res* 12: 71–75
123 Golub LM (2000) *Host-modulation therapy for periodontitis: potential benefits for post-menopausal osteoporosis*. Academic Dissertation held at the 86th Annual Meeting of the American Academy of Periodontology in Conjuction with the Japanese Society of Periodontology, Honolulu, Hawaii
124 Moskow BS, Tannenbaum P (1991) Enhanced repair and regeneration of periodontal lesions in tetracycline-treated patients. Case reports. *J Periodontol* 62: 341–350
125 Beck J, Offenbacher s Williams R, Gibbs P, Garcia R (1998) Periodontitis: A risk factor for coronary heart disease? *Ann Periodontol* 3: 142–150
126 Genco RJ, Offenbacher S, Beck J, Rees T (1999) Cardiovascular disease and oral infections. *In*: LF Rose, RJ Genco, DW Cohen, BL Mealey (eds): *Periodontal Medicine*. B.C. Decker Inc., Ontario
127 Golub LM, Greenwald RA, Thompson RW (1999) Tetracyclines and risk of subsequent first-time acute mycocardial infarction to the editor. *JAMA* 282: 1997–1998
128 Meier CR, Derby LE, Jick SS, Vasilakis C, Jick H (1999) Antibiotics and risk of subsequent and first-time acute myocardial infarction. *JAMA* 251: 427–431
129 Mealy B (1999) Diabetes Mellitis. *In*: LF Rose, RJ Genco, DW Cohen, BL Mealy (eds): *Periodontal Medicine*. B.C. Decker Inc., Ontario
130 Offenbacher S, Jared HL, O'Reilly PG, Wells SR, Salvi GE, Lawrence HP, Socransky SS, Beck JD (1998) Potential pathogenic mechanisms of periodontitis-associated pregnancy complication. *Ann Periodontol* 3: 233–250
131 Lockwood CJ, Kuczynski E (1999) Markers for risk for preterm delivery. *J Perinatal Med* 27: 5–20
132 Gomez R, Romero R, Edwin SS, David C (1997) Pathogenesis of preter labor and preterm premature rupture of membranes associated with intraamniotic infection. Infect. *Dis Clin North Am* 11: 135–176

Tetracyclines in Biology, Chemistry and Medicine
ed. by M. Nelson, W. Hillen and R.A. Greenwald

Inhibition of matrix metalloproteinases (MMPs) by tetracyclines

Roeland Hanemaaijer*, Natascha van Lent, Timo Sorsa, Tuula Salo, Yrjő, T. Konttinen and Jan Lindeman

* *Gaubius Laboratory TNO-PG, P.O. Box 2215, 2301 CE Leiden, The Netherlands*

Summary

Remodelling of the extracellular matrix and destruction of connective tissue are regarded as characteristic features of degenerative and invasive processes such as rheumatoid arthritis, periodontitis, wound healing, tumour growth and metastatic invasion. Matrix remodelling is a complex process in which matrix metalloproteinases (MMPs) play a central role. MMP activity arises from a multi-step process, which is tightly regulated. Overexpression and activation of MMPs are correlated with a number of pathologies.

Tetracyclines and tetracycline derivatives are able to inhibit MMP activity, independent of their anti-microbial action. The various MMPs have a different sensitivity for inhibition by tetracyclines. it is the strongest for MMP-8 and MMP-13, and weak or absent for MMP-1 and MMP-3. Inhibition of *in vitro* activity is strongly dependent on the assay conditions used (pH, kind of sub strate, source of enzyme). Besides activity, the activation step of MMPs and MMP synthesis are also tetracycline-sensitive, varying per MMP. For a number of MMPs (MMP-8, MMP-9 and MMP-13), inhibition occurs at tetracycline levels which are physiologically obtainable upon therapy. In general, MMP inhibition by tetracyclines is observed in cells activated by growth factors or inflammatory mediators, in which MMP expression is induced. No effect of tetracyclines on the synthesis and action of TIMPs, inhibitors of MMPs, was observed.

Use of tetracyclines in animal models and human diseases, in which MMPs play a role, reflects the *in vitro* situation: inhibition of MMP synthesis and reduction or prevention of the pathological condition.

Role of MMPs in pathology

Remodelling of the extracellular matrix is an essential phenomenon in embryogenesis, bone homeostasis, tissue repair and ovulation and gestation [1,

2]. Pathological remodelling of the extracellular matrix is a characteristic feature of tissue-destructive processes such as rheumatoid arthritis and periodontitis, but is also a prerequisite for angiogenesis [3], wound healing [4], tumor growth [5], leukocyte and metastatic invasion [6], necrosis and apoptosis [2]. Matrix remodelling is a complex process involving a variety of proteases, of which matrix metalloproteinases (MMPs) are generally considered the pivot. MMPs comprise a large family of structurally related zinc metallo-endopeptidases, with different substrate specificities, collectively able to degrade most, if not all protein constituents of the extra-cellular matrix (Tab. 1). The first member of the MMP family, a collagenase, was discovered back in 1962 and was identified as the one responsible for resorption of the tadpole tail during metamorphosis and is now referred to as MMP-1 or fibroblast-type collagenase/Collagenase-1. Over the years more than 26 MMPs have been described and probably more are awaiting identification.

MMPs are synthesised and secreted from normal and malignant cells as latent pro-enzymes. MMP activity and overall action represent a tightly regulated process which comprises transcription, pro-enzyme secretion and activation; activity is further controlled by a set of specific inhibitors (tissue inhibitors of metalloproteinases: TIMPs) which are often co-expressed with their respective proteases (Fig. 1).

MMP transcription is controlled by a range of cytokines, pro-inflammatory mediators, growth factors and hormones. These factors may also influence sta-

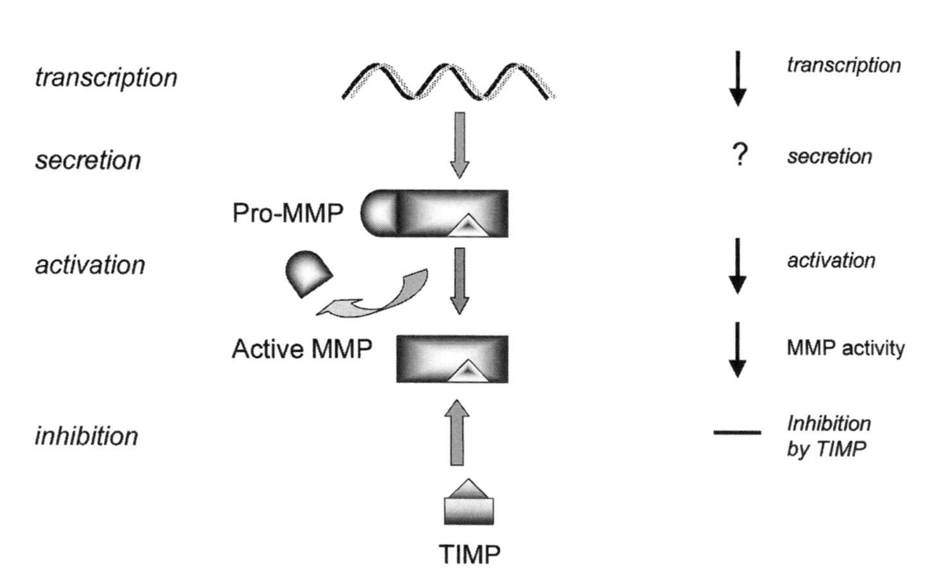

Figure 1. Regulation steps of matrix metalloproteinases and the effect of tetracyclines on MMP regulation. Various regulation steps of MMP-activity formation are given. The effects of tetracyclines are given on the right: ↓ = inhibition; ? = not known; — = no effect.

Table 1. Matrix metalloproteinase family. Names and substrate specificities of the matrix metalloproteinase family

Name	Substrates	Other names	MMP
Collagenases			
Collagenase-1	collagen types I, II, III, VII, VIII and X, gelatin, proteoglycans	interstitial collagenase fibroblast-type collagenase	MMP-1
Collagenase-2	collagen types I, II and III, gelatin, proteoglycans	neutrophil-type collagenase	MMP-8
Collagenase-3	aggrecan, collagen types I, II, III, VII, VIII and X		MMP-13
Gelatinases			
Gelatinase A	collagen types IV, V, VII, X and XII, elastin, gelatin, fibronectin	72 kDa type IV collagenase	MMP-2
Gelatinase B	collagen type IV, V, VII and X, gelatin	92 kDa type IV collagenase	MMP-9
Stromelysins			
Stromelysin-1	collagen types III, IV, IX, and X, gelatin, fibronectin, laminin and proteoglycans	transin, proteoglycanase	MMP-3
Stromelysin-2	collagen types III, IV, X, and XI, gelatin, fibronectin, laminin and proteoglycans	transin-2	MMP-10
Membrane-type MMPs			
MT1-MMP	proMMP-2, proMMP-13, fibrin, vitronectin, collagen types I, II, III, laminin, fibronectin, tenascin		MMP-14
MT2-MMP	gelatin, fibronectin, proMMP-2		MMP-15
MT3-MMP	gelatin, fibronectin, proMMP-2		MMP-16
MT4-MMP			MMP-17
MT5-MMP	proMMP-2		MMP-24
MT6-MMP	proMMP-2	leukolysin	MMP-25
Others			
Matrilysin	elastin, entactin, laminin collagen type IV, fibronectin α1 anti-trypsine	PUMP 1	MMP-7
Stromelysin-3	MMP-11		
Metalloelastase	α1 proteinase inhibitor, elastin, casein	macrophage elastase	MMP-12
Enamelysin	amelogenin	MMP-20	
Not defined	synthetic substrates	MMP-23	
Endometase	gelatin, α1 anti-trypsine	MMP-26	

bility of MMP mRNA. In neutrophils the presynthesised proenzyme is stored in intracellular pools and released and activated upon cellular activation/triggering. Observed directional secretion of MMPs in polar cells [7] (e.g., endothelial cells) implies a controlled secretion mechanism but up till now no data are available.

Following secretion, pro-MMPs are activated by proteolytic splicing of the pro-peptide sequence by membrane-bound MMPs (MT-MMPs) or other proteinases, such as plasmin and cathepsin G [8–10]. Neutrophil-derived MMPs (MMP-8, MMP-9) are efficiently activated *in vitro* by oxygen species, but the physiological role of oxidative activation remains to be established [11]. Proteolytic splicing of the pro-enzyme induces a conformational change that unmasks the active site, followed by autocleavage of the N-terminal pro-peptide. During and upon activation, activity is further regulated by tissue inhibitors of MMP (mainly at tissue level), α_2-macroglobulin (mainly in serum and/or body fluids) and by autodegradation or fragmentation. TIMPs inhibit active forms of MMP by 1:1 non-covalent binding, but TIMPs as well as other MMP-binding molecules, such as neutrophil-gelatinase-associated lipocalin (NGAL), can reduce/retard but not fully abolish MMP-activation by other proteases [12].

Regarding their central role in a large number of pathological processes, inhibition of MMP activity has received considerable interest. A number of specific MMP inhibitors have recently been developed and although pre-clinical studies showed promising results, clinical application is hampered by unacceptable side-effects [13]. Alternatively, tetracycline and its antibiotic derivatives are potent inhibitors of expression as well as activity of a number of MMPs and recently a number of chemically modified tetracyclines (CMT) deprived of their antibiotic action but still inhibiting MMP, have been developed (for a review see [14]). To date, doxycycline with its favourable pharmacokinetics [15] and excellent tolerability is the most extensively studied tetracycline.

Modulation of MMP activity by tetracyclines

Tetracycline antibiotics, which have long been used as adjunct in periodontal therapy based on their antimicrobial effectiveness against a variety of periodontopathogenic micro-organisms, were shown to also inhibit excessive MMP activity (for a review see [16]). Tetracyclines were found to directly inhibit collagenase activity from a wide variety of human and animal sources. However, differences were observed between effectiveness of inhibition against various MMPs: fibroblast collagenase (Collagenase-1; MMP-1) could only be inhibited at high concentrations of doxycycline (IC50 > 200 µM), whereas neutrophil collagenase (Collagenase-2; MMP-8) and the recently discovered Collagenase-3 (MMP-13) were inhibited already at concentrations (IC50 1–10 µM and 5–30 µM, respectively) which are therapeutically attainable upon doxycycline medication [14, 17, 18].

In addition, inhibition studies with pre-activated MMP-13 revealed that MMP-13 activity itself is hardly affected by doxycycline (IC50 > 200 μM), but that *in vitro* activation of latent MMP-13 by APMA is very sensitive for doxycycline [14, 19]. Obviously, besides effects of doxycycline on MMP activity itself as observed for MMP-2, MMP-8 and MMP-9, doxycycline is also able to affect the activation of latent MMPs. Inhibition of *in vitro* activation was not limited to APMA-mediated activation, but was also observed for the activation of MMP-8 by oxygen radicals [20] and activation of MMPs by trypsinogen-2 [21]. No effect was observed on the activation of MMP-9 by MMP-3 [22], which corresponds to the inability of tetracyclines to inhibit MMP-3 activity (unpublished results).

Studies on inhibition of MMP activity by tetracyclines may give contradictory results, because the results greatly depend on the assay conditions of the study:

I) Tetracyclines can affect both the activation step and MMP activity itself (see above) and therefore it is important to be able to discriminate between these processes.

II) Inhibition of MMP activity by tetracyclines is dependent on the substrate used: poor inhibition of MMP-13 activity was observed when peptide-based substrates were used, whereas usage of collagen type II, which is a natural substrate for MMP-13, showed inhibition of activity already at low doxycycline concentrations [23]. On the other hand, no differences were observed for inhibition of MMP-8 using either peptide-based or natural substrates [23], indicating that no general mechanism can be concluded to exist.

III) Inhibition of MMP activity by tetracyclines is dependent on the source of enzyme used: the catalytic domain of recombinant MMP-8 was much less inhibited by doxycycline than the full-length natural MMP-8 [23]. From this study it was concluded that doxycycline not only works via chelation of the Zn^{2+}- and Ca^{2+}-ions, which are needed for enzymatic activity and are present on the catalytic domain, but also has additional interactions with the MMP protein which affect MMP enzyme activity, corresponding with previous data suggesting interaction of doxycycline with Zn^{2+} at a secondary site instead of the catalytic Zn^{2+} [18].

IV) Inhibition of MMP activity by tetracyclines is pH-dependent: At pH > 7.1, inhibition by doxycycline occurs, whereas no inhibition of MMP-8 activity was observed if the assay was performed at pH < 7.1 [24].

In vivo inhibition of MMP activity by tetracyclines is suggested by studies of Golub et al., who showed inhibition of collagen type I degradation in gingival crevicular fluid [25]. However, from these *in vivo* studies it is difficult to conclude whether the observed effects of tetracyclines are due to inhibition of MMP activity or inhibition of MMP synthesis (see below).

Modulation of MMP-synthesis by tetracyclines *in vitro*

In studies on the inhibition of MMP-activity by tetracyclines at cellular conditions, it was observed that treatment with doxycycline not only resulted in inhibition of activity, but also resulted in a decrease in MMP antigen levels. Human endothelial cells treated with 10–50 μg doxycycline showed decreased levels of MMP-9 antigen, as measured by zymography (Fig. 2A). No changes in MMP-2 levels were observed. The reduction in MMP-9 level was not caused by increased degradation of MMP-9 enzyme, but by a decreased MMP-9 expression, as demonstrated by reduction of MMP-9 mRNA upon doxycycline treatment, as determined by Northern blotting using an MMP-9 specific probe (Fig. 2B). Inhibition of MMP-9 expression by doxycycline has been observed in many cell types, such as macrophages (Fig. 3), THP-1 monocytes [26] and to a lesser extent in stimulated chondrocytes [27] and human corneal epithelial cells [22].

Collagenases -1, -2 and -3 (MMP-1, MMP-8 and MMP-13) are interstitial collagenases that are able to digest triple-helical collagens. All three enzymes

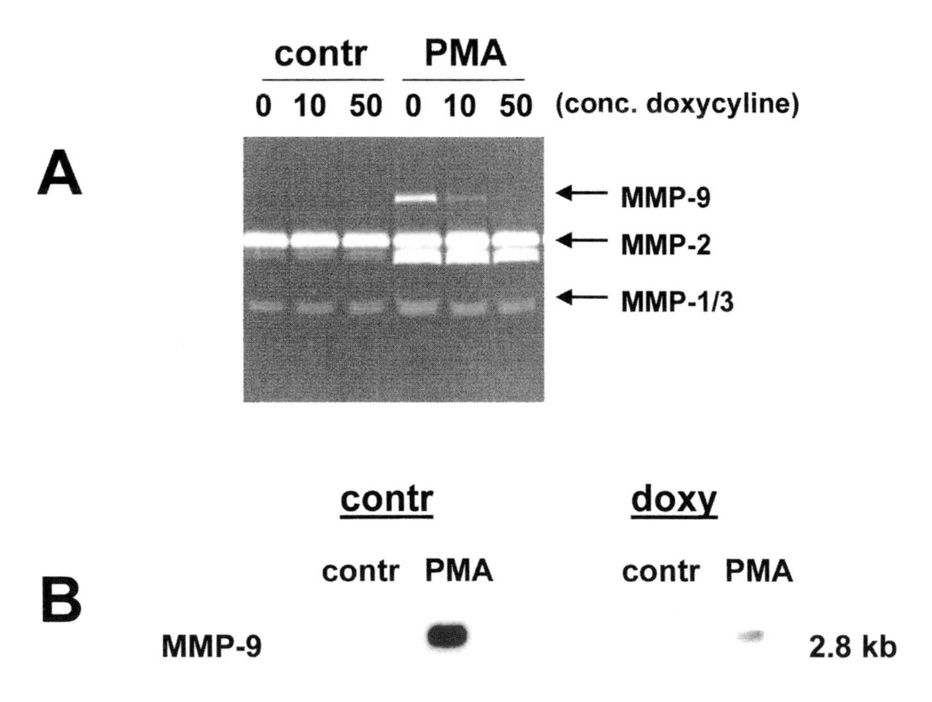

Figure 2. Effect of doxycycline on MMP expression in PMA- and non-stimulated human umbilical vein endothelial cells. (A) Human umbilical vein endothelial cells were stimulated for 24 h with (PMA) or without (contr) 10 nM PMA, in the presence or absence of 10 or 50 μM doxycycline. Aliquots of conditioned medium (10 μl) were analysed by gelatin zymography. The positions of MMP-9, MMP-2 (active [lower band] and pro-form [higher band]) and MMP-1/3 are indicated by arrows. (B) Human umbilical vein endothelial cells were stimulated as described under A). Total RNA was isolated and analysed by Northern blotting, using an MMP-9 specific probe.

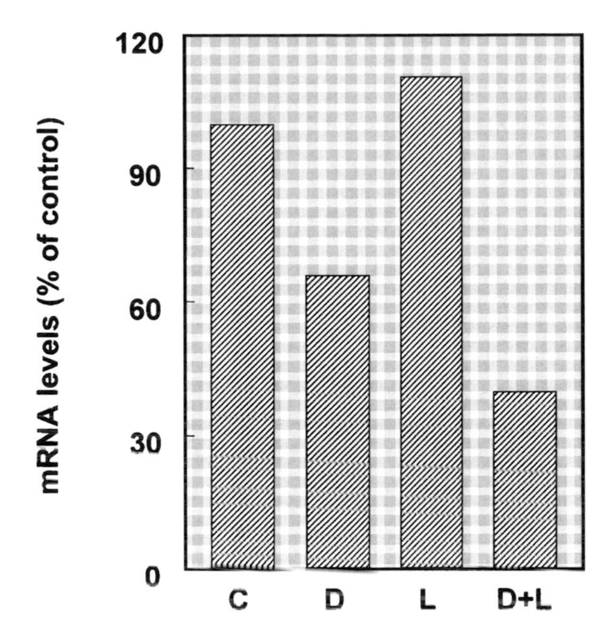

Figure 3. Effect of doxycycline on MMP expression in LPS- and non-stimulated human macrophages. Monocyte-derived macrophages were isolated as described [71]. Cells were stimulated for 24 h with or without 0.1 µg/ml LPS (L) in the presence or absence of 50 µM doxycycline (D). (D+L) represents cells incubated in the presence of both LPS and doxycycline. Total RNA was isolated after 24 h and analysed by Northern blotting, using an MMP-9 specific probe. The blot was scanned, and data are given as percentage of the control.

are associated with bone and cartilage degradation in osteoarthritis and rheumatoid arthritis. Doxycycline at low concentrations (2–20 µM) was able to inhibit the expression of all three collagenases *in vitro* in osteoarthritic chondrocytes [28]. Similar results were obtained in human synovial fibroblasts and endothelial cells, where a doxycycline-mediated inhibition of expression of MMP-8 was observed, both at protein and at mRNA level [29]. No effect of doxycycline was observed on the expression of MMP-2 and the inhibitors of MMPs, TIMP-1 and TIMP-2 [29, 30]. Chemically modified tetracyclines showed similar effects on MMP expression as doxycycline, except for CMT-5, the pyrazole analogue of tetracycline that lacks the Ca^{2+} and the Zn^{2+} binding site, which had lost its MMP inhibitory activity [16].

The mechanism of inhibition of MMP expression was studied in more detail by Jonat et al., who observed tetracycline-mediated inhibition of MMP-3 expression in IL-1-induced fibroblasts and chondrocytes [27, 31]. Using MMP-3-promoter/reporter constructs it was shown that the observed inhibition was transcriptionally regulated [27]. In general concentrations of tetracycline needed for inhibition of expression were lower (<5 µM) than those needed for inhibition of activity (25–100 µM), indicating that these two ways of inhibition act independently.

MMP-2 is the only MMP known at present that is constitutively expressed. No clear effects of doxycycline on MMP-2 expression are known, and only one publication reports that MMP-2 expression is inhibited in keratinocytes *in vitro* by doxycycline, CMT-1 and CMT-8 [32]. However, no clear toxicity studies on keratinocytes were reported, and also no such results have been repeated in other cell types.

In summary, in various cell types following cell activation by, e.g., growth factors and cytokines, MMPs are induced (MMP-1, MMP-3, MMP-8, MMP-9 and MMP-13). Tetracycline and tetracycline analogues are able to prevent induction of MMPs at transcriptional level. Tetracyclines seemingly do not inhibit MMPs and TIMPs that are constitutively expressed. In addition, the ability of tetracyclines to inhibit expression is independent of their ability to inhibit MMP activity. The former usually occurs at clearly lower tetracycline concentrations than concentrations needed for inhibition of MMP activity.

Modulation of MMP synthesis by tetracyclines *in vivo* in animal models

Inhibition of MMP-9 expression by doxycycline has not been limited to *in vitro* cell culture. In a human tissue culture model of vein graft stenosis, doxycycline treatment significantly reduced neointimal thickness, in conjunction with a reduction in MMP-9 expression [33], suggesting a direct role for MMP-9 in stenosis, and a possible therapeutic effect of doxycycline.

In vivo effects of doxycycline were most intensively studied in abdominal aortic aneurysms. In a rat model it was observed that treatment with doxycycline had inhibitory effects on elastin degradation, and in addition inhibited MMP-9 expression [34]. A very elegant study in mice in which aortic aneurysm was induced by transient elastase perfusion, showed suppression of aneurysm development by doxycycline treatment. This treatment also caused suppression of MMP-9 expression in the mononuclear cell infiltration. A similar reduction in aneurysm formation was observed in MMP-9 KO mice, but not in MMP-12 KO mice, suggesting a critical role for MMP-9 in the experimental model of aortic aneurysm disease and a potential therapeutic effect of doxycycline [35]. Besides a reduced MMP-9 expression, tetracycline-mediated inhibition of MMP-8 expression in an animal model has also been described. In a rat wound healing model it was observed that treatment with CMT-8, a modified tetracycline, resulted in a decreased expression of both MMP-8, MMP-9 and MMP-13 [36].

Inhibition of expression by tetracyclines is not limited to MMPs only: *in vitro* at low concentrations (0.1–1 μM), doxycycline inhibited expression of trypsinogen-2, an enzyme able to activate pro-MMPs [21]. This suggests another route of prevention of MMP activity by tetracyclines. In chondrocyte culture, doxycycline also inhibited collagen type II and collagen type X synthesis [37, 38]. iNOS expression, but not iNOS activity was inhibited by doxycycline [39, 40] and CMTs [41]. Very recently minocycline-mediated inhibition of iNOS expression, together with inhibition of caspase-1 and caspase-3, was observed in a mouse model of Huntington disease [42], suggesting a possible therapeutic application for this disease.

Inhibition of MMPs: clinical applications

Matrix metalloproteinases play a central role in 1) migration and invasion of cells, 2) processes related to or dependent on the invasive behaviour of cells such as neointima formation, angiogenesis and tumor metastasis, and 3) degenerative processes, such as degradation of cartilage and bone in rheumatoid arthritis and periodontal diseases. Since many of these processes are dependent on MMP activity, it might be expected that tetracyclines may interfere in these processes (depending on type of MMP involved and whether MMP activity is a prerequisite for the progress of the process). *In vivo* studies with animals underline the possible therapeutic benefit of tetracyclines (see above).

Most studies regarding therapeutic effects of tetracyclines in humans have been carried out in patients with periodontal disease, rheumatoid arthritis or vascular diseases. Some preliminary data are obtained from applications in other diseases.

Adult periodontal disease

Although doxycycline holds many promises as a versatile MMP inhibitor, so far only its use as an adjuvant in treatment of adult periodontal disease has been FDA-approved [43]. Periodontitis is the most common cause of adult tooth loss in Western societies and is characterised by inflammation of the gingivae and destruction of periodontal supporting structures. Although bacterial plaque and infection is essential for initiating periodontitis, persistent neutrophil activation in response to the initial bacterial infection appears primarily responsible for increased neutrophil collagenase (MMP-8) activity in periodontal disease. Increased collagenolysis in response to MMP-8 accumulation leads to destruction of periodontal support structures followed by tooth loss [44].

In patients with periodontitis, bone-degrading activity was reduced in inflammatory exudates (gingival crevicular fluid samples) in low-dose doxycycline-treated subjects (20 mg b.i.d.), whereas no such decrease was observed in non-treated persons. Concomitant with the reduction in bone-degrading activity, a decrease in MMP-8 and MMP-13 levels was observed [25], suggesting that the therapeutic effect of doxycycline was due to inhibition of MMP expression and activity. The clinical relevance of this inhibition has been substantiated in a clinical trial showing prevention of alveolar bone loss and improved attachment levels [43].

Abdominal aortic aneurysms

Doxycycline has been recently proposed as an alternative strategy to preserve the integrity of the aortic wall in abdominal aortic aneurysms (AAA). AAA is a common pathology in the elderly and a major cause of death due to rupture [45]. Despite major advances in medical care, mortality from ruptured AAA is still as high as 80%. AAA is characterised by mononuclear inflammatory cell infiltrate and massive elastolysis [46], which is primarily attributed to excess MMP activity [47]. The pivotal role of MMP-9 in elastase-induced elastolysis has been shown in animal models (see above). Likewise, the efficacy of doxycycline has been confirmed in man: preoperative doxycycline reduced MMP-9 expression (both mRNA and protein) and activation in the aortic wall [26, 48]. No effects on MMP-2 levels were observed, which corresponds with *in vivo* data. Consequently, doxycycline has been proposed as an alternative strategy, stabilising small AAA and thus eliminating the need for invasive treatment, although this hypothesis awaits confirmation in a large clinical trial.

Osteoarthritis and rheumatoid arthritis

Osteoarthritis and rheumatoid arthritis are common and debilitating disorders characterised by loss of collagen and proteoglycans from the cartilage, leading to loss of its physical properties, which results in joint dysfunction. Tetracyclines have been shown to reduce the severity of the arthritis in an animal model of osteoarthritis [49].

Preliminary results in man suggest that doxycycline may have favourable effects in rheumatoid arthritis. It was observed that long-term doxycycline treatment resulted in decreased levels of MMP-8 in serum and saliva samples [50] and a suppression of salivary collagenase activity of both the fibroblast (non-glycosylated) and neutrophil-type (glycosylated) MMP-8 [29, 51]. Doxycycline-mediated reduction of salivary MMP-8 was shown to be reversible; after the cessation of medication, salivary MMP-8 activity returned to levels before medication [52]. Recently, positive effects of minocycline have been reported in a controlled study on minocycline in early rheumatoid arthritis [53].

Neoplasms

A large number of MMPs have been identified in tumor cells and surrounding stromal cells and are associated with (tumor) angiogenesis, tumor growth and migration of tumor cells [5, 54]. Observations in animal models suggested that inhibition of MMP activity reduced tumor growth and metastases [55–57].

Inhibition of MMP activity has thus been proposed as a promising adjuvant cancer treatment. Over the last few years, a number of specific MMP inhibitors have been developed and 3 of them have entered phase III evaluation. Only one has been formally reported so far. This study suggested that marimastat, a synthetic MMP inhibitor, had no survival advantage when compared to chemotherapy in advanced pancreatic carcinoma [58], whereas other trials with marimastat had to be abrogated due to profound musculoskeletal toxicity.

Although doxycycline has been shown to diminish growth rates and to reduce metastases in a number of animal models [59–61], no human studies are available so far.

Cardio-vascular disease

Cardio-vascular disease is the leading cause of morbidity and mortality in Western societies; many forms of cardiac disease as well as many cases of localised cerebral ischemia are secondary to atherosclerosis. Atherosclerosis is a progressive disease, characterised by the formation of a fibrous plaque consisting of smooth muscle cells, collagen, elastin, macrophages and T lymphocytes. Progressive disruption of the plaque promotes thrombosis, embolus formation and vascular spasm, which subsequently cause clinical symptoms. Progression, morphogenesis and stability of the atherosclerotic plaque is highly dependent on extracellular matrix remodelling in which MMPs are considered to be key elements [62–64]. Consequently, modulation of MMP activity holds many promises. Animal studies suggest that inhibition of MMP activity, either by overexpression of TIMP [65, 66] or by MMP inhibitors [67, 68], may alleviate progression of the disease.

Other diseases

In animal models, minocycline was able to delay the disease process in an *in vivo* mouse model of Huntington disease [42] and to prevent lung damage in an ARDS model in the rat [69]. In rats CMT (Metastat) reduced dental caries progression [70].

It has not been shown yet that the observed effects on these disease processes are due to inhibition of MMP activity or expression. However, the data give clue for further study on the effect of tetracycline therapy on a number of diseases in which matrix remodelling, cell migration and cell invasion play a central role.

In conclusion, doxycycline holds many promises as a well-tolerated MMP inhibitor with a favourable pharmacological profile. MMPs are inhibited at various levels: synthesis, activation of the pro-enzyme and MMP activity itself. Expression of MMP-8, MMP-9, and MMP-13 are most sensitive for cell-specific inhibition by tetracyclines. However, possible therapeutic effects have mainly been proven *in vitro* or in animal models, and therefore all promises for therapeutic application await clinical confirmation.

References

1 Woessner JF (1993) Introduction to serial reviews – the extracellular matrix. *FASEB J* 7: 1191–1191
2 Damjanovski S, Ishizuya-Oka A, Shi YB (1999) Spatial and temporal regulation of collagenases-3, -4, and stromelysin-3 implicates distinct functions in apoptosis and tissue remodeling during frog metamorphosis. *Cell Res* 9: 91–105
3 Stetler Stevenson WG (1999) Matrix metalloproteinases in angiogenesis: a moving target for therapeutic intervention. *J Clin Invest* 103: 1237–1241
4 Soo C, Shaw WW, Zhang X, Longaker MT, Howard EW, Ting K (2000) Differential expression of matrix metalloproteinases and their tissue-derived inhibitors in cutaneous wound repair. *Plast Reconstr Surg* 105: 638–647
5 Kahari VM, Saarialhokere U (1999) Matrix metalloproteinases and their inhibitors in tumour growth and invasion. *Ann Med* 31: 34–45
6 Delclaux C, Delacourt C, Dortho MP, Boyer V, Lafuma C, Harf A (1996) Role of gelatinase b and elastase in human polymorphonuclear neutrophil migration across basement membrane. *Amer J Respir Cell Mol Biol* 14: 288–295
7 Unemori EN, Bouhana KS, Werb Z (1990) Vectorial secretion of extracellular matrix proteins, matrix-degrading proteinases, and tissue inhibitor of metalloproteinases by endothelial cells. *J Biol Chem* 265: 445–451
8 Sato H, Takino T, Okada Y, Cao J, Shinagawa A, Yamamoto E, Seiki M (1994) A matrix metalloproteinase expressed on the surface of invasive tumour cells. *Nature* 370: 61–65
9 Heymans S, Luttun A, Nuyens D, Theilmeier G, Creemers E, Moons L, Dyspersin GD, Cleutjens JPM, Shipley M, Angellilo A et al (1999) Inhibition of plasminogen activators or matrix metalloproteinases prevents cardiac rupture but impairs therapeutic angiogenesis and causes cardiac failure. *Nat Med* 5: 1135–1142
10 Sepper R, Konttinen YT, Buo L, Eklund KK, Lauhio A, Sorsa T, Tschesche H, Aasen AO, Sillastu H (1997) Potentiative effects of neutral proteinases in an inflamed lung: relationship of neutrophil procollagenase (proMMP-8) to plasmin, cathepsin g and tryptase in bronchiectasis *in vivo*. *Eur Resp J* 10: 2788–2793
11 Capodici C, Berg RA (1991) Neutrophil Collagenase Activation – the role of oxidants and cathepsin-G. *Agents Actions* 34: 8–10
12 Sorsa T, Salo T, Koivunen E, Tyynela J, Konttinen YT, Bergmann U, Tuuttila A, Niemi E, Teronen O, Heikkila P et al (1997) Activation of type IV procollagenases by human tumor-associated trypsin-2. *J Biol Chem* 272: 21 067–21 074
13 Brown PD (1999) Clinical studies with matrix metalloproteinase inhibitors. *APMIS* 107: 174–180
14 Greenwald RA, Golub LM, Ramamurthy NS, Chowdhury M, Moak SA, Sorsa T (1998) *In vitro* sensitivity of the three mammalian collagenases to tetracycline inhibition: relationship to bone and cartilage degradation. *Bone* 22: 33–38
15 Saivin S, Houin G (1988) Clinical pharmakokinetics of doxycylcine and minocycline. *Clin Pharmacokinet* 15: 355–366
16 Golub LM, Ramamurthy NS, McNamara TF, Greenwald RA, Rifkin BR (1991) Tetracyclines inhibit connective tissue breakdown: new therapeutic implications for an old family of drugs. *Crit Rev Oral Biol Med* 2: 297–322
17 Golub LM, Sorsa T, Lee HM, Ciancio S, Sorbi D, Ramamurthy NS, Gruber B, Salo T, Konttinen YT (1995) Doxycycline inhibits neutrophil (PMN)-type matrix metalloproteinases in human adult periodontitis gingiva. *J Clin Periodontol* 22: 100–109
18 Sorsa T, Ding Y, Salo T, Lauhio A, Teronen O, Ingman T, Ohtani H, Andoh N, Takeha S, Konttinen YT (1994) Effects of tetracyclines on neutrophil, gingival, and salivary collagenases. *Ann N Y Acad Sci* 732: 112–131
19 Lindy O, Konttinen YT, Sorsa T, Ding YL, Santavirta S, Ceponis A, Lopezotin C (1997) Matrix metalloproteinase 13 (collagenase 3) in human rheumatoid synovium. *Arthritis Rheum* 40: 1391–1399
20 Sorsa T, Ramamurthy NS, Vernillo AT, Zhang X, Konttinen YT, Rifkin BR, Golub LM (1998) Functional sites of chemically modified tetracyclines: inhibition of the oxidative activation of human neutrophil and chicken osteoclast pro-matrix metalloproteinases. *J Rheumatol* 25: 975–982
21 Lukkonen A, Sorsa T, Salo T, Tervahartiala T, Koivunen E, Golub L, Simon S, Stenman U H (2000) Down-regulation of trypsinogen-2 expression by chemically modified tetracyclines: asso-

ciation with reduced cancer cell migration. *Int J Cancer* 86: 577–581
22 Sobrin L, Liu Z, Monroy DC, Solomon A, Selzer MG, Lokeshwar BL, Pflugfelder SC (2000) Regulation of MMP-9 activity in human tear fluid and corneal epithelial culture supernatant. *Invest Ophthalmol Visual Sci* 41: 1703–1709
23 Smith GN, Mickler EA, Hasty KA, Brandt KD (1999) Specificity of inhibition of matrix metalloproteinase activity by doxycycline – relationship to structure of the enzyme. *Arthritis Rheum* 42: 1140–1146
24 Smith GN, Brandt KD, Mickler EA, Hasty KA (1997) Inhibition of recombinant human neutrophil collagenase by doxycycline is pH dependent. *J Rheumatol* 24: 1769–1773
25 Golub LM, Lee HM, Greenwald RA, Ryan ME, Sorsa T, Salo T, Giannobile WV (1997) A matrix metalloproteinase inhibitor reduces bone-type collagen degradation fragments and specific collagenases in gingival crevicular fluid during adult periodontitis. *Inflamm Res* 46: 310–319
26 Curci JA, Mao DL, Bohner DG, Allen BT, Rubin BG, Reilly JM, Sicard GA, Thompson RW (2000) Preoperative treatment with doxycycline reduces aortic wall expression and activation of matrix metalloproteinases in patients with abdominal aortic aneurysms. *J Vasc Surg* 31: 325–341
27 Baragi VM, Jonat C, Renkiewicz R, Qui L, Man CF (1998) Effects of tetracycline on cartilage degradation and matrix metalloproteinase activity/expression. *Adv Dent Res* 12: 68–70
28 Shlopov BV, Smith GN, Cole AA, Hasty KA (1999) Differential patterns of response to doxycycline and transforming growth factor beta 1 in the down-regulation of collagenases in osteoarthritic and normal human chondrocytes. *Arthritis Rheum* 42: 719–727
29 Hanemaaijer R, Sorsa T, Konttinen YT, Ding YL, Sutinen M, Visser H, Vanhinsbergh VWM, Helaakoski T, Kainulainen T, Ronka H (1997) Matrix metalloproteinase-8 is expressed in rheumatoid synovial fibroblasts and endothelial cells – regulation by tumor necrosis factor alpha and doxycycline. *J Biol Chem* 272: 31 504–31 509
30 Hanemaaijer R, Visser H, Koolwijk P, Sorsa T, Salo T, Golub LM, van Hinsberg VWM (1998) Inhibition of MMP synthesis by doxycycline and chemically modified tetracyclines (CMTs) in human endothelial cells. *Adv Dent Res* 12: 114–118
31 Jonat C, Chung FZ, Baragi VM (1996) Transcriptional downregulation of stromelysin by tetracycline. *J Cell Biochem* 60: 341–347
32 Uitto VJ, Firth JD, Nip L, Golub LM (1994) Doxycycline and chemically modified tetracyclines inhibit gelatinase A (MMP-2) gene expression in human skin keratinocytes. *Ann N Y Acad Sci* 732: 140–151
33 Porter KE, Thompson MM, Loftus IM, Mcdermott E, Jones L, Crowther M, Bell PR, London NJ (1999) Production and inhibition of the gelatinolytic matrix metalloproteinases in a human model of vein graft stenosis. *Eur J Vasc Endovasc Surg* 17: 404–412
34 Petrinec D, Liao SX, Holmes DR, Reilly JM, Parks WC, Thompson RW (1996) Doxycycline inhibition of aneurysmal degeneration in an elastase-induced rat model of abdominal aortic aneurysm: preservation of aortic elastin associated with suppressed production of 92 kd gelatinase. *J Vasc Surg* 23: 336–346
35 Pyo R, Lee JK, Shipley JM, Curci JA, Mao D, Ziporin SJ, Ennis TL, Shapiro SD, Senior RM, Thompson RW (2000) Targeted gene disruption of matrix metalloproteinase-9 (Gelatinase B) suppresses development of experimental abdominal aortic aneurysms. *J Clin Invest* 105: 1641–1649
36 Ramamurthy NS, McClain SA, Pirila E, Maisi P, Salo T, Kucine A, Sorsa T, Vishram F, Golub LM (1999) Wound healing in aged normal and ovariectomized rats: effects of chemically modified doxycycline (CMT-8) on MMP expression and collagen synthesis. *Ann N Y Acad Sci* 878: 720–723
37 Beekman B, Verzijl N, Deroos JADM, Koopman JL, Tekoppele JM (1997) Doxycycline inhibits collagen synthesis by bovine chondrocytes cultured in alginate. *J Paediat Endocrinol MeTab* 237: 107–110
38 Davies SR, Cole AA, Schmid TM (1996) Doxycycline inhibits type x collagen synthesis in avian hypertrophic chondrocyte cultures. *J Biol Chem* 271: 25 966–25 970
39 Amin AR, Attur MG, Thakker GD, Patel PD, Vyas PR, Patel RN, Patel IR, Abramson SB (1996) A novel mechanism of action of tetracyclines: effects on nitric oxide synthases. *Proc Natl Acad Sci USA* 93: 14 014–14 019
40 Trachtman H, Futterweit S, Greenwald R, Moak S, Singhal P, Franki N, Amin AR (1996) Chemically modified tetracyclines inhibit inducible nitric oxide synthase expression and nitric oxide production in cultured rat mesangial cells. *J Paediat Endocrinol Metab* 229: 243–248
41 Cillari E, Milano S, D'Agostino P, Di Bella G, La Rosa M, Barbera C, Ferlazzo V, Cammarata G,

Grimaudo S, Tolomeo M et al (1998) Modulation of nitric oxide production by tetracyclines and chemically modified tetracyclines. *Adv Dent Res* 12: 126–130

42 Chen M, Ona VO, Li M, Ferrante RJ, Fink KB, Zhu S, Bian J, Guo L, Farrel LA, Hersch SM et al (2000) Minocycline inhibits caspase-1 and caspase-3 expression and delays mortality in a transgenic mouse model of Huntington disease. *Nat Med* 6: 797–801

43 Ashley RA (1999) Clinical trials of a matrix metalloproteinase inhibitor in human periodontal disease. SDD Clinical Research Team. *Ann N Y Acad Sci* 878: 335–346

44 Mancini S, Romanelli R, Laschinger CA, Overall CM, Sodek J, Mcculloch CAG (1999) Assessment of a novel screening test for neutrophil collagenase activity in the diagnosis of periodontal diseases. *J Periodontol* 70: 1292–1302

45 Anonymus (1998) Mortality results for randomised controlled trial of early elective surgery or ultrasonographic surveillance for small abdominal aortic aneurysms. The UK Small Aneurysm Trial Participants. *Lancet* 352: 1649–1655

46 Satta J, Laurila A, Paakko P, Haukipuro K, Sormunen R, Parkkila S, Juvonen T (1998) Chronic inflammation and elastin degradation in abdominal aortic aneurysm disease: an immunohistochemical and electron microscopic study. *Eur J Vasc Endovasc Surg* 15: 313–319

47 Knox JB, Sukhova GK, Whittemore AD, Libby P (1997) Evidence for altered balance between matrix metalloproteinases and their inhibitors in human aortic diseases. *Circulation* 95: 205–212

48 Franklin IJ, Harley SL, Greenhalgh RM, Powell JT (1999) Uptake of tetracycline by aortic aneurysm wall and its effect on inflammation and proteolysis. *Brit J Surg* 86: 771–775

49 de Bri E, Lei W, Svensson O, Chowdhury M, Moak SA, Greenwald RA (1998) Effect of an inhibitor of matrix metalloproteinases on spontaneous osteoarthritis in guinea pigs. *Adv Dent Res* 12: 82–85

50 Lauhio A, Salo T, Ding Y, Konttinen YT, Nordstrom D, Tschesche H, Lahdevirta J, Golub LM, Sorsa T (1994) *In vivo* inhibition of human neutrophil collagenase (MMP-8) activity during long-term combination therapy of doxycycline and non-steroidal anti-inflammatory drugs (NSAID) in acute reactive arthritis. *Clin Exp Immunol* 98: 21–28

51 Nordstrom D, Lindy O, Lauhio A, Sorsa T, Santavirta S, Konttinen YT (1998) Anti-collagenolytic mechanism of action of doxycycline treatment in rheumatoid arthritis. *Rheumatol Int* 17: 175–180

52 Lauhio A, Salo T, Tjaderhane L, Lahdevirta J, Golub LM, Sorsa T (1995) Tetracyclines in treatment of rheumatoid arthritis. *Lancet* 346: 645–646

53 O'Dell JR, Paulsen G, Haire CE, Blakely K, Palmer W, Wees S, Eckhoff PJ, Klassen LW, Churchill M, Doud D et al (1999) Treatment of early seropositive rheumatoid arthritis with minocycline: four-year followup of a double-blind, placebo-controlled trial. *Arthritis Rheum* 42: 1691–1695

54 Mccawley LJ, Matrisian LM (2000) Matrix metalloproteinases: multifunctional contributors to tumor progression. *Mol Med Today* 6: 149–156

55 Aparicio T, Kermorgant S, Dessirier V, Lewin MJ, Lehy T (1999) Matrix metalloproteinase inhibition prevents colon cancer peritoneal carcinomatosis development and prolongs survival in rats. *Carcinogenesis* 20: 1445–1451

56 Price A, Shi Q, Morris D, Wilcox ME, Brasher PM, Rewcastle NB, Shalinsky D, Zou H, Appelt K, Johnston RN et al (1999) Marked inhibition of tumor growth in a malignant glioma tumor model by a novel synthetic matrix metalloproteinase inhibitor AG3340. *Clin Cancer Res* 5: 845–854

57 An ZL, Wang XE, Willmott N, Chander SK, Tickle S, Docherty AJP, Mountain A, Millican AT, Morphy R, Porter JR, Epemolu RO, Kubota T, Moossa AR, Hoffman RM (1997) Conversion of highly malignant colon cancer from an aggressive to a controlled disease by oral administration of a metalloproteinase inhibitor. *Clin Exp Metastasis* 15: 184–195

58 Rosemurgy A, Harris J, Langleben A, Casper E, Goode S, Rasmussen H (1999) Marimastat in patients with advanced pancreatic cancer: a dose-finding study. *Amer J Clin Oncol* 22: 247–252

59 Cakir Y, Hahn KA (1999) Direct action by doxycycline against canine osteosarcoma cell proliferation and collagenase (MMP-1) activity *in vitro*. *In Vivo* 13: 327–331

60 Lokeshwar BL (1999) MMP inhibition in prostate cancer. *Ann N Y Acad Sci* 878: 271–289

61 Seftor REB, Seftor EA, Delarco JE, Kleiner DE, Leferson J, Stetler Stevenson WG, McNamara TF, Golub LM, Hendrix MJC (1998) Chemically modified tetracyclines inhibit human melanoma cell invasion and metastasis. *Clin Exp Metastasis* 16: 217–225

62 Sukhova GK, Schonbeck U, Rabkin E, Schoen FJ, Poole AR, Billinghurst RC, Libby P (1999)

Evidence for increased collagenolysis by interstitial collagenases-1 and-3 in vulnerable human atheromatous plaques. *Circulation* 99: 2503–2509
63 Li ZH, Li L, Zielke R, Cheng L, Xiao RP, Crow MT, Stetler Stevenson G, Froehlich J, Lakatta EG (1996) Increased expression of 72-kd type IV collagenase (MMP-2) in human aortic atherosclerotic lesions. *Amer J Pathol* 148: 121–128
64 Galis ZS, Sukhova GK, Lark MW, Libby P (1994) Increased expression of matrix metalloproteinases and matrix degrading activity in vulnerable regions of human atherosclerotic plaques. *J Clin Invest* 94: 2493–2503
65 Lijnen HR, Soloway P, Collen D (1999) Tissue inhibitor of matrix metalloproteinases-1 impairs arterial neointima formation after vascular injury in mice. *Circ Res* 85: 1186–1191
66 Rouis M, Adamy C, Duverger N, Lesnik P, Horellou P, Moreau M, Emmanuel F, Caillaud JM, Laplaud PM, Dachet C et al (1999) Adenovirus-mediated overexpression of tissue inhibitor of metalloproteinase-1 reduces atherosclerotic lesions in apolipoprotein e-deficient mice. *Circulation* 100: 533–540
67 de Smet BJ, de Kleijn D, Hanemaaijer R, Verheijen JH, Robertus L, van Der H, Borst C, Post MJ (2000) Metalloproteinase inhibition reduces constrictive arterial remodeling after balloon angioplasty: A study in the atherosclerotic yucatan micropig. *Circulation* 101: 2962–2967
68 Porter KE, Loftus IM, Peterson M, Bell PR, London NJ, Thompson MM (1998) Marimastat inhibits neointimal thickening in a model of human vein graft stenosis. *Brit J Surg* 85: 1373–1377
69 Carney DE, Lutz CJ, Picone AL, Gatto LA, Ramamurthy NS, Golub LM, Simon SR, Searles B, Paskanik A, Snyder K et al (1999) Matrix metalloproteinase inhibitor prevents acute lung injury after cardiopulmonary bypass. *Circulation* 100: 400–406
70 Tjaderhane L, Sulkala M, Sorsa T, Teronen O, Larmas M, Salo T (1999) The effect of MMP inhibitor metastat on fissure caries progression in rats. *Ann N Y Acad Sci* 878: 686–688
71 Noorman F, Braat EA, Rijken DC (1995) Degradation of tissue-type plasminogen activator by human monocyte-derived macrophages is mediated by the mannose receptor and by the low-density lipoprotein receptor-related protein. *Blood* 86: 3421–3427

Tetracyclines in Biology, Chemistry and Medicine
ed. by M. Nelson, W. Hillen and R.A. Greenwald

Structure/function studies of doxycycline effects on matrix metalloproteinase activity and cartilage degeneration

Gerald N. Smith Jr. and Karen A. Hasty*

* *Department of Orthopaedic Surgery/Campbell Clinic, University of Tennessee Health Science Center, VA Medical Center, 1030 Jefferson Ave., Memphis, TN 38104, USA*

Studies of inhibition of MMP activity *in vivo*

The observation that doxycycline and other tetracyclines reduced the level of matrix metalloproteinase (MMP) activity in periodontal disease, even in germ-free rats [1, 2] prompted the testing of these compounds for treatment of osteoarthritis (OA) [3, 4], rheumatoid arthritis (for extensive review of this topic see [5–9]) and other diseases in which MMP-mediated destruction of connective tissues is prominent, such as abdominal aortic aneurysms [10] and wound healing in diabetes [11]. Doxycycline and chemically modified tetracyclines have been tested in animal models of OA, a disease in which the degradation of joint connective tissues is thought to depend, at least partially, on MMP activity. In our laboratory, we have employed doxycycline to treat canine experimental OA. We have examined the effect of treatment with oral doxycycline on the MMP of tissues from patients with rheumatoid arthritis, see [5–9], and patients with OA at the time of joint replacement surgery [3]. Similar studies have been performed using minocycline for treatment of adjuvant arthritis in rats [12], and for chemically modified tetracycline-7 (cmt-7) in a spontaneous model of OA in Taft-Hartley guinea pigs [13]. The guinea pig model resembles human disease in that the development of OA is weight- and age-dependent. Treatment with doxycycline slowed disease progression in the animal models, while MMP activity was reduced in both animal and human tissues.

Oral administration of doxycycline dramatically slowed the progression of articular cartilage degradation in an accelerated model of canine OA [14]. In this model, the dogs undergo unilateral L4-S1 dorsal root ganglionectomy to block sensory input from the hind limb. Two weeks later, the ipsilateral anterior cruciate ligament is transected, following which the dogs develop rapidly progressive OA in the operated knee. In all of the six untreated dogs, most of the weight-bearing cartilage on the femoral condyle was destroyed, and eburnation of bone was apparent 8 weeks after ligament transection. Over the same time period, among the five dogs that were treated with doxycycline (50 mg, twice daily, approximately 3.5 mg/kg) beginning the day after anterior cruciate ligament transection, two exhibited no damage at the articular surface while the remaining three showed only cartilage pitting or partial thinning [14]. In a similar experiment, administration of doxycycline was delayed until the fourth week after anterior cruciate ligament transection when cartilage destruction was already in progress. Again, dogs receiving doxycycline showed less severe changes than the control dogs [15].

This slowing of disease progression was associated with much lower levels of collagenase and gelatinase activity in the articular cartilage, as measured in guanidine extracts of the articular cartilage. The levels of collagenase and gelatinase activity of both active and total enzyme were reduced significantly. Specifically, active collagenase (collagenolytic activity measured after destruction of TIMP by reduction and alkylation) and total collagenase (collagenase measured after destruction of TIMP and activation of latent enzyme) were, respectively, 2-fold and 5-fold higher in the cartilage from knees of untreated dogs than that from treated dogs. Active gelatinase was 3-fold higher, while total gelatinase was 4-fold higher. Chondroprotection was noted also in experiments where doxycycline administration was delayed until the destruction of cartilage was already underway. In a system of explant culture of chicken embryo cartilages, doxycycline blocked spontaneous resorption [16, 17]. Reduced collagenase and gelatinase activities and decreased proteoglycan loss were seen in cultures while cell death was decreased.

These data indicate substantial changes in the production of MMP that go far beyond inhibition of the active enzyme at the site of cartilage degradation. The methods that were used to measure the collagenase and gelatinase activity involved extraction of the cartilage and extensive dialysis of the tissue extracts over several days. Any soluble doxycycline that was in the tissue when the cartilage was harvested would be expected to be removed by these procedures so that, at the time the MMP activity in the extracts was assayed, only irreversibly bound doxycycline would be present.

Additional evidence that differences in the amount of enzyme were measured, not simply acute inhibition of active enzyme, came from an experiment performed to determine the appropriate dose of doxycycline for a clinical study of OA treatment [3]. Patients that were scheduled to undergo hip replacement surgery because of OA were given 100 mg/kg doxycycline twice daily for 5 days prior to surgery. At surgery, articular cartilage from the hip was isolated, extracted and the supernatants assayed for collagenase activity. In extracts of cartilage from the head of the femur there was a significant reduction in the level of total collagenase ($p = 0.02$) and gelatinase ($p = 0.04$) in the samples from the treated patients compared to the levels in extracts of cartilage from OA patients who were given placebo capsules [3]. Again, only irreversibly bound doxycycline would still be present after the extraction of the enzyme and destruction of TIMP.

Although enzyme levels tend to be higher in tissues with severe lesions, it is unlikely that the differences in enzyme level reflected differences in the extent of cartilage pathology in the randomly assigned patients. Similar results were obtained in a study of synovial collagenase in patients with bilateral rheumatoid arthritis who underwent unilateral joint replacement, were treated with minocycline for 10 days, then had the contralateral joint replaced [2]. Tissue extracts and culture media from minced synovial tissue were analyzed for collagenase activity, and the data from the untreated samples were compared with those from the treated samples from the same patient. Again, significant reduction of collagenase activity was noted, although only irreversibly bound minocycline would be present at assay.

In vivo experiments in other systems also indicate significant reduction for MMP protein present rather than simply inhibition of enzyme activities. For example, cartilage from rachitic rats treated with tetracyclines had significantly reduced MMP-3 activity, even though MMP-3 enzymatic activity could not be inhibited *in vitro* [18]. In the elastase-induced model of aortic aneurysm, both 92 and 72 kDa gelatinase (MMP-9 and –2) proteins were reduced significantly when doxycycline was given prior to the induction of aortic aneurysm [19]. These *in vivo* studies suggest that doxycycline and other cmts affect the levels of MMP that exist in the tissue through a mechanism that affects the synthesis and/or stability of MMP.

Studies of inhibition of MMP activity *in vitro*

Inhibition of MMP activity *in vitro* by tetracycline and chemically modified tetracyclines (cmt) was first demonstrated in diabetic rat tissues [1], and the observation was quickly extended to a variety of MMPs [20–23]. Initially, it was proposed that the ability of tetracyclines to chelate calcium and zinc ions was the mechanism of action. Work from a number of laboratories confirmed the ability of various tetracyclines to inhibit collagenase and gelatinase and that high calcium and zinc concentrations blocked the inhibitory activity [1–3, 24]. In addition, cmts which have no antibiotic activity, but which retain metal binding capacity, are effective inhibitors of MMPs. In contrast, cmt-5, in which the oxygen atoms involved with metal binding have been replaced with nitrogen atoms to produce a cmt that does not bind divalent cations, is totally inactive as an inhibitor of MMP activity. Occasional studies have reported that excess calcium and zinc do not prevent inhibition of enzyme activity [2, 25]. Our experience has been that the calcium requirement varies with the particular MMP tested, but that inhibition can be prevented with sufficiently high concentrations of calcium and/or zinc.

Studies of different enzyme preparations and studies utilizing different cmts indicate differences in the sensitivity of specific MMPs to inhibition (for review see [24, 26]). Inhibition of the collagenases MMP-1, MMP-8 and MMP-13, the gelatinases MMP-2 and MMP-9, and stromelysin, or MMP-3, have been characterized using a variety of substrates, enzyme preparations and specific cmts [25, 27–30]. In general, there is agreement on the relative sensitivity of most MMPs, although the particular preparation of enzyme employed, the substrate and the assay conditions generate significant differences in the IC50s and inhibition constants reported.

Two mechanisms have been suggested to explain both inhibition and the differential sensitivity of various MMPs to inhibition by tetracycline and its derivatives. Although there is general agreement that calcium and/or zinc binding by tetracyclines mediates MMP inhibition, it is not clear whether the drug acts on the catalytic zinc or on structural calcium and/or zinc. It has been suggested that the inhibitory action of the tetracyclines is due to binding of the catalytic zinc ion at the active site of the enzyme. If this is the mechanism of inhibition, then the shape of the substrate pocket of the enzyme could be the primary determinant of susceptibility to inhibition. Evidence marshalled in support of this hypothesis includes the enhanced susceptibility of those enzymes with relatively high catalytic rates and broader substrate profiles, such as MMP-8 and MMP-13 [31]. It is argued that the accessibility of the catalytic zinc in these enzymes to the inhibitor would be enhanced, because these MMPs have very large substrate pockets. The ability of zinc to reverse inhibition, and the fact that the active concentrations of inhibitor are much lower than buffer calcium concentrations, also support a role for zinc binding in enzyme inhibition.

The second proposed mechanism is alteration of enzyme structure by binding enzyme-associated calcium and/or zinc. Our studies with the specific collagenases (MMP-1, MMP-8, and MMP-13) and with structurally modified recombinant forms of these enzymes support this mechanism [4]. The collagenolytic activity of MMP-8 and MMP-13 is inhibited by doxycycline at concentrations which can be obtained in the tissues by oral dosing, while that of MMP-1 is not inhibited, even at very high concentrations of the drug. The inhibition is non-competitive and evidence from recent studies of recombinant human MMP-8 and MMP-13 suggests that it is due to the modification of enzyme conformation, presumably due to the binding of enzyme-associated calcium ions. For MMP-8, an effect on enzyme conformation was suggested by studies of the effect of doxycycline on the activation of latent enzyme [32]. The initial observation was that proenzyme activated in the presence of doxycycline had much lower activity than a similar amount of enzyme which was first activated, then mixed with doxycycline. When these mixtures were examined by Western blotting with antisera raised against MMP-8, the reason for the difference became apparent. As expected, upon activation with either trypsin or aminophenyl mercuric acetate, most of the reactive protein was reduced from the latent proenzyme form with a molecular weight of ~55 kDA to an active form of ~46 kDa. In contrast, in the presence of doxycycline and either activator, the enzyme was reduced to smaller fragments. At high concentrations of trypsin in the presence of 30 μM doxycycline, the entire enzyme preparation was reduced to fragments of 30 kDA or less; no pro-MMP-8 latent enzyme or 46 kDa active enzyme survived activation. We interpreted these results as an indication that the configuration of MMP-8 was altered in the presence of doxycycline, resulting in increased susceptibility of the MMP to proteolysis [28, 32].

Further evidence that the effects of doxycycline on MMP *in vitro* are partially due to calcium-induced conformation changes was obtained from kinetic studies of specific and structurally modified MMPs [4]. In our initial studies, we employed a small substrate, which is cleaved at a similar rate by both activated full-length enzyme (the catalytic domain and the hemopexin-like domain), and a truncated, activated form (the isolated catalytic domain). In these studies, we found that the activity of each enzyme preparation was inhibited reversibly in a noncompetitive fashion. We found that MMP-1 and MMP-13 were highly resistant to doxycycline inhibition, while the activity of MMP-8 was strongly inhibited. In the presence of 90 μM doxycycline, the activity of full-length MMP-1 and MMP-13 for the small substrate was reduced by 14% and 10%, respectively, while the activity of truncated MMP-13, which lacked the hemopexin-like domain, was increased by 8%. In contrast, the activity of full-length MMP-8 was reduced by 50% in the presence of doxycycline, while that of truncated MMP-8 required slightly higher concentrations of doxycycline to achieve the same level of inhibition. Since cleavage of the small substrate does not require an intact hemopexin-like domain, we interpreted these data as a demonstration that MMP-1 and MMP-13 were not susceptible to doxycycline inhibition, while MMP-8 was strongly inhibited.

However, when we sought to confirm these observations in an assay against native, type II collagen fibers – the natural substrate for collagenolytic MMPs in the OA joint – the results were both unexpected and revealing. Because truncated MMPs do not cleave the triple helix of the native collagen molecule at a significant rate, only full-length enzyme was tested in this system. When we tested the full-length enzymes against type II collagen, the results with MMP-1 and MMP-8 were similar to those using the small substrate, but the collagenolytic activity of MMP-13 was highly susceptible to inhibition. In the presence of five μM doxycycline, cleavage of type II collagen by MMP-13 was reduced by 55%. We interpreted these data as a demonstration that doxycycline affects the function of the hemopexin-like domain in MMP-13 at concentrations that fail to affect the function of the catalytic domain.

For MMP-8, the activity against the small substrate and the activity against type II collagen are inhibited to a similar extent. Since activity against the small substrate requires only the catalytic domain, while activity against native collagen requires both the catalytic domain and the hemopexin-like domain, doxycycline appears to affect the catalytic region of MMP-8. No effect of doxycycline on the function of the hemopexin-like domain of this enzyme is suggested by these data. In contrast, MMP-1 was not inhibited in either the small substrate assay or the native collagen substrate assay, suggesting that neither the catalytic domain nor the hemopexin-like domain of MMP-1was affected by doxycycline. Similarly, MMP-3 has been reported to be highly resistant to inhibition by a variety of tetracyclines.

Studies of inhibition of MMP synthesis

Regardless of the ultimate consensus on MMP inhibition *in vitro*, it is apparent that direct inhibition of enzyme activity is an unlikely mechanism to explain the results observed *in vivo*. The concentration of antibiotic necessary for inhibition of most MMPs *in vitro* was significantly higher than the concentrations normally achieved in tissues. Even at concentrations an order of magnitude greater than those that could reasonably be achieved by oral dosing, the activity of some MMPs was not inhibited. Both fibroblast collagenase (MMP-1) and stromelysin (MMP-3), which are highly expressed in a number of pathological conditions, were identified as inhibition-resistant enzymes. Nevertheless, MMP-3 production by rachitic rat growth plate chondrocytes was significantly reduced by minocycline [18].

Amin et al reported that doxycycline and cmt-3 augmented inducible nitric oxide synthetase (iNOS) mRNA degradation in LPS-stimulated murine macrophages [33]. Reduced iNOS resulted in decreased nitric oxide (NO) production by the macrophages and suggests a mechanism for regulation of the inflammatory process. The reduction in NO was significantly less when treated after 6 h following stimulation and absent after 12 h, indicating the interference of an early event [34, 35].

Since many different factors trigger activation of macrophages and NO production, doxycycline might be expected to affect inflammation mediated by a variety of pathways. Indeed, tetracyclines have been shown to affect bone metabolism via its effects on osteoclast function (for review see [36]). In models of brain ischemia and reperfusion, minocyline inhibits microglial activation and induction of interleukin 1 beta-converting enzyme and it reduces cyclooxygenase-2 expression and production of PGE_2 [37–40]. Decreased infiltration of inflammatory cells in lung tissues was seen with minocycline treatment in mice injected intraperitoneally with lipopolysaccharide [41].

These studies suggest that MMP production might be decreased indirectly through inhibiting macrophage activation and cytokine release. However, doxycycline also reduces metalloproteinase activity by interfering in its induction by inflammatory cytokines. Jonat et al previously reported that tetracycline reduced interleukin-1 beta transcription and translation of stromelysin-1 [42]. The synthesis of MMP-8 by rheumatoid synovial fibroblasts and endothelial cells that is stimulated by tumor necrosis factor alpha or phorbol myristate acetate has been shown to be decreased at both protein and mRNA levels [43]. Understanding the intracellular pathways that might be relevant to metalloproteinase synthesis is complicated by the numerous intracellular events that occur in response to inflammatory cytokine stimulus. Increased synthesis of MMP-1 in response to interleukin-1 beta involves phosphorylation triggered by binding of the interleukin receptor to its ligand and initiation of downstream pathways [44, 45]. The response seen for mMMP-1 mRNA occurs relatively late, appearing at 8 h, and is due to increased transcription and stabilization of the mRNA [46]. Interference by doxycycline could have an impact at one or more of these steps.

Table 1. Cytokine and cytokine receptor mRNA production by the osteoarthritic chondrocytes in response to treatment with doxycycline

Cytokine	Treated[1]/control × 100	Cytokine receptor	Treated[1]/control × 100
IL-6	60%	IL-1 RI	200% *
IL-1α	50%*	IL-1 RII	190% *
IL-1β	80%	TNF RI	100%
INFγ	70%	TNF RII	100%
TGFβ1	100%	IL-6R	100%
TGFβ3	220*	TGFβ RI	200% *
IL-1Ra	140%*	TGFβ RII	170% *

[1] mRNA was measured by RNase protection assay.
Results presented as a percent of that seen in cultured, untreated osteoarthritic chondrocytes (n = 9). $P \leq 0.05$ for values labeled with an asterisk (*).

Recently, we have found that treatment with doxycycline not only effectively inhibits the level of MMP-1 and MMP-13 mRNA and protein production in chondrocytes isolated from around the OA lesion [47], but also regulates autocrine production of cytokines and cytokine receptors (Tab. 1.) [48]. OA chondrocytes from around the lesion constitutively produce mRNA for both proinflammatory (IL-1α, IL-1β, IL-6, TNFα) and antiinflammatory cytokines (IL-1Ra, TGFβ1 and TGFβ3). The balance between these two groups of cytokines is very important and may determine the extent of cartilage destruction. The collective activity of doxycycline may be to decrease proinflammatory cytokines in conjunction with an upregulation of the protective cytokine TGFβ3. We have shown that TGFβ1 decreases MMP-1 and MMP-13 enzyme production in OA lesional chondrocytes and presumably TGFβ3 would act similarly [49]. In our latest experiments, doxycycline also increased the level of mRNA for TGFβ RI and TGFβ RII receptors. These data also explain and support previous observations, demonstrating that tetracycline analogs increase collagen synthesis, osteoblast activity and bone formation [50–52]. The fact that doxycycline is capable of modulating cytokine levels provides evidence that it may act indirectly to modulate MMPs and may partially explain its inhibitory properties on synthesis. Thus, doxycycline regulation of collagenase production through modulation of cytokine and cytokine receptors would also serve to protect articular cartilage from proteolytic degradation.

Recently, we have obtained an immortalized human chondrocyte line from Dr. Mary Goldring. These cells were isolated from an articular chondrocyte and immortalized using the SV40 T-antigen which induces propagation at 31 °C but is inactivated at 37 °C. We have found that MMP-1 mRNA and protein are induced upon stimulation with TNFα and have initiated studies into the inhibition of MMP synthesis by doxycycline using these cells as a prototypic model (Fig. 1). As we have seen with OA chondrocytes, doxycycline effectively inhibits the amount of collagenase synthesized by immortalized

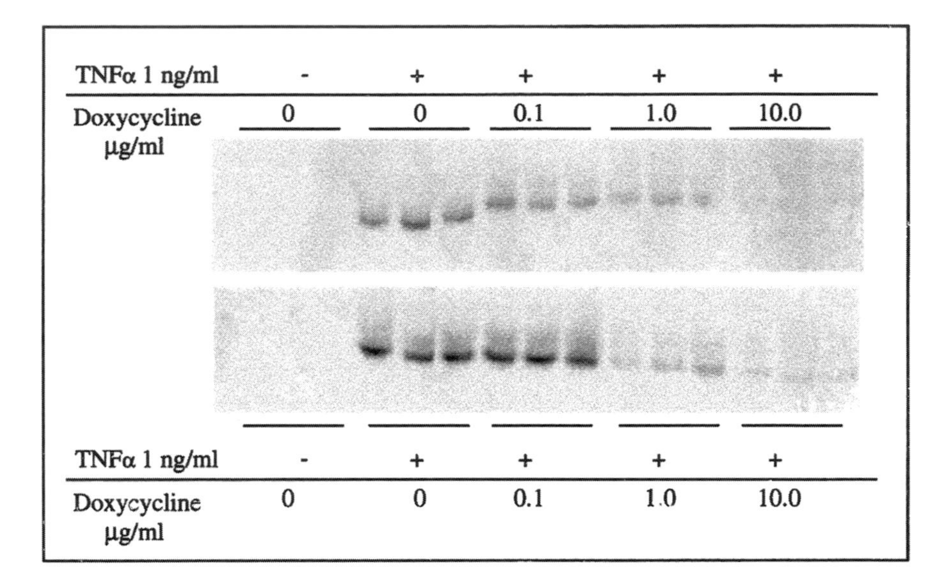

Figure 1. Western blot analysis of the effect of treatment with 0.1, 1.0 and 10 mg/ml of doxycycline on MMP-1 enzyme production by immortalized human chondrocytes stimulated with 1 or 10 ng/ml of TNFa. Triplicate samples are shown for each of the conditions used.

human chondrocytes stimulated with TNFα in a dose-dependent manner. These cells are extremely sensitive to this inhibition, showing decreased MMP-1 production at doxycycline concentrations of 0.1 μg/ml. This cell line will be an ideal tool to study the mechanism of doxycycline inhibition of collagenase synthesis, as large numbers of cells can be prepared routinely.

Our studies of the kinetics of inhibition of different metalloenzymes by doxycycline prompted additional studies of metalloproteinase synthesis. The relative ineffectiveness of doxycycline on MMP-1 activity *in vitro* conflicted with our data showing that oral doxycycline reduced MMP-1 levels in tissues taken from patients at joint replacement. Recent findings by other investigators and us suggest that MMP proteins are regulated at both transcriptional and translational levels by the tetracycline family as well as by direct inhibition of enzyme activity. We have identified at least four potential mechanisms by which doxycycline could slow matrix degradation. First, direct inhibition of MMP-13 or of an unidentified MMP remains a viable mechanism. However, to have the effect on total MMP protein observed in human and animal studies, this mechanism would require that a cascade of events initiated by an MMP is essential to drive matrix degradation. Alternatively, net reduction of total enzyme protein could be achieved through enhanced degradation of MMPs, as seen in our studies of MMP-8 *in vitro* or through reduced transcription of MMP message, as seen in the studies on MMP expression from this and other laboratories [43] or through inhibition of translation of enzyme message as reported for type X collagen [53].

Acknowledgements
This material is based upon work supported by the Office of Research and Development (R&D), Medical Research Service), Department of Veterans Affairs (KAH) and a Biomedical Grant from the Arthritis Foundation (KAH) and in part by NIH grant AR 20582 to the Indiana University Multipurpose Arthritis and Musculoskeletal Diseases Center (GNS) and by grants from Procter & Gamble and the Showalter Trust (GNS).

References

1 Golub LM, Lee HM, Lehrer G, Nemiroff A, McNamara TF, Kaplan R, Ramamurthy NS (1983) Minocycline reduces gingival collagenolytic activity during diabetes. Preliminary observations and a proposed new mechanism of action. *J Periodont Res* 18: 516–526
2 Greenwald RA, Golub LM, Lavietes B, Ramamurthy NS, Gruber B, Laskin RS, McNamara TF (1987) Tetracyclines inhibit human synovial collagenase *in vivo* and *in vitro*. *J Rheumatol* 14: 28–32
3 Smith GN Jr, Yu LP Jr, Brandt KD, Capello WN (1998) Oral administration of doxycycline reduces collagenase and gelatinase activities in extracts of human osteoarthritic cartilage. *J Rheumatol* 25: 532–535
4 Smith GN Jr, Mickler EA, Hasty KA, Brandt KD (1999) Specificity of inhibition of matrix metalloproteinase activity by doxycycline: Relationship to structure of the enzyme. *Arthritis Rheum* 42: 1140–1146
5 Alarcon GS (2000) Tetracyclines for the treatment of rheumatoid arthritis. Expert Opin Investig Drugs 9: 1491–1498
6 Langevitz P, Livenah A, Bank I, Pras M (2000) Benefits and risks of minocycline in rheumatoid arthritis. *Drug Safety* 22: 405–414
7 Alarcon GS, Bartolucci AA (2000) Radiographic assessment of disease progression in rheumatoid arthritis patients treated with methotrexate or minocycline. *J Rheumatol* 27: 530–534
8 Eichenfield AH (1999) Minocycline and autoimmunity. *Curr Opin Pediatr* 11: 447–456
9 O'Dell JR, Paulsen G, Haire CE, Blakely K, Palmer W, Wees S, Eckhoff PJ, Klassen LW, Churchill M, Doud D et al (1999) Treatment of early seropositive rheumatoid arthritis with minocycline: four-year followup of a double-blind, placebo-controlled trial. *Arthritis Rheum* 42: 1691–1695
10 Curci JA, Mao D, Bohner DG, Allen BT, Rubin BG, Reilly JM, Sicard GA, Thompson RW (2000) Preoperative treatment with doxycycline reduces aortic wall expression and activation of matrix metalloproteinases in patients with abdominal aortic aneurysms. *J Vasc Surg* 31: 325–342
11 Ramamurthy NS, Kucine AJ, McClain SA, McNamara TF, Golub LM (1998) Topically applied CMT-2 enhances wound healing in streptozotocin diabetic rat skin. *Adv Dent Res* 12: 144–148
12 Zernicke RF, Wohl GR, Greenwald RA, Moak SA, Leng W, Golub LM (1997) Administration of systemic matrix metalloproteinase inhibitors maintains bone mechanical integrity in adjuvant arthritis. *J Rheumatol* 24: 1324–1331
13 de Bri E, Lei W, Svensson O, Chowdhury M, Moak SA, Greenwald RA (1998) Effect of an inhibitor of matrix metalloproteinases on spontaneous osteoarthritis in guinea pigs. *Adv Dent Res* 12: 82–85
14 Yu LP Jr, Smith GN Jr, Brandt KD, Myers SL, O'Connor BL, Brandt DA (1992) Reduction of the severity of canine osteoarthritis by prophylactic treatment with oral doxycycline. *Arthritis Rheum* 35: 1150–1159
15 Yu LP Jr, Smith GN Jr, Brandt KD, O'Connor BL, Myers SL (1993) Therapeutic administration of doxycycline slows the progression of cartilage destruction in canine osteoarthritis. *Trans Orthop Res Soc* 18: 724
16 Cole AA, Chubinskaya S, Luchene LJ, Chlebek K, Orth MW, Greenwald RA, Kuettner KE, Schmid TM (1994) Doxycycline disrupts chondrocyte differentiation and inhibits cartilage matrix degradation. *Arthritis Rheum* 37: 1727–1734
17 Cole AA, Chubinskaya S, Chlebek K, Orth MW, Luchene LL, Schmid TM (1994) Doxycycline inhibition of cartilage matrix degradation. *Ann N Y Acad Sci* 732: 414–415
18 Arsenis C, Moak SA, Greenwald RA (1992) Tetracyclines (TETs) inhibit the synthesis and/or activity of cartilage proteinases *in vivo* and *in vitro*. *Matrix Suppl* 11: 314

19 Curci JA, Petrinec D, Liao S, Golub LM, Thompson RW (1998) Pharmacologic suppression of experimental abdominal aortic aneurysms: a comparison of doxycycline and four chemically modified tetracyclines. *J Vasc Surg* 28: 1082–1093
20 Ryan ME, Ramamurthy NS, Sorsa T, Golub LM (1999) MMP-mediated events in diabetes. *Ann N Y Acad Sci* 878: 311–334
21 Rifkin BR, Vernillo AT, Golub LM (1993) Blocking periodontal disease progression by inhibiting tissue-destructive enzymes: a potential therapeutic role for tetracyclines and their chemically-modified analogs. *J Periodontol* 64 (8 Suppl): 819–827
22 Sasaki T, Kaneko H, Ramamurthy NS, Golub LM (1991) Tetracycline administration restores osteoblast structure and function during experimental diabetes. *Anat Rec* 231: 25–34
23 Golub LM, McNamara TF, D'Angelo G, Greenwald RA, Ramamurthy NS (1987) A non-antibacterial chemically-modified tetracycline inhibits mammalian collagenase activity. *J Dent Res* 66: 1310–1314
24 Golub LM, Lee HM, Ryan ME, Giannobile WV, Payne J, Sorsa T (1998) Tetracyclines inhibit connective tissue breakdown by multiple non-antimicrobial mechanisms. *Adv Dent Res* 12: 12–26
25 Burns FR, Stack MS, Gray RD, Paterson CA (1989) Inhibition of purified collagenase from alkali-burned rabbit corneas. *Invest Ophthalmol Visual Sci* 30: 1569–1575
26 Ryan ME, Greenwald RA, Golub LM (1996) Potential of tetracyclines to modify cartilage breakdown in osteoarthritis. *Curr Opin Rheumatol* 8: 238–247
27 Suomalainen K, Sorsa T, Ingman T, Lindy O, Golub LM (1992) Tetracycline inhibition identifies the cellular origin of interstitial collagenases in human periodontal diseases *in vivo*. *Oral Microbiol Immunol* 7: 121–123
28 Smith GN Jr, Brandt KD, Hasty KA (1996) Activation of recombinant human neutrophil procollagenase in the presence of doxycycline results in fragmentation of the enzyme and loss of enzyme activity. *Arthritis Rheum* 39: 235–244
29 Yu LP Jr, Smith GN Jr, Hasty KA, Brandt KD (1991) Doxycycline inhibits type XI collagenolytic activity in human osteoarthritic cartilage and of gelatinase. *J Rheumatol* 18: 1450–1452
30 Nip LH, Uitto V-J, Golub LM (1993) Inhibition of epithelial cell matrix metalloproteinases by tetracyclines. *J Periodont Res* 28: 379–385
31 Greenwald RA, Golub LM, Ramamurthy NS, Chowdhury M, Moak SA, Sorsa T (1998) *In vitro* sensitivity of the three mammalian collagenases to tetracycline inhibition: relationship to bone and cartilage degradation. *Bone* 22: 33–38
32 Smith GN Jr, Brandt KD, Hasty KA (1994) Procollagenase is reduced to inactive fragments upon activation in the presence of doxycycline. *Ann N Y Acad Sci* 732: 436–438
33 Amin AR, Attur MG, Thakker GD, Patel PD, Vyas PR, Patel RN, Patel IR, Abramson SB (1996) A novel mechanism of action of tetracyclines: effects on nitric oxide synthases. *Proc Natl Acad Sci USA* 93: 14 014–14 0149
34 Cillari E, Milano S, D'Agostino P, Di Bella G, La Rosa M, Barbera C, Ferlazzo V, Cammarata G, Grimaudo S, Tolomeo M et al (1998) Modulation of nitric oxide production by tetracyclines and chemically modified tetracyclines. *Adv Dent Res* 12: 126–130
35 D'Agostino P, Arcoleo F, Barbera C, Di Bella G, La Rosa M, Misiano G, Milano S, Brai M, Cammarata G et al (1998) Tetracycline inhibits the nitric oxide synthase activity induced by endotoxin in cultured murine macrophages. *Eur J Pharmacol* 346: 283–290
36 Vernillo AT, Rifkin BR (1998) Effects of tetracyclines on bone metabolism. *Adv Dent Res* 12: 56–62
37 Yrjanheikki J, Tikka T, Keinanen R, Goldsteins G, Chan PH, Koistinaho J (1999) A tetracycline derivative, minocycline, reduces inflammation and protects against focal cerebral ischemia with a wide therapeutic window. *Proc Natl Acad Sci USA* 96: 13 496–13 500
38 Yrjanheikki J, Keinanen R, Pellikka M, Hokfelt T, Koistinaho J (1998) Tetracyclines inhibit microglical activation and are neuroprotective in global brain ischemia. *Proc Natl Acad Sci USA* 95: 15 769–15 774
39 Weingart JD, Sipos EP, Brem H (1995) The role of minocycline in the treatment of intracranial 9L glioma. *J Neurosurg* 82: 635–640
40 Clark WM (1997) Cytokines and reperfusion injury. *Neurology* 49 (Suppl 4): S10–4
41 Yamaki K, Yoshida N, Kimura T, Ohbayashi H, Takagi K (1998) Effects of cytokines and minocycline on subacute lung injuries induced by repeated injection of lipopolysaccharide. *Kansenshogaku Zasshi* 72: 75–82
42 Jonat C, Chung FZ, Baragi VM (1996) Transcriptional downregulation of stromelysin by tetracy-

cline. *J Cell Bioch* 60: 341–347

43 Hanemaaijer R, Visser H, Koolwijk P, Sorsa T, Salo T, Golub LM, van Hinsbergh VW (1998) Inhibition of MMP synthesis by doxycycline and chemically modified tetracyclines (CMTs) in human endothelial cells. *Adv Dent Res* 12: 114–118

44 Mengshol JA, Vincenti MP, Coon CI, Barchowsky A, Brinckerhoff CE (2000) Interleukin-1 induction of collagenase 3 (matrix metalloproteinase 13) gene expression in chondrocytes requires p38, c-Jun N-terminal kinase, and nuclear factor kappaB: differential regulation of collagenase 1 and collagenase 3. *Arthritis Rheum* 43: 801–811

45 Vincenti MP, Coon CI, Brinckerhoff CE (1998) Nuclear factor kappaB/p50 activates an element in the distal matrix metalloproteinase 1 promoter in interleukin-1beta-stimulated synovial fibroblasts. *Arthritis Rheum* 41: 1987–1994

46 Vincenti MP, Coon CI, Lee O, Brinckerhoff CE (1994) Regulation of collagenase gene expression by IL-1 beta requires transcriptional and post-transcriptional mechanisms. *Nucleic Acids Res* 22: 4818–4827

47 Shlopov BV, Lie WR, Mainardi CL, Cole AA, Chubinskaya S, Hasty KA (1997) Osteoarthritic lesions: involvement of three different collagenases. *Arthritis Rheum* 40: 2065–2074

48 Shlopov BV, Stuart JM, Gumanovskaya ML, Hasty KA (2001) Regulation of cartilage collagenase by doxycycline. *J Rheumatol* 28: 835–842

49 Shlopov BV, Smith GN Jr, Cole AA, Hasty KA (1999) Differential response patterns to doxycycline and tgfβ1 in down-regulation of collagenases in osteoarthritic and normal human chondrocytes. *Arthritis Rheum* 42: 719–727

50 Craig RG, Yu Z, Xu L, Barr R, Ramamurthy N, Boland J, Schneir M, Golub LM (1998) A chemically modified tetracycline inhibits streptozotocin-induced diabetic depression of skin collagen synthesis and steady-state type I procollagen mRNA. *Biochim Biophys Acta* 1402: 250–260

51 Schneir M, Ramamurthy N, Golub L (1990) Minocycline-treatment of diabetic rats normalizes skin collagen production and mass: possible causative mechanisms. *Matrix* 10: 112–123

52 Sasaki T, Ramamurthy NS, Yu Z, Golub LM (1992) Tetracycline administration increases protein (presumably procollagen) synthesis and secretion in periodontal ligament fibroblasts of streptozotocin-induced diabetic rats. *J Periodont Res* 27: 631–639

53 Davies SR, Cole AA, Schmid TM (1996) Doxycycline inhibits type X collagen synthesis in avian hypertrophic chondrocyte cultures. *J Biol Chem* 271: 25 966–25 970

Tetracyclines in Biology, Chemistry and Medicine
ed. by M. Nelson, W. Hillen and R.A. Greenwald

Regulation of inflammatory mediators by tetracyclines

Mukundan G. Attur*, Mandar N. Dave, Nirupama Mohandas, Indravadan R. Patel, Steven B. Abramson and Ashok R. Amin

*Department of Rheumatology, Hospital for Joint Diseases, NYU School of Medicine, 301 E. 17th St., Room 1600, New York, NY 10003, USA

Abstract

Tetracyclines (TCs) are a group of antibiotic compounds which possess anti-inflammatory activity both *in vivo* and *in vitro* at pharmacological concentrations. They inhibit the activity of pleiotropic inflammatory mediators such as nitric oxide, prostaglandins, cytokines and MMPs. The mechanism of action of this group of compounds, to inhibit various inflammatory mediators is distinctive. TCs and chemically modified TCs (CMTs) both inhibit nitric oxide synthase (mRNA level), TNFα convertase (enzyme level), which subsequently downregulates nitric oxide, and TNFα production, respectively. TCs and CMTs modulate anti inflammatory mediators such as IL-10 (mRNA level) and type II IL-1β decoy receptor (secretory enzyme level). Doxycycline and minocycline augment COX-2 (mRNA level)-mediated PGE_2 production, whereas CMTs preferentially inhibit COX-2 both at mRNA and protein level. These pleiotropic properties of TCs and CMTs make them candidate drugs for complex multifactoral inflammatory diseases.

Introduction

Doxycycline and minocycline (a semisynthetic tetracycline) are members of the tetracycline family of broad-spectrum antibiotics. During recent years, it has been established that TCs, which are rapidly absorbed and have a prolonged half-life, exert biological effects independent of their antimicrobial activity [1–3]. Such effects include inhibition of matrix metalloproteases [(e.g., collagenase (MMP-1), gelatinase (MMP-2) and stromelysin (MMP-3) activities)] and prevention of pathogenic tissue destruction [1]. Furthermore, TCs and inhibitors of metalloproteases have been reported to inhibit tumor progression [4], bone resorption [5], angiogenesis [6] and inflammation [7].

Regulation of nitric oxide in inflammation

Nitric oxide (NO) is a multifunctional pro-inflammatory mediator that is readily detected at sites of inflammation [8]. It is now recognized that NO plays a vital role in host defense and immunity, including the modulation of inflammatory responses. Upregulation of iNOS generates excessive NO production during the course of a variety of diseases, including sepsis, ulcerative colitis, psoriasis, multiple sclerosis, type I diabetes, giant cell arthritis, systemic lupus erythematosus, Sjögren's syndrome, rheumatoid arthritis (RA) and osteoarthritis (OA) [8]. Although the inhibition of NO synthesis improves outcome in experimental models of systemic lupus erythematosus and arthritis, the role of this highly reactive free radical in the pathogenesis of these diseases is unclear [9].

Regulation of NO by TCs

There is a growing body of literature that demonstrates the capacity of TCs to inhibit NO production in experimental animal models in association with amelioration of disease. For example, intraperitoneal injections of TCs protect mice from lethal endotoxemia, downregulating iNOS in various organs and cytokine and NO secretion in blood [10]. TCs have also been shown to exert protective effects in animal models of inflammatory arthritides, such as type II collagen-induced arthritis [11]. Yu et al. have shown that prophylactic administration of doxycycline markedly reduced the severity of OA in dog models, where it is postulated that excess NO production results in inhibition of proteoglycan and collagen type II synthesis [12] and upregulation of metalloprotease activity [12, 13]. In human OA- or RA-affected cartilage (but not normal cartilage), there is upregulation of NOS and increased NO production in quantities sufficient to cause cartilage damage [14–16]. Minocycline has been demonstrated to be superior to placebo in the treatment of rheumatoid arthritis [17]. In view of these observations, the effects of TCs were evaluated in human OA-affected cartilage and rodent macrophage cell lines (RAW 264.7). Among the tetracycline group of compounds, doxycycline > minocycline blocked and reversed both spontaneous and IL-1β-induced OA-NOS activity in *ex vivo* conditions. Similarly, minocycline and doxycycline inhibited both LPS- and IFNγ-stimulated iNOS in RAW 264.7 cells *in vitro*, as assessed by nitrite accumulation. Although doxycycline and minocycline (20–40 μg/ml) could inhibit both these enzyme isoforms, their susceptibility to each of these drugs was distinct. Unlike acetylating agents (e.g., aspirin), or competitive inhibitors, such as L-NMMA, that directly inhibit the specific activity of NOS, doxycycline or minocycline had no significant effect on the specific activity of iNOS in cell-free extracts. The mechanism of action of these drugs on murine iNOS expression was found to be, at least in part, at the level of RNA expression and protein synthesis of the enzyme, which would account for the decreased iNOS protein and specific activity of the enzyme. TCs had no significant effect on the levels of mRNA for β-actin and glyceraldehyde-3-PO_4 dehydrogenase (GAPDH), nor on levels of β-actin and total COX-2 protein. These studies indicate that a novel mechanism of action of a group of antibiotics, *i.e.,* TCs, is to inhibit the expression of NOS (Fig. 1).

Chemically modified TCs [CMT-3 (IC_{50} ~ 6–13 μM = ~2.5–5 μg/ml) and CMT-8 (IC_{50} ~ 26 μM = 10 μg/ml), but not CMT-1, -2 or -5], which lack antimicrobial activity, inhibited nitrite production in LPS-stimulated macrophage and mesengial cells [18, 19]. Unlike competitive inhibitors of L-arginine, which inhibited the specific activity of iNOS in cell-free extracts, CMTs exerted no such direct effect on the enzyme. However, CMT-3 inhibits both iNOS mRNA accumulation and protein expression in LPS-stimulated cells [18]. TCs and CMT-3 (unlike hydrocortisone) had no significant effect on murine macrophages transfected with iNOS promoter (tagged to a luciferase reporter gene) and exposed to LPS. However, doxycycline and CMT-3 augmented iNOS mRNA degradation in the presence of actinomycin D, in LPS-stimulated murine macrophages. These studies show a novel mechanism of action by TCs and CMT-3 which harbors properties to increase iNOS mRNA degradation and decrease iNOS protein synthesis and NO production in macrophages

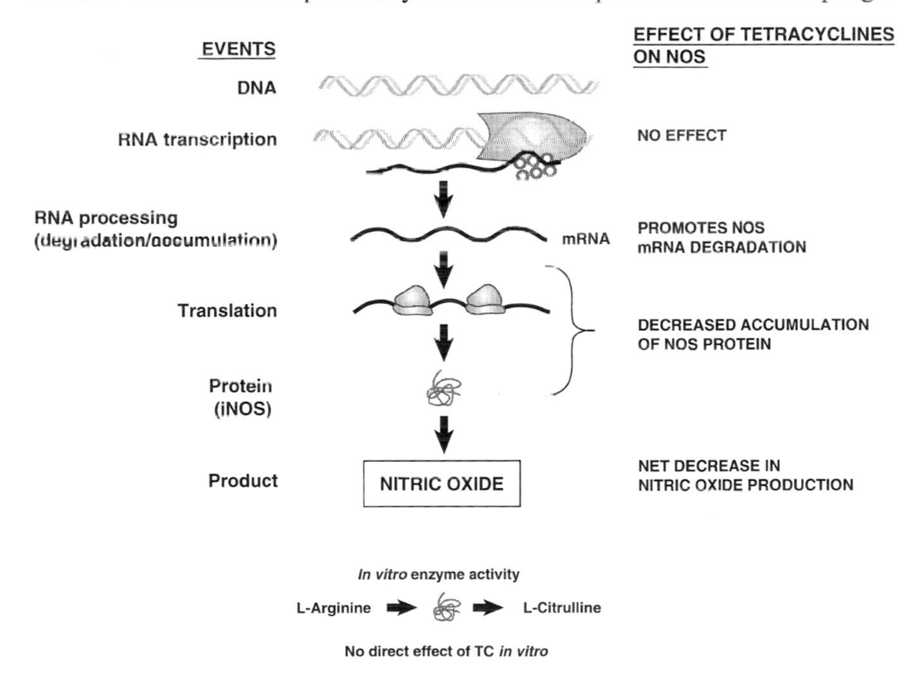

Figure 1. Regulation of NOS by TCs. Cells treated with doxycycline and minocycline (IC_{50} 20–40 μg/ml) CMT-3 and CMT-8 (IC_{50} 5–10 μg/ml) inhibit iNOS mRNA accumulation and have no significant effect on the transcription of iNOS mRNA. The decrease in RNA accumulation (due to decrease in the half-life of the mRNA) leads to decrease in protein expression and the subsequent production of NO. TCs (unlike aspirin or competitive substrate inhibitors, NMA of NOS) do not significantly inhibit the enzyme activity of NOS *in vitro* assays. The data are compiled from [18, 19].

[18]. Recent studies by D'Agostino et al., 1998, have suggested that the effects of TCs on NOS (*in vivo*) are independent of IL-10 [20]. This property of TCs may have beneficial effects in the treatment of various diseases where excess NO has been implicated.

Regulation of serum NO by TCs in human arthritis

Farrell et al have reported significantly increased nitrite levels in the sera of patients with OA and RA as compared to normal individuals [21]. We performed a preliminary study to examine the effect of TCs in patients with arthritis. Patients with OA (n = 5) were treated with doxycycline (200 mg) daily for a period of 10 days and the levels of nitrite were examined before and after the treatment. These OA patients showed an increase in nitrite levels in their serum as compared to normal individuals, similar to that observed by Farrell et al. [21]. Treatment with doxycycline showed significant decrease in nitrite levels in these patients as compared to normal individuals. This preliminary study, which requires tissue corroboration, suggests that doxycycline can affect NO production *in vivo* by cartilage in patients with arthritis (Fig. 2). Were this to be the case, these results would raise the possibility that serum nitrate/nitrite could serve as a surrogate marker for tissue disease activity of OA; this type of assessment of effects of drug therapy on disease markers would be of value, comparable to the situation in RA patients treated with anti-TNF therapy, who show a decrease in serum nitrite levels with decreased expression of iNOS in

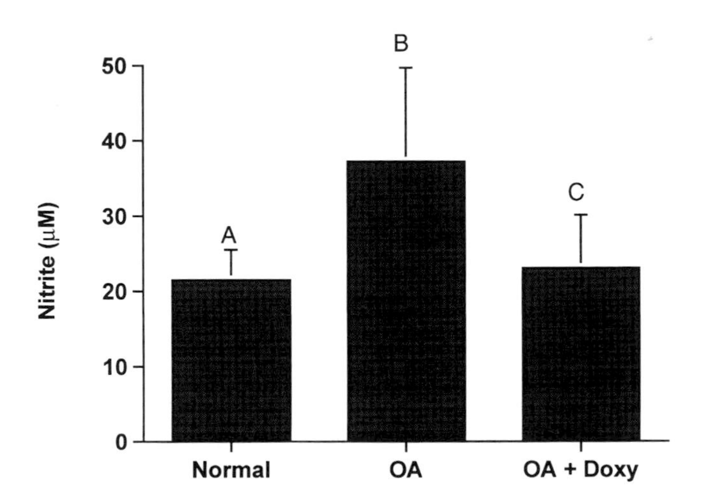

Figure 2. Regulation of nitric oxide by TCs in human serum of arthritis-affected individuals. Five normal and OA-affected individuals were examined for accumulation of nitrite in the serum. The OA-affected individuals were treated with 200 mg/day BID of doxycline for 10 days and then examined again for nitrite accumulation in the serum. n = 5. The p values between different parameters were: A/B $p \leq 0.007$, B/C $p \leq 0.006$, A/C $p \leq 0.4$

peripheral blood cells [22]. Since the overproduction of NO has been implicated in the pathogenesis of arthritis, as well as other inflammatory diseases, these observations suggest that TCs should be evaluated as potential therapeutic modulators of NO for various pathological conditions.

Regulation of prostaglandins in inflammation

Major sources of prostaglandins in acute inflammation include endothelial cells, which produce large amounts of PGE_2 and PGI_2, and mononuclear phagocytes, which produce a wide spectrum of prostaglandins (including PGE_2 and PGI_2), thromboxanes and leukotrienes. Platelets are a major source of thromboxane A (TXA_2), whereas mast cells synthesize leukotrienes and prostaglandins, notably PGD_2. Neutrophils are a poor source of prostaglandins and thromboxanes; they are, however, a source of leukotriene production, particularly LTB4 [23].

Cyclooxygenases (COX-1 and -2) are key enzymes in the synthesis of PGE from arachidonic acid. COX-1 is constitutively expressed in many tissues and cell types, but in some cases it increases during differentiation [24]. By contrast, the expression of COX-2 is upregulated by mitogen, cytokines and tumor promoters and is involved in pathological conditions, notably inflammation. In experimental acute and chronic inflammation animal models, enhanced COX-2 expression parallels the degree of tissue inflammation. In addition, prostanoids such as $PGF_{2\alpha}$, PGI_2 and PGE_2 generated by COX-2 may stimulate COX-2 expression, forming a positive loop for COX-2 induction [25].

Prostaglandins are produced at elevated levels in inflamed tissues such as rheumatoid synovium [26, 27]. PGE_1 and PGE_2 contribute to synovial inflammation by increasing local blood flow and potentiating the effects of mediators such as bradykinin that induce vasopermeability [23]. PGE_2 has been shown to induce apoptosis in chondrocytes, reverse proteoglycan degradation by IL-1, sensitize cells to IGF-1, decrease alpha1 collagen synthesis, increase DNA synthesis and growth of chondrocytes, enhance aggregan and collagen synthesis and trigger osteoclastic bone resorption [28, 29], suggesting that although this molecule may also contribute to the pathophysiology of joint erosion in arthritis, it may also have some chondroprotective functions during the complex process of cartilage homeostasis.

Regulation of COX-2 by TCs

TCs [doxycycline and minocycline (5–40 µg/ml)] augmented (1–2-fold) PGE_2 production in human OA-affected cartilage (in the presence or absence of cytokines and endotoxin) in *ex vivo* conditions. Similarly, bovine chondrocytes stimulated with LPS showed an (1–5-fold) increase in PGE_2 accumulation in the presence of doxycycline. This effect was observed at drug concentrations which did not affect NO production. In murine macrophages (RAW 264.7) stimulated with LPS, TCs inhibited NO release and increased PGE_2 production. Tetracycline(s) and L-NMMA (NOS inhibitor) showed an additive effect on inhibition of NO and PGE_2 accumulation, thereby uncoupling the effects of TCs on NO and PGE_2 production. The enhancement of PGE_2 production in RAW 264.7 cells by TCs was accompanied by the accumulation of both COX-2 mRNA and soluble cytosolic COX-2 protein, but not membrane COX-2 (which represents ~85% of the enzyme in the cell).

In contrast to TCs, L-NMMA at low concentrations (100 μM) inhibited the spontaneous release of NO in OA-affected cartilage explants and LPS-stimulated macrophages, but had no significant effect on the PGE_2 production. At higher concentrations, L-NMMA (500 μM) inhibited NO release but augmented PGE_2 production. This study indicates a novel mechanism of action by TCs to augment the expression of soluble cytosolic COX-2 and PGE_2 production, an effect which is independent of the endogenous concentration of NO (Fig. 3a).

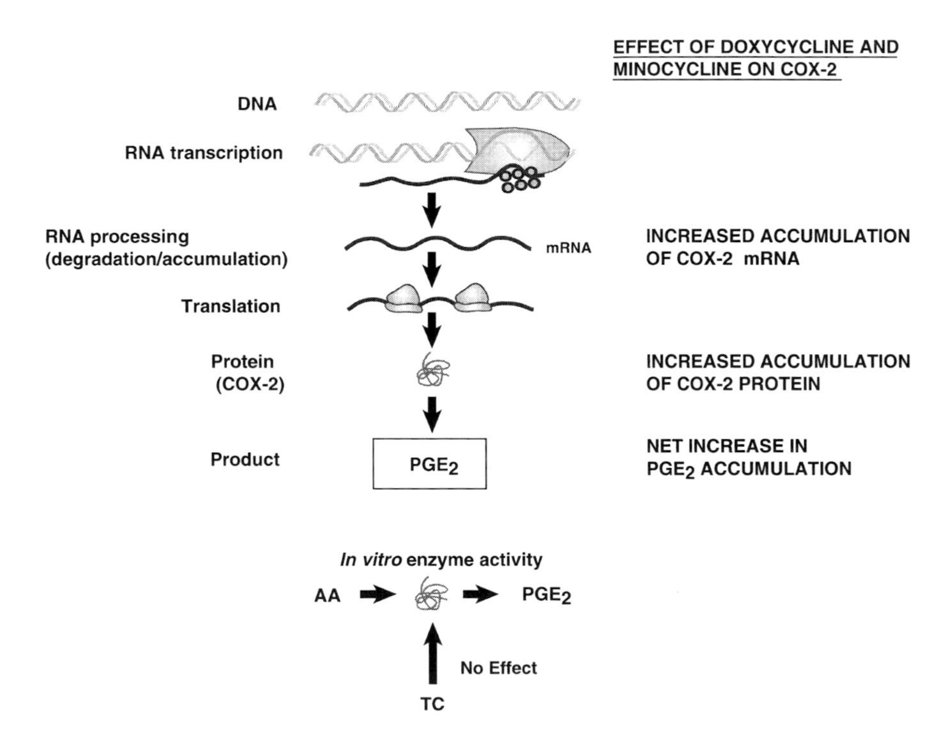

Figure 3a. Regulation of COX-2 by generic TCs (doxycycline [DOX] and minocycline [MINO]). The scheme shows the action of TCs on COX-2 and PGE_2 production. The action of TC is shown on the right-hand side of the scheme. Cells treated with DOX and MINO increase in the accumulation of COX-2 mRNA, and show an increase in the accumulation of cytosolic COX-2 (but not total COX-1). There is a net increase in the accumulation of PGE_2 in cells treated with DOX or MINO. DOX and MINO have no significance on the specific activity of COX-2 *in vitro*.

CMTs have recently been shown to be potent inhibitors of tissue and matrix break-down and to attain a longer half-life in serum than TCs [30]. Recent studies have shown that CMT-3 and CMT-8 inhibit tumor metastasis and arthritis-affected synoviocyte invasion in animal models [30, 31].

In contrast to doxycycline and minocycline, CMT-1 and -8 (1 to 10 μg/ml), but not CMT-1, -2 and -5, inhibit both NO and PGE_2 production in LPS-stimulated murine macrophages, bovine chondrocytes and human OA-affected cartilage (Fig. 3b). Furthermore, CMT-3 augments COX-2 protein expression but inhibits net PGE_2 accumulation [32]. This coincides with the ability of CMT-3 and -8 to inhibit COX-2 enzyme activity *in vitro*. The action of CMTs is distinct from those observed with doxycycline and minocyclines because (a) CMT-3-mediated inhibition of PGE_2 production coincides with modification of COX-2 protein, which is distinct from the non-glycosylated COX-2 protein generated in the presence of tunicamycin, as observed by western blot analysis, (b) CMT-3 and -8 have no significant effect on COX-2 mRNA accumulation, (c) CMT-3 and -8 do not inhibit COX-1 expression at the level of protein and mRNA accumulation, (d) CMT-3 and -8 inhibit the specific activity of

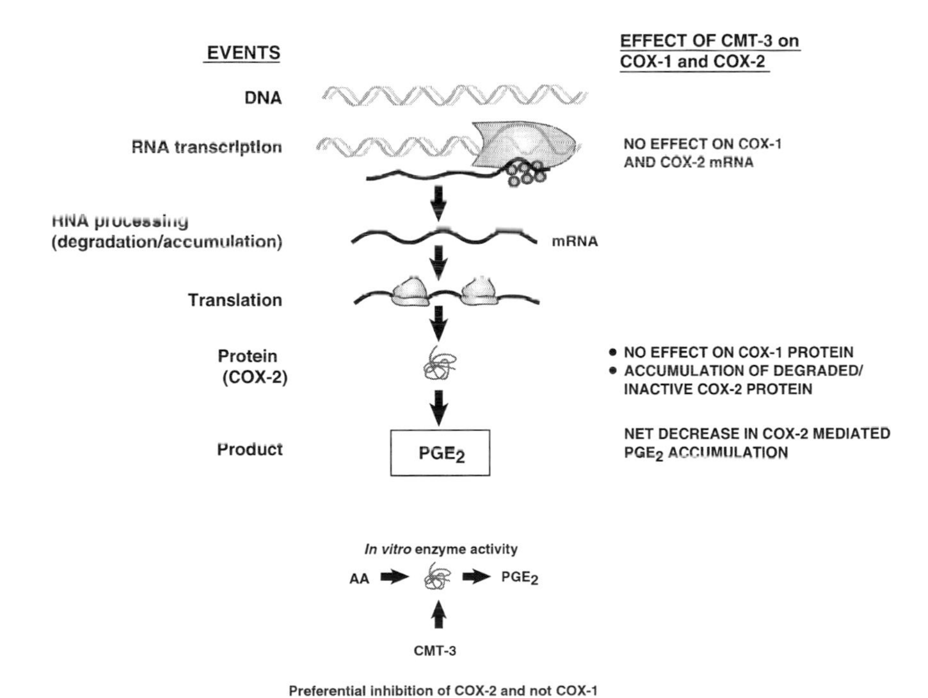

Figure 3b. Regulation of COX-2 by CMTs. The scheme shows the action of CMTs on COX-1, COX-2 and PGE_2 production. Cells treated with CMT-3 have no effect on COX-1 and COX-2 mRNA accumulation but there is a net decrease in PGE_2 production. CMT-3 accumulates the inactive form of COX-2 protein and preferentially inhibits the activity of COX-2 (but not COX-1) *in vitro* enzyme assays similar to NS293 and Celebrex. Data compiled from Amin et al. [52], Attur et al. [53].

Table 1. Effect of CMTs and COX-2 specific inhibitors (NS398 and Celecoxib) on COX-1 and COX-2 specific activity *in vitro*

Treatment		COX-1 specific activity (PGE_2 ng/mg protein)	$p \geq$	COX-2 specific activity (PGE_2 ng/mg protein)	$p \leq$
Control		92.5 ± 1.3		188.9 ± 18.9	
CMT-3	(5 µg)	79.2 ± 1.1	≥ 0.05	92.0 ± 1.6	≤ 0.0005
CMT-5	(5 µg)	90.2 ± 0.8	≥ 0.05	105.1 ± 3.0	≤ 0.0008
CMT-8	(5 µg)	91.1 ± 0.5	≥ 0.05	107.3 ± 2.3	≤ 0.0009
NS398	(1 µg)	80.0 ± 0.7	≥ 0.05	79.3 ± 0.8	≤ 0.0001
Celecoxib	(1 µg)	86.6 ± 0.9	≥ 0.05	110.7 ± 0.6	≤ 0.001

Cell-free extracts were prepared from serum-starved (24 h) unstimulated and IL-1β (10 ng/ml) stimulated A549 human epithelial cells for 24 h. The extracts were incubated with different modulators for 20 min at 37 °C before adding the substrate to initiate the enzyme reaction which was terminated after 30 min by heat inactivation. The data represent one of the two similar experiments, the PGE_2 analysis from each sample was performed in triplicate n = 3. The *p* values are comparison between control and experimentals.

COX-2 (but not COX-1) in cell-free extracts (Tab. 1). These results demonstrate differential action of CMT-3 on COX-1 and -2 enzyme activity, which is distinct from other generic TCs such as doxycyclines and minocyclines. The action of TCs and CMTs on NO and PGE_2 are summarized in Table 2.

Regulation of TNFα and TACE

TNFα (tumor necrosis factor α), a pleiotropic cytokine, produces a broad spectrum of injurious effects which makes it an important target for therapeutic intervention. TNFα is involved in the pathophysiology of arthritis, AIDS, cancer, autoimmune diseases (immune complex diseases), lung fibrosis, multiple sclerosis, skin delayed-type hypersensitivity reactions, bacterial and parasitic infections [33–36].

The synthesis of TNFα is known to be regulated at transcriptional, translational and post-translational levels [33]. Three independent groups have demonstrated that broad-spectrum inhibitors of matrix metalloproteases can specifically inhibit the release of membrane pro-TNFα from various cell surfaces, including RA synovial cell cultures [37–40]. This inhibitor of pro-TNFα processing could protect mice against a lethal dose of endotoxin administered to them [38].

The TNFα converting enzyme (TACE/ADAM-17) is a metalloprotease-disintegrin that releases soluble TNFα from cells by cleaving within the extracellular domain of membrane-bound pro-TNFα [39, 40]. TACE (an 80 kDa protein) cleaves the Gln-Ala-↓-Val-Arg sequence of pro-TNFα [39–41]. TACE also contributes to the "shedding" of several other transmembrane proteins, including the p75 TNFα receptors, transforming growth factor-α, L-selectin

Table 2. Summary of the action of tetracyclines on NOS/COX expression

	COX-1 Expression				COX-2 Expression				NOS Expression			
	PGE_2	Protein	Enzyme activity	mRNA	PGE_2	Protein	Enzyme activity	mRNA	NO	Protein	Enzyme activity	mRNA
Dox/Mino 5–10 μg/ml	ND	ND	ND	ND	↑	^ change	No	↑	←— No change —→			
Dox/Mino 40 μg/ml	ND	ND	ND	ND	↑	↑	No	↑ change	↓	↓	↓	↓
CMT-3 5–10 μg/ml	↓	No change	ND	No change	↓	↑	↓	No change	↓	↓	↓	↓
CMT-8 5–10 μg/ml	↓	No change	ND	No change	↓	No change	↓	No change	↓	↓	↓	↓
CMT-1,-2,-5	←— ND —→				←— No change —→				←— No change —→			

The table summarizes the effect of Doxycycline (Dox), Minocycline (Mino) and CMTs on COX-2 and NOS expression in activated murine macrophages, bovine chondrocytes and human OA-affected cartilage. The data are compiled from Amin et al., 1997 [18], Tractman et al., 1996 [19] and Patel et al., 1999 [32]. ND denotes "not done".

and the amyloid precursor protein [42, 43]. In view of these observations, we and others developed a cell line to assay for inhibitors of TACE [41, 44].

Regulation of TACE and TNFα in pro-TNFα transfected cells

Wild-type pro-TNFα and mutant pro-TNFα cDNAs, both containing a carboxyl terminal FLAG Tag, were used to generate stable cell lines in the human embryonic kidney cell line (HEK 293) and mouse T cells (BW5147). In contrast to the mouse BW5147 T cells (stably transfected with human pro-TNFα), the pro-TNFα$^+$ HEK 293 cells did not show significant staining using anti-TNFα antibodies of the cell surface pro-TNFα, but did show significantly high amounts (~600 ng/ml) of TNFα in the culture medium when compared to the mouse BW5147 cells transfected with pro-TNFα: (~20 pg/ml) (Fig. 4a). Furthermore, a clone (pro-TNFα mut$^+$ HEK 293) harboring the mutated pro-TNFα (at the cleavage site) showed, as expected, low amounts of TNFα (~20 pg/ml) in the medium similar to that seen with the non-transfected cells. These experiments indicated that the pro-TNFα$^+$ in human cells is rapidly cleaved by the endogenous TACE from the cell surface, whereas the mouse cells TNFα convertase do not seem to recognize human pro-TNFα as a substrate.

TACE inhibitors inhibit release of soluble TNFα in pro-TNFα$^+$ HEK 293 cells and human OA-affected cartilage

The effects of various MMP and ADAM inhibitors were tested in pro-TNFα$^+$ HEK 293 cells. Addition of cycloheximide inhibited, as expected, the sponta-

---

Figure 4. Regulation of TNFα in Pro-TNFα transfected cells. Role of TACE. The cDNA encoding the pro-TNFα(wild) was isolated by PCR amplification of first strand cDNA generated from 1 μg of total RNA from THP-1 cells. A sense primer (5' GGA ATT CAT ATG AGC ACT GAA AGC) containing a NdeI site and antisense primer (CGG GAT CCT CAC TTG TCA TCG TCG TCC TTG TAG TCC AGG GCA TTG A) containing FLAG +ochre termination codon and a BamHI site were used. PCR amplification was carried according to Patel et al., 1998 [41]. The mutant pro-TNFα cDNA was constructed as follows. Sense and antisense primers were generated (sense 5' GCTAGC TCA TCT TCT CGA ACC and antisense 5' AGA TGA GCT AGC CAG AGG GCT GAT) for the sequence at and near the cleavage site Ser–Pro–Leu–Ala–Gln–**Ala–↓–Val**–Arg–Ser–Ser. The oligos looped out 6 amino acids (Ala-Gln-Ala-Val-Arg-Ser) from the wild type [pro-TNFα (wt)] and looped in (Ala-Ser, GCT AGC) resulted in [pro-TNFα (mut)] clone. Two clones were separately cloned into the eukaryotic expression vectors. (a) The pro-TNFα (wild) and pro-TNFα (mut) expressing HEK293 cell lines were selected by Bector Dickenson cell sorter. Control represents untransfected cells. TNFα, which was spontaneously produced in the medium, was measured using a sandwich ELISA technique. (b) Different modulators of TACE were tested in these cells, and assayed for the levels of soluble TNFα in the medium. Control: untreated cells were considered as the 100% value which represents 300 ng/ml TNFα accumulated by 7×10^6 cells at 72 h. n = 3. CMT-5 and CMT-3 (10 μg/ml): doxycycline (DOXY) 20 μg/ml; cycloheximide (CHX) 10 μg/ml; and BB-16 (10 μg/ml). (c) proTNF$^+$HEK293 cells which were treated with CMT-3 (1 μg/ml) for 24 h were also stained for membrane-bound pro-TNFα with anti-TNFα antibodies and FACScan.

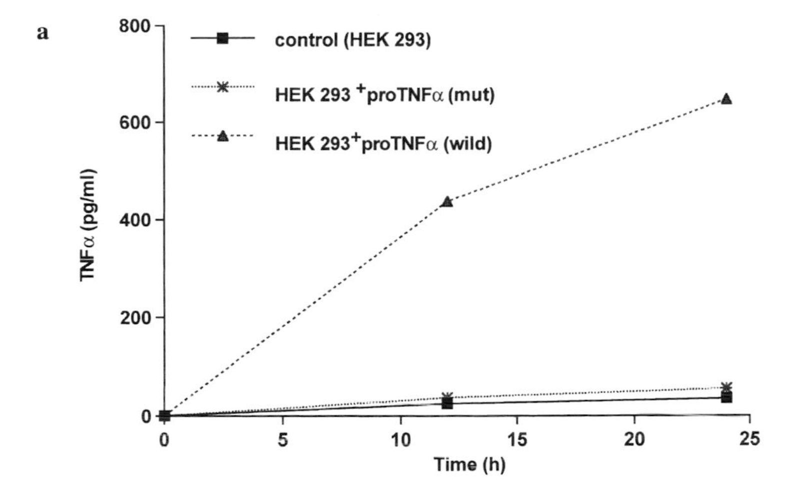
a
800
600
400
200
0
TNFα (pg/ml)
control (HEK 293)
HEK 293 +proTNFα (mut)
HEK 293+proTNFα (wild)
0
5
10
15
20
25
Time (h)

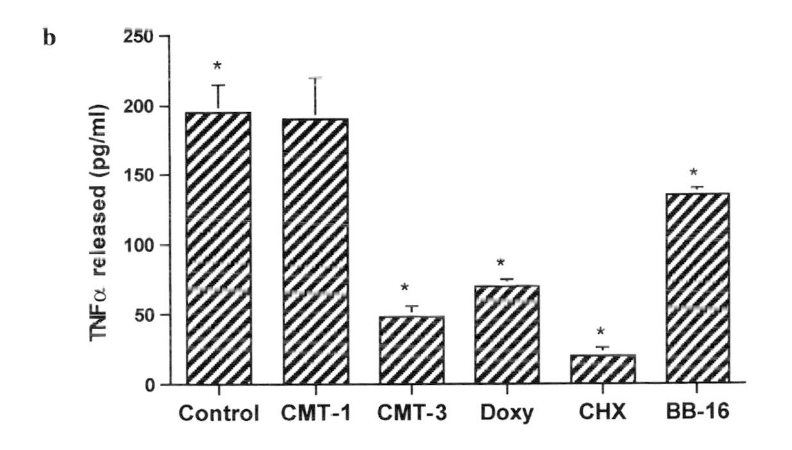
b
250
200
150
100
50
0
TNFα released (pg/ml)
Control
CMT-1
CMT-3
Doxy
CHX
BB-16

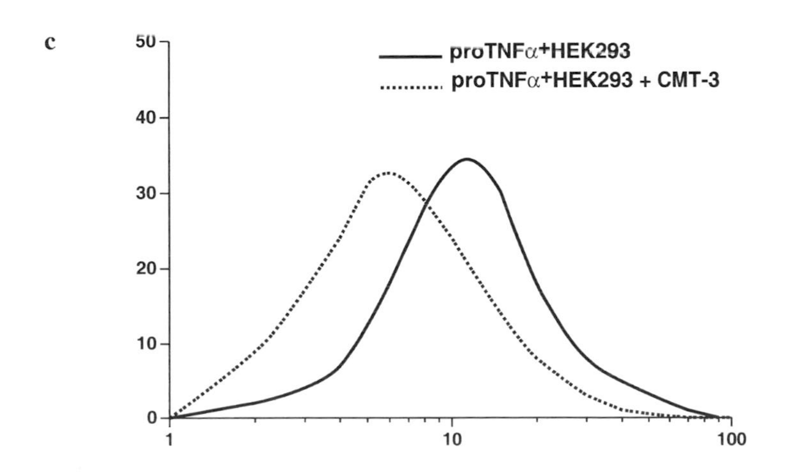
c
50
40
30
20
10
0
proTNFα+HEK293
proTNFα+HEK293 + CMT-3
1
10
100

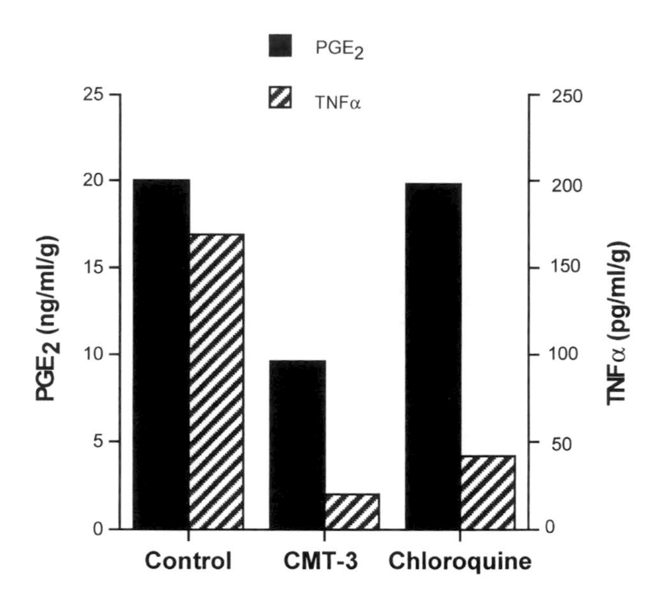

Figure 5. Regulation of TNFα in human OA-affected cartilage by CMT-3. Human OA-affected cartilage was incubated in *ex vivo* as described [41]. The OA-affected cartilage spontaneously released PGE_2 (dark bars) and TNFα (hutch bars) as shown in control. CMT-3 and chloroquine were added at 0 time period and the mediators (TNFα and PGE_2) were analyzed at 72 h from the medium.

neous release of the TNFα in pro-TNFα$^+$ HEK 293 cells. BB-16 (an inhibitor of TACE/MMPs), CMT-3 (but not CMT-1) and TCs significantly blocked the release of TNFα in pro-TNFα$^+$ HEK 293 cells in the medium (Fig. 4b).

FACScan analysis of pro-TNFα$^+$ HEK 293 cells treated with CMT-3 and stained with TNFα antibody showed increased accumulation of surface pro-TNFα. These experiments demonstrated that inhibitors of TACE could block the spontaneous release of TNFα in these cells. BB-16 and CMT-3 share the property to inhibit TNFα release by blocking TNFα convertase activity. These studies show that the tetracycline group of compounds (which inhibits iNOS and COX-2) [18] also inhibited TNFα production and TACE activity in this whole-cell assay.

We also examined the effect of CMT-3 on TACE in human OA-affected cartilage. We took advantage of the observation that human OA-affected cartilage spontaneously released TNFα in the medium in *ex vivo* conditions [41]. Chloroquine has recently been reported to inhibit processing of pro-TNFα (26 kDa) in murine macrophage cells (RAW 264.7) [45]. Addition of chloroquine or CMT-3 (but not CMT-1) significantly blocked TNFα production in OA-affected cartilage (Fig. 6). We also monitored the release of PGE_2 [18] in the same experiment (Fig. 6). CMT-3 blocked, as expected, the release of both TNFα and PGE_2 but chloroquine inhibited only the release of TNFα but not PGE_2. These experiments suggest that human OA-affected cartilage has TACE activity which is responsible for the production of functional native soluble

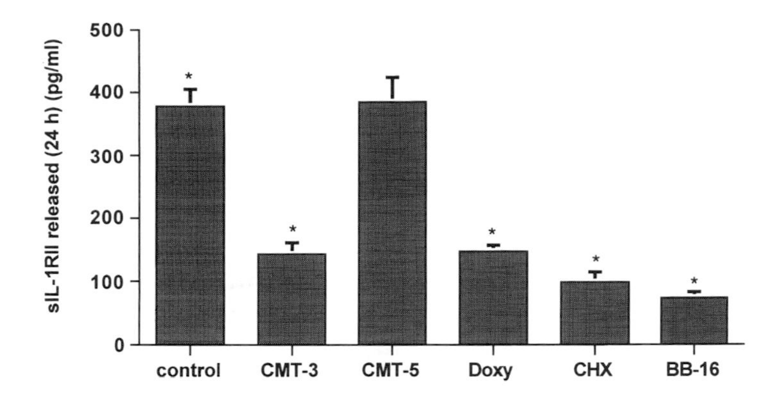

Figure 6. Regulation of Type II IL-1β decoy receptor by CMT-3. HEK293 cells were transfected with full-length IL-1RII cDNA in an expression vector as shown previously [49]. The stably transfected cells (IL-1RII+HEK293) spontaneously released soluble IL-1RII in the medium. CMT-3 (10 μg/ml), CMT-5 (10 μg/ml), doxycycline (20 μg/ml), cycloheximide (10 μg/ml) and BB-16 (available from Dupont Merck, USA: Wilmington, DE) were added to the medium. The release of soluble IL-1RII was assayed at 72 h. * $p \leq 0.05$

TNFα. In summary, CMT-3, like TCs, can also inhibit the activity of TACE and TNFα production [46].

Upregulation of anti-inflammatory mediators by TCs

IL-10 is an anti-inflammatory cytokine which has been reported to inhibit inflammatory mediators such as NO, PGE_2, IL-1β, TNFα and IL-6 [47]. Recent studies by Ritchlin et al. [48] have shown upregulation (25 to 45%) of IL-10 mRNA and protein in IL-1 induced synovial cells and LPS-stimulated PBMC in the presence of CMTs and minocyclines (5 μg/ml) but not CMT-1 and doxycycline. The upregulation of a pleiotropic anti-inflammatory cytokine such as IL-10 by TC gives TCs an added anti-inflammatory activity independent of their direct effect on NO, PGE_2 and TNFα production.

The soluble type II IL-1β decoy receptor has potent IL-1 neutralizing activity. For example, 100 pg/ml of exogenously added soluble type II IL-1β receptor can block 50% of NO production (as compared to 1 mg of soluble IL-1 type I receptor) in human OA-affected cartilage [49]. However, transfection of the type II decoy receptor and its expression on the cell surface (in IL-1RII+HEK293 cells as compared to HEK293) render the cells more resistant to the insults of exogenous IL-1. IL-1RII+HEK293 not only produces significantly high amounts of sIL-1RII in the medium, but also blocks the activity of IL-1 transcription and type I IL-1β receptor signaling within the cell [49]. CMT-3 and doxycycline (but not CMT-5) inhibit the spontaneous shedding of type II IL-1 receptor in cells that are permanently transfected with the IL-1RII receptor as shown in Figure 6. These experiments suggest that the functional activity of the type II IL-1β decoy receptor may be enhanced by CMT-3 against the insults of IL-1 by accumulating the receptor on the cell surface.

In summary, TCs and their derivatives exert significant effects on inflammatory mediators such as NO, PGE_2 and MMP activity (matrixins and ADAMs) *in vitro* and *in vivo*. They may also upregulate endogenous anti-inflammatory molecules such as IL-10. Experimental data in both animals and humans suggest that these pleiotropic anti-inflammatory properties may account for beneficial therapeutic effects of the TCs in diverse medical conditions such as acne, adult respiratory distress syndrome (ARDS) and arthritis [50, 51]. Thus, based upon *in vitro* properties which have been corroborated by a limited number of *in vivo* observations, TCs are potential disease-modifying agents worthy of further study.

Acknowledgments
We would like to thank CollaGenex Corporation (Newtown, PA) for their generous support during the course of these studies performed at the Hospital for Joint Diseases. We would like to thank Dr. Christopher Ritchlin (University of Rochester School of Medicine and Dentistry, Rochester, NY) and Dr. Nieman for sharing some of their unpublished data during the preparation of this manuscript. We are grateful to the Journal of Immunology for allowing us to reproduce some of the tables and figures used in this article. We would also like to thank Ms. Andrea L. Barrett for preparing the manuscript.

Abbreviations
NO: nitric oxide; PGE_2: Prostaglandin E2; NOS: nitric oxide synthase; OA: Osteoarthritis; RA: Rheumatoid arthritis; OA-NOS: Osteoarthritis-affected NOS; CMT: chemically modified tetracycline; IL-1RII: interleukin 1 type II decoy receptor; COX: cyclooxygenase; TACE: TNFα convertase enzyme; MMP: matrix metalloproteases

References

1 Golub LM, Ramamurthy NS, McNamara TF, Greenwald RA, Rifkin B (1991) Tetracyclines inhibit connective-tissue breakdown – new therapeutic implications for an old family of drugs. *Crit Rev Oral Biol Med* 2: 297–322
2 Golub LM, Suomalainen K, Sorsa T (1992) Host modulation with tetracyclines and their chemically modified analogues. *Curr Opin Dent* 2: 80–90
3 Uitto VJ, Firth JD, Nip L, Golub LM (1994) Doxycycline and chemically modified tetracyclines inhibit gelatinase A (MMP-2) gene-expression in human skin keratinocytes. *Ann N Y Acad Sci* 732: 140–151
4 DeClerck YA, Shimada H, Taylor SM, Langley KE (1994) Matrix metalloproteinases and their inhibitors in tumor progression. *Ann N Y Acad Sci* 732: 222–232
5 Rifkin B, Vernillo AT, Golub LM, Ramamurthy NS (1994) Modulation of bone resorption by tetracyclines. *Ann N Y Acad Sci* 732: 165–180
6 Maragoudakis ME, Peristeris P, Missirlis E, Aletras A, Andriopoulou P, Haralabopoulos G (1994) Inhibition of angiogenesis by anthracyclines and titanocene dichloride. *Ann N Y Acad Sci* 732: 280–293
7 Amin AR, Attur MG, Abramson SB (1999) Regulation of nitric oxide and inflammatory mediators in human osteoarthritis-affected cartilage: implication for pharmacological intervention. *In*: GM Rubanyi (ed.): *The Pathophysiology and Clinical Applications of Nitric Oxide*. Harwood Academic Publishers, Richmond, CA, 397–412
8 Nathan C (1997) Inducible nitric oxide synthase: what difference does it make? *J Clin Invest* 100: 2417–2423

9 Belmont HM, Amin AR, Abramson SB (1997) Nitric oxide in systemic lupus erythematosus. *In*: GM Kammer and GC Tsokos (eds): *Lupus: Molecular and Cellular Pathogenesis*. Humana Press, Totowa, NJ, 21–42

10 Milano S, Arcoleo F, D'Agostino P, Cillari E (1997) Intraperitoneal injection of tetracyclines protects mice from lethal endotoxemia downregulating inducible nitric oxide synthase in various organs and cytokine and nitrate secretion in blood. *Antimicrob Agents Chemother* 41: 117–121

11 Hanyu T, Chotanaphuti T, Arai K, Tanaka T, Takahashi HE (1999) Histomorphometric assessment of bone changes in rats with type II collagen-induced arthritis. *Bone* 24: 485–490

12 Yu LPJ, Smith GNJ, Brandt KD, Myers SL, O'Connor BL, Brandt DA (1992) Reduction of the severity of canine osteoarthritis by prophylactic treatment with oral doxycycline. *Arthritis Rheum* 35: 1150–1159

13 TeKoppele JM, Beekman B, Verzijl N, Koopman JL, DeGroot J, Bank RA (1998) Doxycycline inhibits collagen synthesis by differentiated articular chondrocytes. *Adv Dent Res* 12: 63–67

14 Amin AR, Di Cesare P, Vyas P, Attur MG, Tzeng E, Billiar TR, Stuchin SA, Abramson SB (1995) The expression and regulation of nitric oxide synthase in human osteoarthritis-affected chondrocytes: Evidence for up-regulated neuronal nitric oxide synthase. *J Exp Med* 182: 2097–2102

15 Amin AR, Abramson SB (1998) The role of nitric oxide in articular cartilage breakdown in osteoarthrtis. *Curr Opin Rheumatol* 10: 263–268

16 Amin AR, Attur M, Abramson SB (1999) Nitric oxide synthase and cyclooxygenases: distribution, regulation, and intervention in arthritis. *Curr Opin Rheumatol* 11: 202–209

17 Tilley BC, Trentham DE, Neuner R, Kaplan DA, Clegg DO, Leisen JCC, Buckley L, Cooper SM, Duncan H, Pillemer SR et al (1995) Minocycline in rheumatoid arthritis. *Ann Intern Med* 122: 81–89

18 Amin AR, Patel RN, Thakker GD, Lowenstein CJ, Attur MG, Abramson SB (1997) Post-transcriptional regulation of inducible nitric oxide synthase mRNA in murine macrophages by doxycyclines and chemically modified tetracyclines. *FEBS Lett* 410: 259–264

19 Tractman H, Futterweit S, Greenwald RA, Moak SA, Singhal P, Franki N, Amin AR (1996) Chemically modified tetracyclines inhibit inducible nitric oxide synthase expression and nitric oxide production in cultured rat mesengial cells. *Biochem Biophys Res Commun* 229: 243–248

20 D'Agostino P, La RM, Barbera C, Arcoleo F, Di BG, Milano S, Cillari E (1998) Doxycycline reduces mortality to lethal endotoxemia by reducing nitric oxide synthesis via an interleukin-10-independent mechanism. *J Infect Dis* 177: 489–492

21 Farrell AJ, Blake DR, Palmer RM, Moncada S (1992) Increased concentrations of nitrite in synovial fluid and serum samples suggest increased nitric oxide synthesis in rheumatic diseases. *Ann Rheum Dis* 51: 1219–1222

22 Perkins DJ, St Misukonis MA, Weinberg JB (1998) Reduction of NOS2 overexpression in rheumatoid arthritis patients treated with anti-tumor necrosis factor alpha monoclonal antibody (cA2). *Arthritis Rheum* 41: 2205–2210

23 Davies P, MacIntyre DE (1992) Prostaglandins and inflammation. *In*: JI Gallin, IM Goldstein, R Snyderman (eds): *Basic principles and clinical correlates*. Raven Press, New York, 123–138

24 Crofford LJ (1997) COX-1 and COX-2 tissue expression: implications and predictions. *J Rheumatol* 24 Suppl 49: 15–19

25 Wu KK (1995) Inducible cyclooxygenase and nitric oxide synthase. *Adv Pharmacol* 33: 179–207

26 Bombardier S, Cattani P, Giabattoni G, Di Munno O, Paero G, Patrono C, Pinca E, Pugliese F (1981) The synovial prostaglandin system in chronic inflammatory arthritis: differential effects of steroidal and non-steroidal anti-inflammatory drugs. *Brit J Pharmacol* 73: 893–901

27 Davies P, Bailey PJ, Goldenberg MM, Ford-Hutchinson AW (1984) The role of arachidonic acid oxygenation products in pain and inflammation. *Annu Rev Immunol* 2: 335–357

28 Robinson DR, Tashjian AH, Levine L (1975) Prostaglandin-stimulated bone resportion by rheumatoid synovia. A possible mechanism for bone destruction in rheumatoid arthritis. *J Clin Invest* 56: 1181–1188

29 Amin AR, Dave M, Attur M, Abramson B (2000) COX-2, NO and cartilage damage and repair. *Curr Rheumatoll Rep* 2: 447–453

30 Seftor EA, Seftor RE, Nieva DR, Hendrix MJ (1998) Application of chemically modified tetracyclines (CMTs) in experimental models of cancer and arthritis. *Adv Dent Res* 12: 103–110

31 Seftor RE, Seftor EA, De LJ, Kleiner DE, Leferson J, Stetler-Stevenson WG, McNamara TF, Golub LM, Hendrix MJ (1998) Chemically modified tetracyclines inhibit human melanoma cell invasion and metastasis. *Clin Exp Metastasis* 16: 217–225

32 Patel RN, Attur MG, Dave MN, Patel IV, Stuchin SA, Abramson SB, Amin AR (1999) A novel mechanism of action of chemically modified tetracyclines: inhibition of COX-2-mediated prostaglandin E2 production. *J Immunol* 163: 3459–3467

33 Vassalli P (1992) The pathophysiology of tumor necrosis factors. *Annu Rev Immunol* 10: 411–452

34 Tracey KJ, Cerami A (1993) Tumor necrosis factor, other cytokines and disease. *Annu Rev Cell Virol* 9: 317–343

35 Feldmann M, Brennan FM, Williams RO, Elliott MJ, Maini RN (1996) Cytokine expression and networks in rheumatoid arthritis: rationale for anti-TNFα antibody therapy and its mechanism of action. *J Inflamm* 47: 90–96

36 Tracey KJ, Cerami A (1994) Tumor necrosis factor: a pleiotropic cytokine and therapeutic target. *Annu Rev Med* 45: 491–503

37 Gearing AJH, Beckett P, Christodoulo M, Churchill M, Clements J, Davidson AH, Drummond AH, Galloway WA, Gilbert R, Gordon JL et al (1994) Processing of tumor necrosis factor-α precursor metalloproteinases. *Nature* 370: 555–557

38 Mohler KM, Sleath PR, Fitzner JN, Cerretti D, Alderson MR, Kerwar SS, Torrance DS, Otten-Evans C, Greenstreet T, Weerawarna K et al (1994) Protection against a lethal dose of endotoxin by an inhibitor of tumor necrosis factor processing. *Nature* 370: 218–220

39 Moss M, Catherine Jin S-L, Milla ME, Burkhart W, Luke Carter H, Chen W-J, Clay WC, Didsbury JR, Hassler D, Hoffman CR et al (1997) Cloning of a disintegrin metalloproteinase that processes precursor tumor-necrosis factor-a. *Nature* 385: 733–736

40 Black RA, Rauch CT, Kozlosky CJ, Peschon JJ, Slack JL, Wolfson MF, Castner B, Stocking KL, Reddy P, Srinivasan S et al (1997) A metalloproteinase disintegrin that releases tumor-necrosis factor-a from cells. *Nature* 385: 729–733

41 Patel IR, Attur MG, Patel RN, Stuchin SA, Abagyan RA, Abramson SB, Amin AR (1998) TNF-α convertase from human arthritis-affected cartilage: Isolation of cDNA by differential display, expression of the active enzyme, and regulation of TNF-α. *J Immunol* 160: 4570–4579

42 Doedens JR, Black RA (2000) Stimulation-induced down-regulation of tumor necrosis factor-alpha converting enzyme. *J Biol Chem* 275: 14 598–14 607

43 Peschon JJ, Slack JL, Reddy P, Stocking KL, Sunnarborg SW, Lee DC, Russell WE, Castner BJ, Johnson RS, Fitzner JN et al (1998) An essential role for ectodomain shedding in mammalian development. *Science* 282: 1281–1284

44 Glaser KB, Falduto M, Metzer R, Pederson T, Pease L, Shiosaki K, Morgan DW (1997) Expression, release, and regulation of human TNFα from stable transfectants of HEK-293 cells. *Inflamm Res* 46 (Suppl 2): S127-S128

45 Jeong J-Y, Jue D-M (1997) Chloroquine inhibits processing of tumor necrosis factor in lipopolysaccharide-stimulated RAW 264.7 macrophages. *J Immunol* 158: 4901–4907

46 Shapira L, Houri Y, Barak V, Soskolne WA, Halabi A, Stabholz A (1997) Tetracycline inhibits porphyromonas gingivalis lipopolysaccharide-induced lesions *in vivo* and TNF alpha processing *in vitro*. *J Periodont Res* 32 (1 Pt 2): 183–188

47 Hawkins D, MacKay R, MacKay S, Moldawer L (1998) Human interleukin-10 suppresses production of inflammatory mediators by LPS-stimulated equine peritoneal macrophages. *Vet Immunol Immunopathol* 66: 1–10

48 Ritchlin C, Haas-Smith S, Schwarz E, Greenwald R (2000) Minocycline but not doxycycline upregulates IL-10 production in human synoviocytes, mononuclear cells and synovial explants. ACR 64th Annual Scientific Meeting Abstract

49 Attur M, Dave M, Cipolletta C, Kang P, Goldring M, Patel I, Abramson S, Amin A (2000) Reversal of autocrine and paracrine effects of IL-1 in human arthritis by type II IL-1 decoy receptor: Potential for pharmacological intervention. *J Biol Chem* 275: 40307–40315

50 Nieman G, Searles B, Carney D, McCann U, Schiller H, Lutz C, Finck C, Gatto LA, Hodell M, Picone A (1999) Systemic inflammation induced by cardiopulmonary bypass: A review of pathogenesis and treatment. *J Extra-Corp Technol* 31: 202–210

51 Shah PK (1999) Targeting the proteolytic arsenal of neutrophils. A promising approach for post-pump syndrome and ARDS. *Circulation* 100: 333–334

52 Amin AR, Attur MG, Patel RN, Thakker GD, Marshall PJ, Rediske J, Abramson SB (1997) Superinduction of cyclooxygenase-2 activity in human osteoarthritis-affected cartilage: Influence of nitric oxide. *J Clin Invest* 99: 1231–1237

53 Attur M, Patel R, Patel P, Abramson S, Amin A (1999) Tetracycline up-regulates COX-2 expression and prostaglandin E_2 production independent of its effect on nitric oxide. *J Immunol* 162: 3160–3167

Tetracyclines in Biology, Chemistry and Medicine
ed. by M. Nelson, W. Hillen and R.A. Greenwald

Tetracyclines and autoimmunity

Helena Marzo-Ortega, Roland M Strauss and Paul Emery*

* *University of Leeds, 36 Clarendon Road, Leeds LS2 9NZ, UK*

Introduction

Tetracyclines (TCs) are pharmaceutical compounds derived from streptomyces which have been widely used as broad-spectrum antibiotics since their introduction in 1948.

In the last few years there has been a renewed interest in these drugs due to the discovery of their non-antimicrobial properties. The first reports came from the odontological literature after Golub and colleagues identified in a seminal experiment an unexpected property of the TCs independent of their anti-microbial actions. They demonstrated the ability of these drugs to inhibit metalloproteinases, enzymes known to play a major role in the breakdown of cartilage and bone. This finding opened up new avenues for research directed at novel therapeutic options for a variety of dental and medical diseases characterized by excessive collagen breakdown, such as periodontitis, arthritis or cancer invasion.

In the field of rheumatology, double-blind trials have confirmed the role of certain new generations of semisynthetic tetracyclines as disease-modifying agents in inflammatory arthritides. Studies in man of osteoarthritis and osteoporosis are currently awaited. By contrast, a new intriguing property of these drugs is their capacity to trigger autoimmune syndromes, in particular the development of lupus erythematosus. Much attention has been paid in recent years to minocycline, both for its potential as a disease-modifying anti-rheumatic agent and for its link with autoimmunity. Increasing awareness of the unwanted side-effects, which on occasions have proven fatal, has resulted in a large number of reports in the medical literature.

Tetracyclines in arthritis

Rheumatoid arthritis (RA), the commonest inflammatory arthritis, is a disease of unknown aetiology characterized by persistent joint swelling and tenderness, leading to joint destruction, functional disability and increased mortality. With a prevalence of 1%, it is estimated that it affects about 2 million people

Table 1. Non-antibiotic properties of the tetracyclines

Connective tissue breakdown inhibition
Inhibition of phospholipase A2
Inhibition of bone resorption
Inhibition of prostaglandin synthesis
Modification of chemotaxis
Scavenging of superoxide radicals

in the USA; no curative therapy exists at present. The available therapeutic drugs, commonly referred to as disease-modifying anti-rheumatic drugs (DMARDs), aim to suppress inflammation and attenuate the progression of the disease process.

With the description of their anti-inflammatory properties, TCs became an obvious choice to consider in the treatment of arthritis for several reasons. First is their ability, especially doxycycline and minocycline, to directly inhibit MMPs, responsible for cartilage and bone breakdown, in synovial tissue, synovial fluid [1–4], chondrocytes and osteoblasts [5, 6]. TCs can also retard excessive connective tissue breakdown and bone resorption, and therefore prevent articular damage, which is the main cause of functional disability in RA and osteoarthritis sufferers. Tetracycline derivatives have been shown to downregulate different inflammatory pathways. They can, for example, inhibit phospholipase A2, a pro-inflammatory enzyme able to induce synovitis in diarthrodial joints [7–9] and have also been shown to be scavengers of oxygen radicals, important mediators of inflammation [10, 11] (Tab. 1).

However, most of these actions have been demonstrated *in vitro.* Some of the work done in adjuvant arthritis suggests that the mode of action of the tetracyclines might be based on their immunomodulatory T cell effect, with alteration of T cell calcium flux and of the expression of T cell-derived collagen-binding protein [12]. Since RA is a T-cell -mediated disease, these results could explain part of the anti-inflammatory effect of tetracyclines in RA. Kloppenburg et al. published in 1995 results on human T cell clones derived from the synovium of a patient with RA and showed that minocycline could produce a dose-dependent inhibition of T cell proliferation and reduction in production of IL-2, interferon gamma and tumour necrosis alpha when T cells were activated via the T cell receptor/CD3 complex. Besides an inhibition in IL-2 production, minocycline could exert its effect on T cell proliferation by induction of a decreased IL-2 responsiveness [13] (Tab. 2).

Evidence-based use of tetracyclines in inflammatory arthritides

Sporadic use of tetracyclines in arthritis dates back almost forty years. The initial rationale for the use of these drugs was the finding of mycoplasmas in the

Table 2. Mechanisms of action of tetracyclines in RA

Immunomodulatory activity:
– Reduced leukotaxis
– Alteration of phagocytic function of human neutrophils
– Alteration of proliferative response of human lymphocytes
– Delayed erythroid and lymphoid cell proliferation and differentiation
– Decreased synovial T-cell proliferation and cytokine production
Anti-inflammatory activity:
– Inhibition of metalloproteinases
– Decrease production of phospholipase A2
– Reduction in production of leukotrienes and prostaglandins
– Scavengers of oxygen radicals

synovial fluid of patients with different inflammatory arthritis [14]. A number of small open label studies followed in patients with rheumatoid and reactive arthritis using different tetracycline compounds. In reactive arthritis, preliminary uncontrolled trials using doxycycline [15] and minocycline [16] were also inconclusive. The first controlled study of short-term tetracycline therapy in Reiter's disease, a form of reactive arthritis, failed to show any benefit [17]. The largest group of patients (21) was reported by Lauhio et al. [18] in a double-blind, placebo-controlled, 3-month treatment study. No significant difference was found in time for remission between groups, although this took longer than 3 months in both groups. Only the patients with documented chlamydia infection remitted in a significantly shorter period of time (15 weeks *versus* 39.5 weeks for controls). There was, however, no information given about use of any other anti-inflammatory agents, in particular corticosteroids.

In RA, the first open label studies were performed in the early nineties in small numbers of patients and were statistically underpowered [19–21]. No clear information is given about use of concomitant anti-inflammatory therapies such as non-steroidal anti-inflammatory drugs (NSAIDs) or corticosteroids, making conclusions difficult. Two independent double-blind randomized controlled trials in established RA followed [22, 23]. In both studies patients were treated with 100 mg oral minocycline or placebo and were followed up for 26 weeks [22] and 48 weeks respectively [23]. The first study included 80 patients with a mean disease duration of 13 years who remained on other DMARDs whilst on the study. The second trial had 219 patients with a mean disease duration of 8 years and who had stopped other DMARDs at least 4 weeks before starting the study period. In both studies, the intergroup differences were statistically significant for standard clinical parameters. During treatment with minocycline, a marked decrease in serum levels of C-reactive protein (CRP) occurred, which correlated with a decrease in serum IL-6 and suppression of serum Ig M rheumatoid factor levels, independent of albumin levels, thus suggesting modulation of local synovial inflammation by minocycline [24]. One of the main measures of outcome in therapeutic studies for RA is the evaluation of radiographic evidence of bone erosions, as in the long term these correlate with poor physical function and mean irreversible disease progression. When the radiographic data from the MIRA study were analyzed, no difference in the rate of disease progression was observed within both groups, although erosions or bone damage tended to occur in a larger proportion of patients in the placebo than the minocycline-treated groups [25].

There is now a growing body of evidence to support the use of disease-modifying therapy earlier on in the disease process in RA, in order to minimize disease progression and irreversible joint damage [26, 27]. The first trial in early RA using minocycline was a multicentre, double-blind, placebo-controlled study in patients with a disease duration of less than 12 months [28]. This was a selected group in that all patients were seropositive for rheumatoid factor, one of the markers of poor prognosis in RA. Data from this study show a favorable effect of minocycline in patients with early disease with follow-up at 4 years, confirming that remissions were more frequent and further need for DMARDs was less in the patients treated early in the disease course with minocycline [29]. It is assumed that such an effect would be seen because of the effect on CRP, generally a good marker of cytokine production, and the theoretical effect on MMPs and TIMPs. No mention is made, however, about the presence of other known poor prognostic markers such as the shared epitope or radiographic erosions at baseline. In fact, no radiographic data are available from this study, so the potential role of minocycline as a true DMARD, namely reducing structural damage, cannot be established from these results.

New insights into the pathogenesis of RA have led to the successful targeting of pro-inflammatory cytokines known to play a central role in RA. Current data show TNF alpha blockade with chimeric and humanized compounds such as infliximab [30] and etanercept [31] to be superior to conventional DMARDs in clinical and radiological outcomes. Larger studies are needed which compare the efficacy of tetracyclines *versus* other DMARDs to allow for definite conclusions to be drawn with regard to their place in the therapeutic pyramid of RA.

Tetracycline-induced autoimmune disease

Despite growing research on and knowledge of the TCs, the most enigmatic aspect remains their relationship with autoimmunity. On one side, as discussed above, is the potential use of these drugs in the treatment of autoimmune processes such as inflammatory arthritis, on the other is the recently described association between these drugs, in particular minocycline with the development of autoimmune syndromes. Minocycline is a semisynthetic tetracycline currently used for a variety of disorders, in particular for the treatment of acne vulgaris; it is the most widely prescribed antibiotic for this condition in the United States, Canada and the United Kingdom [32]. It is very lipophilic and like doxycycline it is nearly completely absorbed after oral administration. Like the rest of the natural or semisynthetic tetracyclines, minocycline has a wide range of biological activity dependant on its C4-dimethylamino group. It is administered in a once- or twice-daily dosage because of its long half-life. This has been associated with good compliance and lack of resistance [33].

Despite having been generally well tolerated [34], a large number of adverse side-effects have been reported with minocycline. These include: vestibular symptoms [35], photosensitivity [36], skin pigmentation [37], eosinophilia, hypersensitivity pneumonitis, toxic hepatitis, etc. [33]. More recently interest has been fuelled about a relationship between minocycline and the development of autoimmune syndromes.

The first written account on the exacerbation of systemic lupus erythematosus after the use of TCs appeared in the late fifties [38]. Since then, there have been an increasing number of reports about the rheumatic manifestations associated with the use of minocycline, most of them in the last decade and nearly all reported in acne patients, reflecting the population exposed to the drug. Although acne itself can predispose to joint pains or arthralgia, the fact that the symptoms started only after exposure to the drug, with resolution after discontinuation and re-appearance on rechallenge, favors the association between the drug and development of the phenomenon [11]. In general the observed patterns of autoimmune reactions to TC antibiotics can be divided into early reactions such as serum-sickness-like and hypersensitivity reactions or late reactions, occurring on average 2 years after initiation of therapy. These include: vasculitis, drug-induced lupus erythematosus and auto-immune hepatitits (Tab. 3).

Table 3. Autoimmune syndromes associated with tetracycline use

1. Serum-sickness-like syndrome and hypersensitivity reactions
2. Vasculitis
3. Drug-induced lupus
4. Autoimmune hepatitis

Serum-sickness-like and hypersensitivity reactions

Serum-sickness-like reactions have been reported with minocycline, including fever, arthralgia, lymphadenopathy and urticarial rash [39–42]. Generalized hypersensitivity reactions, including Stevens-Johnson syndrome, have also been described [43–47]. These manifestations have in common that they tend to occur within a few weeks after initiation of minocycline therapy, are short-lived, resolve after discontinuation of the drug and in general, although suggestive of an antibody-mediated event, are not associated with antinuclear antibody (ANA) positivity [32]. Tetracycline and doxycycline have also been associated with these reactions; however, minocycline has been the implicated agent in over 80% of the cases [48].

Minocycline-induced vasculitis

Several cases have been reported of minocycline-induced vasculitis, usually associated with clinical manifestations such as fever and arthralgia or arthritis. Livedo reticularis has been reported in three series [49–51]. There have been single reports, including one of acute febrile neutrophilic dermatosis [52] and another recent case of a 15-year-old girl who developed a purpuric rash associated with fever, myalgia and cervical lymphadenopathy. Biopsy of the rash showed changes consistent with polyarteritis nodosa [53]. It is likely that the prevalence of this problem is underestimated, as histological diagnosis is rarely pursued if resolution of the clinical signs occurs on discontinuation of therapy.

Minocycline-induced lupus

Drug-induced lupus (DIL) has been so far the commonest autoimmune condition reported with minocycline. Since the first case appeared in the Lancet in 1992 [54], reports have increased in the last few years, perhaps reflecting awareness of this problem. No established consensus criteria exist for the diagnosis of DIL, but the accepted definition is that of a syndrome presenting with at least one clinical feature of systemic lupus erythematosus [55] (Tab. 4) with positive ANA and circumstantial association between the use of the drug and the development of clinical and serological signs [56, 57] (Tab. 5). There should be no previous history of systemic lupus erythematosus and cessation of the syndrome should occur on withdrawal of the drug, with reappearance on rechallenge. More than 80 cases have been described so far and it is suspected that this might be an underestimation of the frequency of this condition [58].

Three recent reviews [32, 58, 59] describe joint pains or arthralgia to be the commonest reported symptoms in minocycline-induced lupus, with synovitis described in about a quarter of the patients [58]. Constitutional symptoms such as fever, malaise, fatigue, anorexia or weight loss have been reported as associated with the musculoskeletal manifestations. In the most recent report, 57 cases of minocycline-induced lupus were reviewed [59]. Asymptomatic elevation of liver enzymes or symptomatic liver involvement with chronic hepatitis or autoimmune hepatitis were the second predominant manifestations [59]. Only a minority had any other clinical criteria of lupus such as dermatological (12 patients), pleuropulmonary (8 patients), haematological (5 patients) or neurological abnormalities (4 cases) and reassuringly only 1 case had renal involvement presenting with nephritis. Of interest is the fact that the majority

Table 4. The 1982 Revised ACR criteria for the diagnosis of systemic lupus erythematosus

Criterion	Definition
1. Malar rash	Fixed erythema, flat or raised, over the malar eminences, sparing the nasolabial folds
2. Discoid rash	Erythematosus raised patches with adherent keratotic scaling and follicular plugging
3. Photosensitivity	Skin rash as a result of abnormal reaction to sunlight, by patient history or physician observation
4. Oral ulcers	Oral or nasopharyngeal ulceration, usually painless observed by a physician
5. Arthritis	Nonerosive arthritis involving two or more peripheral joints, characterized by tenderness, swelling or effusion
6. Serositis	a) Pleuritis – convincing history or pleuritic pain or rub heard by a physician or evidence of pleural effusion; or b) Pericarditis – documented by ECG, or rub or evidence of pericardial effusion
7. Renal disorder	a) Persistent proteinuria greater than 0.5 g per day or greater than 3+ if quantification not performed b) Cellular casts – may be red cell, hemoglobin, granular, tubular or mixed
8. Neurologic disorder	a) Seizures – in the absence of offending drugs or known metabolic derangement b) psychosis
9. Hematologic disorder	a) Hemolytic anaemia with reticulocytosis or b) Leukopenia – less than 1500/mm^3 on more than two occasions c) Thrombocytopenia – less than 100,000/mm^3 in the absence of offending drugs
10. Immunologic disorder	a) Positive LE cell preparation; or b) Anti-DNA: antibody native to DNA in abnormal titre; or c) Anti-Sm: presence of antibody to Sm nuclear antigen; or d) False positive serologic test for syphilis known to be positive for at least 6 months and confirmed by *Treponema pallidum* immobilization or fluorescent treponemal antibody test
11. Antinuclear antibody	An abnormal titre of antinuclear antibody by immunofluorescence or an equivalent assay at any point of time in the absence of drug

Table 5. Guidelines for the identification of drug-induced lupus

- Absence of history suggestive of systemic lupus erythematosus before drug therapy
- Continuous treatment with a suspected drug for at least 1 month (usually much longer)
- Positive antinuclear antibody (ANA)
- At least one clinical feature of systemic lupus erythematosus during sustained treatment
- Rapid resolution in clinical symptoms and gradual fall in antinuclear antibodies after drug withdrawal

of the cases were reported in female patients, as is the case in autoimmune systemic lupus erythematosus. This sex ratio is perhaps surprising, as DIL normally follows the epidemiology of the underlying condition and acne has no sex predilection [32, 60]. However, in common with DIL linked to other drugs, all cases appeared after prolonged exposure to the drug, with a median onset of 19 months [59].

Serologically all the patients had a positive ANA. Anti-histone antibodies known to be associated with DIL, were present only in a small minority of the cases. These are not thought to play a role in the development of minocycline-induced lupus. Recent reports have noted a striking association with perinuclear anti-neutrophil cytoplasmic antibodies (p-ANCA), usually present in high titers in systemic vasculitides such as microscopic polyangiitis and Churg-Strauss vasculitis.

The p-ANCA antibodies are directed against several cytoplasmic components such as myeloperoxidase, lactoferrin and elastase [61]. Reports on the cytotoxic mechanism of action of other lupus-inducing drugs associated with p-ANCA, such as hydralazine [62], have shown that these were mediated through the myeloperoxidase enzyme pathway [63]. A similar mechanism has been proposed for minocycline [64, 65]; however, further research is awaited.

A recent report [65] on 29 patients that developed a lupus-like illness whilst taking minocycline for acne showed that all the symptomatic patients (14 in total) were p-ANCA positive, the majority with anti-myeloperoxidase (anti-MPO) specificity, whereas none of the well minocycline users were. This correlates with our own findings. To establish the prevalence of minocycline-induced autoimmune phenomena, we examined 252 acne sufferers, of which 69% had been exposed to minocycline at some point during their illness. We found that 6.9% of the exposed group were ANCA-positive, the majority with a p-ANCA pattern, whereas no ANCA positivity was found in the non-exposed group. Four patients became symptomatic and discontinued the drug on their own because of new onset of arthritis; interestingly all of them were ANA and p-ANCA positive [66]. Perhaps because of the small number studied, we did not find a statistically significant relationship between clinical symptoms and the presence of anti-MPO antibodies. A larger, longitudinal study is now under way to try and establish the incidence and true prevalence of these phenomena.

Interestingly, despite the growing use of minocycline in RA, there does not seem to be an increased incidence of minocycline-induced lupus in these patients. Our group has recently reported on a patient who developed a lupus-like syndrome after prolonged exposure to minocycline for the treatment of RA [67], which was associated to new ANA and p-ANCA positivity with anti-MPO specificity. Clinical and serological abnormalities resolved promptly on discontinuation of the drug. At 18 months follow-up off minocycline, the patient remains asymptomatic with a decreasing p-ANCA titre. The expected presence of overlapping autoimmune syndromes in rheumatological conditions might account for the apparent lack of reported drug-induced lupus in this group of patients. Also new presentations of polyarthralgia or polyarthritis, the most common presenting symptom of minocycline-induced lupus, are easily misinterpreted in the context of RA as a new flare-up of disease [68].

Autoimmune hepatitis

High-dose tetracyclines are known to cause a degree of liver toxicity that can range from hepatic steatosis to liver failure. Minocycline-induced hepatotoxicity was first reported after high-dose intravenous administration [69–71], suggesting a dose-dependent toxic effect. Around 30 cases of autoimmune hepatitis (AIH) induced by minocycline have been reported [58], ranging from mildly deranged liver serology to one case of fulminant hepatitis leading to liver transplant in a teenage girl [72] and two deaths due to hepatitis and pancytopenia [73].

The clinical picture of AIH in the reported cases was similar to that in patients with systemic lupus erythematosus. The presence of ANA is a common feature and there have been reports of the simultaneous occurrence of both conditions [73–75]. The main differences were the higher prevalence of males and the longer time for serum transaminases to return to normal after discontinuation of the drug [58] in AIH. Some cases required further therapy with corticosteroids or azathioprine [64, 76]. The time of exposure seems to vary, with some cases of acute hepatitis developing within days or weeks of receiving minocycline [32] to as long as months or years into therapy [73]. When liver biopsies have been performed, changes consistent with chronic active hepatitis and lobular hepatitis have been reported [73, 64].

Laboratory findings have shown patients to have a positive serum ANA, with anti-smooth muscle antibodies present only in a small proportion [32]. Perinuclear antineutrophil cytoplasmic antibodies were sought in only three cases but were positive in all [32].

Why minocycline?

So far there are more severe side-effects associated with the use of minocycline than with the rest of the TCs. DIL has only been reported in association with minocycline, but not with doxycycline or tetracycline. The pathogenesis of this phenomenon is far from understood, although several mechanisms have been proposed by different authors. Factors that have been associated with DIL include the use of the drug for long-term therapy, dose dependency and the presence of a reactive metabolite [48]. Minocycline is likely to be prescribed for long-term therapy as is the case for the treatment of acne, whereas the other TC compounds are usually prescribed for acute infections requiring relatively shorter courses [48]. It is not known, however, whether minocycline metabolism produces any specifically reactive metabolites.

Some authors have suggested the possibility of predisposing genetic factors. A recent paper reported on preliminary results in 14 symptomatic acne patients who had developed minocycline-induced lupus after long-term exposure to the drug (median 3.8 years) and suggested an association with the MHC class II genes HLA-DRB1 1104*, HLA-DR4 and HLA-DR2 [65]. Larger cohorts of patients with minocycline-induced lupus need to be studied to confirm these results.

Risks associated with exposure to tetracyclines

In a comparative review of the safety of tetracycline, minocycline and doxycycline prescriptions, Shapiro et al. reported retrospectively on data from local and national databases in Canada and a computer-based Medline search. Although they did not specify the number of patients reviewed in the study, it is estimated that in 1994 there were 1,866,000 prescriptions for tetracycline antibiotics in Canada. In total, more adverse reactions including hypersensitivity reactions, drug-induced lupus and single organ dysfunction were reported on the minocycline-exposed group. The authors concluded that these effects are still rare, bearing in mind that in the United Kingdom 28 million minocycline tablets were taken annually when this study was published [48].

A recent nested case-control study, using computerized records, looked at the risk of developing lupus-like symptoms in British patients receiving minocycline for acne vulgaris, compared to other tetracyclines or no tetracyclines at all. They reported in a cohort of 27,688 patients with acne, an 8.5-fold risk of developing a lupus-like syndrome with current use of minocycline that increased to 16-fold with long-term use when compared with non-users and past users of tetracyclines combined [77]. Females were at a 14-times increased risk of developing minocycline-induced lupus when compared with males, interesting when we bear in mind that 60% of the total of all minocycline prescriptions were given to male acne patients.

Summary

Tetracyclines are widely used, broad-spectrum antibiotics. The recently described non-antibiotic properties of these drugs have opened a new era for research into the aetiopathogenesis and therapy of a wide number of disorders in both the dental and medical fields. Due to their anti-inflammatory and immunomodulatory effects, they have been shown to be effective as disease-modifying agents in the treatment of RA. However, like any chemical compound, tetracyclines are not free from side-effects, and despite being generally well tolerated, these adverse effects can on occasions be severe. Of particular concern is the role of these drugs in the development of autoimmune disorders. Further knowledge of the newly designed chemically modified tetracyclines should, it is hoped, minimize these problems.

References

1 Golub LM, Lee HM, Lehrer G, Nemiroff A, Mc Namara TF, Kaplan R, Ramamurthy NS (1983) Minocycline reduces gingival collagenolytic activity during diabetes: preliminary observations and a proposed new mechanism of action. *J Periodont Res* 18: 516–526

2 Golub LM, Wolff M, Lee HM, Mc Namara TF, Ramamurthy NS, Zambon J, Ciancio S (1985) Further evidence that tetracyclines inhibit collagenase activity in human crevicular fluid and form other mammalian sources. *J Periodont Res* 20: 12–23

3 Golub LM, Lee HM, Ryan ME (1998) Tetracyclines inhibit connective tissue breakdown by multiple non-antimicrobial mechanisms. *Adv Dent Res* 12: 12–26

4 Greenwald R, Golub L, Lavietes B, Ramamurthy NS, Gruber B, Laslin RS, Mc Namara TF (1987) Tetracyclines inhibit human synovial collagenase *in vivo* and *in vitro*. *J Rheumatol* 13: 28–32

5 Rifkin BR, Vernillo A, Golub LM, Ramamurthy N (1994) Modulation of bone resorption by tetracyclines. *Ann N Y Acad Sci* 732: 165–180

6 Gomes BC, Golub LM, Ramamurthy NS (1984) Tetracyclines inhibit parathyroid hormone-induced bone resorption in organ culture. *Experientia* 40: 1273–1275

7 Vadas P, Pruzanski W, Kim J, Fornassier V (1989) The proinflammatory effect of the intraarticular injection of soluble venom and human phospholipase A2. *Amer J Pathol* 134: 807–811

8 Bomalaski JS, Lawton P, Browning JL (1991) Human extracellular recombinant phospholipase A2 induces an inflammatory response in rabbit joints. *J Immunol* 146: 3904–3910

9 Pruzanski W, Greenwald RA, Street IP, Laliberte F, Stefanski E, Vadas P (1992) Inhibition of the enzymatic activity of phospholipasese A2 by minocycline and doxycycline. *Biochem Pharmacol* 44: 1165–1170

10 Van Barr HM, van de Kerkhop PC, Mier PD, Happle R (1987) Tetracyclines are potent scavengers of the superoxide radical. *Br J Dermatol* 117: 131–132

11 Alarcon GS (1998) Minocycline for the treatment of rheumatoid arthritis. *Rheum Dis Clin N Amer* 24: 489–499

12 Sewell KL, Breedveld F, Furrie E, O'Brien J, Brinckerhoff C, Dynesius-Trentham R, Nosaka Y, Trentham DE (1996) The effect of minocycline in rat models of inflammatory arhtritis: correlation of arthritis suppression with enhanced T cell calcium influx. *Cell Immunol* 167: 195–204

13 Kloppenburg M, Verweij CL, Miltenburg AM, Verhoeven AJ, Daha MR, Dijkmans BA, Breedveld FC (1995) The influence of tetracyclines on T cell activation. *Clin Exp Immunol* 102: 635–641

14 Bartholomew LE (1965) Isolation and characterization of mycoplasma from patients with rheumatoid arthritis, systemic lupus erythematosus and Reiter's syndrome. *Arthritis Rheum* 8: 376–388

15 Pott HG, Wittenborg A, Junge-Hulsing G (1988) Long-term antibiotic treatment in reactive arthritis. *Lancet* 1(8579): 245–246

16 Panayi GS, Clark B (1989) Minocycline in the treatment of patients with Reiter's syndrome. *Clin Exp Rheumatol* 7: 100–101

17 Popert AJ, Gill AJ, Laird SM (1964) A prospective study of Reiter's syndrome: an interim report on the first 82 cases. *Br J Veneral Dis* 40: 160–165

18 Lauhio A, Lerisalo-Repo M, Lahdevirta J, Saikku P, Repo H (1991) Double-blind, placebo controlled study of three-month treatment with lymecycline in reactive arthritis, with special reference to chlamydia arthritis. *Arthritis Rheum* 34: 6–14

19 Skinner M, Cathcart ES, Mills JA, Pinals RS (1971) Tetracyclines in the treatment of rheumatoid arthritis. *Arthritis Rheum* 14: 727–732

20 Breedveld FC, Dijkmans BAC, Mattie H (1990) Minocycline treatment for rheumatoid arthritis. An open dose finding study. *J Rheumatol* 17: 43–46

21 Langevitz P, Bak I, Zemer D, Book M, Pras M (1992) Treatment of resistant rheumatoid arthritis with minocycline. An open study. *J Rheumatol* 19: 1502–1504

22 Kloppenburg M, Breedveld FC, Terwiel JP, Mallee C, Dijkmans BAC (1994) Minocycline in active rheumatoid arthritis. *Arthritis Rheum* 37: 629–636

23 Tilley BC, Alarcon GS, Heyse SP, Trentham DE, Neuner R, Kaplan DA, Clegg DO, Leisen JC, Buckley L, Cooper SM et al (1995) Minocycline in rheumatoid arthritis: a 48-week, double-blind, placebo-controlled trial. *Ann Intern Med* 122: 81–89
24 Kloppenburg M, Dijkmans BA, Verweik CL, Breedveld FC (1996) Inflammatory and immunological parameters of disease activity in rheumatoid arthritis patients treated with minocycline. *Immunopharmacology* 31: 163–169
25 Bluhm GB, Sharp JT, Tilley B, Alarcon GS, Cooper SM, Pillemer SR, Clegg DO, Heyse SP, Trentham DE, Neuner R et al (1997) Radiographic results from the minocycline in rheumatoid arthritis trial (MIRA). *J Rheumatol* 24: 1295–1302
26 Emery P (1994) The optimal management of early rheumatoid disease: the key to preventing disability. The Roche rheumatology prize lecture. *Br J Rheumatol* 33: 765–768
27 Van der Heide A, Jacobs JW, Bijlsma JWJ, Heurkens AH, van Booma-Frankfort C, van der Veen MJ, Haanen HC, Hofman DM, van Albada-kuipers GA, ter Borg EJ et al (1996) The effectiveness of early treatment with "second-line" antirheumatic drugs: a randomized controlled trial. *Ann Intern Med* 124: 699–707
28 O'Dell JR, Haire CR, Palmer W, Drymalski W, Wees S, Blakely K, Churchill M, Eckhoff PJ, Weaver A, Doud D et al (1997) Treatment of early rheumatoid arthritis with minocycline or placebo: results of a randomized, double-blind, placebo-controlled trial. *Arthritis Rheum* 40: 842–848
29 O'Dell JR, Paulsen G, Haire C, Blakely K, Palmer W, Wees S, Eckhoff J, Klassen LW, Churchill M, Doud D et al (1999) Treatment of early seropositive rheumatoid arthritis with minocycline. Four-year follow up of a double-blind, placebo-controlled trial. *Arthritis Rheum* 42: 1691–1695
30 Maini R, St Clair E, Breedveld F, Furst D, Kaldem J, Weisman M, Smolen J, Emery P, Harriman G, Feldmann M et al (1999) Infliximab (chimeric anti-tumor necrosis factor α monoclonal antibody) *versus* placebo in rheumatoid arthritis patients receiving concomitant methotrexate: a randomised phase III trial. *Lancet* 354: 1932–1939
31 Weinblatt M, Kremer J, Bankhurst A, Bulpitt K, Fleischmann R, Fox R, Jackson C, Lange M, Burge D (1999) A trial of etanercept, a recombinant tumor necrosis factor receptor:Fc fusion protein, in patients with rheumatoid arthritis receiving methotrexate. *N Engl J Med* 340: 253–259
32 Eichenfield AH (1999) Minocycline and autoimmunity. *Curr Opin Pediatr* 11: 447–456
33 Mitscher LA (ed) (1978) *The chemistry of the tetracycline antibiotics*. Marcel Dekker, New York
34 Goulden V, Glass D, Cunliffe WJ (1996) Safety of long-term high-dose minocycline in the treatment of acne. *Br J Dermatol* 134: 693–695
35 Allen JC (1976) Minocycline. *Ann Intern Med* 85: 482–487
36 Frost P, Weinstein GD, Gomez EC (1972) Phototoxic potential of minocycline and doxycycline. *Arch Dermatol* 105: 681–683
37 Ridway HA, Sonnex TS, Kennedy CTC, Millard PR, Henderson WJ, Gold SC (1982) Hyperpigmentation associated with oral minocycline. *Br J Dermatol* 107: 95–102
38 Domz C, Mc Namara D, Holzpfel H (1959) Tetracycline provocation in lupus erythematosus. *Ann Intern Med* 50: 1217–1226
39 Puyana J, Urena V, Quirce S, Fernandez-Rivas Cuevas M, Fraj J (1990) Serum sickness-like syndrome associated with minocycline therapy. *Allergy* 45: 313–315
40 Levenson T, Masood D, Patterson R (1996) Minocycline-induced serum sickness. *Allergy Asthma Proc* 17: 79–81
41 Knowles SR, Shapiro L, Shear NH (1996) Serious adverse reactions induced by minocycline: report of 13 patients and review of the literature. *Arch Dermatol* 132: 934–939
42 Harel L, Amir J, Livni E, Straussberg R, Varsano I (1996) Serum sickness-like reaction with minocycline therapy in adolescents. *Ann Pharmacol* 30: 481–489
43 Rosin MA (1984) Viral-like syndrome associated with minocycline (letter). *Arch Dermatol* 120: 575
44 Shoji A, Someda Y, Hamada T (1987) Stevens-Johnson syndrome due to minocycline therapy (letter). *Arch Dermatol* 123: 18–20
45 Gorard DA (1990) Late onset drug fever associated with minocycline. *Postgrad Med* 66: 404–405
46 Kaufmann D, Pichler W, Beer JH (1994) Severe episode of high fever with rash, lymphadenopathy, neutropenia and eosinophilia after minocycline therapy for acne. *Arch Intern Med* 154: 1983–1984
47 Parneix-Spake A, Bastuji-Garin S, Lobut JB, Erner J, Guyet-Rousset P, Revuz J, Roujeau JC (1995) Minocycline as possible cause of severe and protracted hypersensitivity drug reaction (letter). *Arch Dermatol* 131: 490–491

48 Shapiro LE, Knowles SR, Shear NH (1997) Comparative safety of tetracycline, minocycline and doxycycline. *Arch Dermatol* 133: 1224–1230
49 Elkayam O, Yaron M, Caspi D (1996) Minocycline induced arthritis associated with fever, livedo reticularis and p-ANCA. *Ann Rheum Dis* 55: 769–771
50 Elkayam O, Levartovsky D, Brautbar C, Yaron M, Burke M, Vardihor H, Caspi D (1998) Clinical and immunological study of 7 patients with minocycline induced autoimmune disease. *Amer J Med* 105: 484–487
51 Teitelbaum JE, Perez-Atayde AR, Cohen M, Bousvaros A, Jonas MM (1998) Minocycline-related autoimmune hepatitis: case series and literature review. *Arch Pediat Adolesc Med* 152: 1132–1136
52 Mensing H, Kowalzick L (1991) Acute febrile neutrophilic dematosis (Sweet's syndrome) caused by minocycline. *Dermatologica* 182: 43–46
53 Schrodt BJ, Callen JP (1999) Polyarteritis nodosa attributable to minocycline treatment for acne vulgaris. *Pediatrics* 103: 503–504
54 Matsuura T, Shimuzu Y, Fujimoto H, Miyazaki T, Kano S (1992) Minocycline related lupus. *Lancet* 340: 1553
55 Tan EM, Cohen AS, Fries JF, Masi AT, Mc Shane DJ, Rothfield NF, Schaller JG, Talal N, Winchester RJ (1982) The 1982 revised criteria for the classification of systemic lupus erythematosus. *Arthritis Rheum* 25: 1271–1277
56 Hess EV (1988) Drug-related lupus. *N Engl J Med* 318: 1460–1464
57 Rubin RL (1999) Etiology and mechanisms of drug-induced lupus. *Curr Opin Rheumatol* 11: 357–363
58 Elkayam O, Yaron M, Caspi D (1999) Minocycline-induced autoimmune syndromes: an overview. Sem. *Arthritis Rheum* 28: 292–297
59 Schlienger RG, Bircher AJ, Meier CR (2000) Minocycline-induced lupus. *Dermatology* 200: 230–231
60 Krohn K, Bennett R (1998) Drug-induced autoimmune disorders. *Immunol Allergy Clin N Amer* 18: 897–911
61 Wiik A, Stummann L, Kjeldsen L, Borregaard N, Ullman S, Jacobsen S, Halberg P (1995) The diversity of perinuclear antineutrophilic cytoplasmic antibodies (p-ANCA) antigens. *Clin Exp Immunol* 101: 15–17
62 Nassberger L, Sjoholm AG, Jonsson H, Sturfelt G, Ahesson A (1990) Autoantibodies against neutrophil cytoplasm components in lupus erythematosus and in hydralazine-induced lupus. *Clin Exp Immunol* 81: 380–383
63 Jiang X, Khursigara G, Rubin RL (1994) Transformation of lupus inducing drugs to cytotoxic products by activated neutrophils. *Science* 266: 810–813
64 Elkayam O, Levartovsky D, Brautbar C, Yaron M, Burke M, Vardinon N, Caspi D (1998) Clinical and immunological study of 7 patients with minocycline induced autoimmune disease. *Amer J Med* 105: 484–487
65 Dunphy J, Oliver M, Rands AL, Lovell CR, Mc Hugh NJ (2000) Antineutrophil cytoplasmic antibodies and HLA class II alleles in minocycline-induced lupus-like syndrome. *Br J Dermatol* 142: 461–467
66 Marzo Ortega H, Drysdale S, Griffiths B, Jackson G, Misbah S, Gough A, Cunliffe W, Emery P (1999) Minocycline-induced autoimmune disease: defining the problem. *Arthritis Rheum* 42:S324
67 Marzo-Ortega H, Misbah S, Emery P (2001) Minocycline induced autoimmune disease in rheumatoid arthritis: a missed diagnosis?. *J Rheumatol* 28: 377–378
68 Angulo J, Sigal LH, Espinoza LR (1999) Minocycline induced lupus and autoimmune hepatitis. *J Rheumatol* 26: 1420–1421
69 Schultz JC, Adamson JS, Workman WW, Norman TD (1963) Fatal liver disease and intravenous administration of tetracycline in high dosage. *N Engl J Med* 269: 999–1004
70 Peters RL, Edmonson HA, Mikkelsen WP, Tatter D (1967) Tetracyclines-induced fatty liver in non pregnant patients. *Amer J Surg* 113: 622–632
71 Burette A, Finet C, Prigogine T, De Roy G, Deltenre M (1984) Acute hepatic injury associated with minocycline. *Arch Intern Med* 144: 1491–1492
72 Davies MG, Kersey PJW (1989) Acute hepatitis and exfoliative dermatitis associated with minocycline. *BMJ* 298: 1523–1524
73 Gough A, Chapman S, Wagstaff K, Emery P, Elias E (1996) Minocycline induced autoimmune hepatitis and systemic lupus erythematosus-like syndrome. *BMJ* 312: 169–172

74 Angulo JM, Sigal LH, Espinoza LR (1998) Coexistent minocycline-induced systemic lupus erythematosus and autoimmune hepatitis. Sem. *Arthritis Rheum* 28: 187–192
75 Crosson J, Stillman T (1997) Minocycline-related lupus erythematosus with associated liver disease. *J Amer Acad Dermatol* 36: 867–868
76 Malcolm A, Heap TR, Eckstein RP, Lunzer MR (1996) Minocycline-induced liver injury. *Amer J Gastroenterol* 91: 1641–1643
77 Sturkenboom MCJ, Meier CR, Jick H, Stricker BHC (1999) Minocycline and lupus-like syndrome in acne patients. *Arch Intern Med* 159: 493–497

Tetracyclines in Biology, Chemistry and Medicine
ed. by M. Nelson, W. Hillen and R.A. Greenwald

Tetracycline photodynamics

Brad R. Zerler

CollaGenex Pharmaceuticals, Inc., 41 University Drive, Newtown, PA 18940, USA

Introduction

Tetracyclines are a family of broad-spectrum antibiotics which have been used for decades to treat a variety of infections. Tetracyclines also have numerous non-antimicrobial properties which make them attractive for the treatment of inflammatory disorders and cancer. These include the ability to inhibit matrix metalloproteinases (MMPs), serine proteases, and various inflammatory cytokines such as TNF-α, iNOS, and PLA_2 (reviewed in [1]). The clinical value of the tetracyclines as antibiotics is enhanced by their low toxicity during short-term administration. However, the treatment of chronic disorders such as inflammation and cancer can require long-term, if not life-long treatment. This type of prolonged use of tetracyclines can be associated with certain treatment obstacles such as the emergence of antibiotic-resistant flora, gastrointestinal upset, photosensitivity and related skin disorders, all of which can put constraints on the administered dose. To circumvent obstacles associated with the development of microbial resistance and the normal balance of the GI flora, a series of chemically modified tetracyclines (CMTs) have been developed that lack anti-microbial activity but retain anti-inflammatory and anti-cancer activity. However, the chemical modifications that affect the antimicrobial activities of the tetracycline family do not necessarily translate into reduced photosensitivity. Therefore, phototoxicity may still limit the therapeutic potential of the non-antimicrobial CMTs.

The skin abnormalities associated with tetracycline photosensitivity appear as an exaggerated sunburn reaction. Some cases are associated with edema and blistering and a small cohort of patients develop photo-onycholysis on the fingers and toes. Photoreaction to tetracyclines is dose-dependent; therefore, reports of phototoxicity will vary depending on the dose administered and subsequent serum levels achieved in individuals. For example, Layton and Cunliffe [2] report a dose-dependent incidence of phototoxicity of 20% and 42% in individuals receiving long-term treatment with 150 mg/day and 200 mg/day of doxycycline, respectively. This phototoxic rate is substantially greater than the incidence of 3% reported for individuals receiving 100 mg/day doxycycline.

Phototoxicity assays

The phototoxic potential of the tetracyclines varies greatly among family members. Demethylchlortetracycline (DMTC) is associated with a high incidence of phototoxic reactions in individuals, whereas doxycycline is less phototoxic and methacycline and minocyline are rarely associated with phototoxicity [3]. However, the methods used to compare phototoxicity in individuals varies greatly between investigators and different protocols can lead to dramatically different results. A comparison of DMTC, lymecycline, and doxycycline, at similar serum concentrations, in a double-blind cross-over study in healthy human volunteers, revealed that doxycycline was the most potent photosensitizer [4]. Lymecycline was observed to be less phototoxic than doxycycline and DMTC was reported to be the least phototoxic of the three tetracyclines tested. Photoreactions were assessed 24 h following controlled exposure to long-wave ultraviolet radiation (320 to 400 nm; UVA). Although this ranking reflects the observations from a scientifically well-controlled study, the results do not correlate with the phototoxic ranking observed in individuals taking tetracyclines clinically and receiving only natural sunlight exposure under controlled conditions [3, 5]. In these studies the phototoxicity associated with DMTC was much greater than doxycycline. The absence of concomitant UVB light exposure may explain the observed discrepancies in DMTC phototoxicity in the two studies described above. Bjellerup [6] observed that UVB, although unable to induce tetracycline phototoxicity itself, significantly augments the phototoxic reaction from UVA light.

In the development of more effective tetracycline analogs, the photosensitizing potential of the derivatives should be incorporated into the screening cascade to identify derivatives with undesirable phototoxicity early in the development process. As described below, there are numerous *in vitro* phototoxicity assays suitable for this purpose. In addition to accuracy, reproducibility, simplicity, and a requirement for small amounts of test material, the most important criterion of such *in vitro* assays is to correctly predict the *in vivo* phototoxic potential of the derivatives.

The 3T3 Neutral Red Uptake Phototoxicity Test is one assay that has undergone extensive validation under the direction of the European Union and COLIPA (European Cosmetic Industry Association [7, 8]). The assay uses fibroblasts treated with dilutions of test material in the presence or the absence of a non-cytotoxic dose of UVA and visible light [9, 10]. Cytotoxicity is determined by comparing the extent of neutral red dye uptake, by means of a spectrophotometer, in the presence and absence of light exposure. Phototoxicity is predicted from the relative cytotoxicity observed between the two treatments. In this assay the investigators observed a phototoxicity ranking of doxycycline > tetracycline > minocycline. These results correlate with the *in vivo* phototoxic ranking of these tetracyclines in individuals and, therefore, could be used to predict the phototoxic potential of new tetracycline derivatives.

There is also a similar tissue culture assay based on the extent of photohemolysis of red blood cells (RBC) freshly isolated from human volunteers [11]. RBC are quantified by measuring hemoglobin content through the use of Drabkin's solution, which converts hemoglobin to the stable pigment cyanmethemoglobin which is detected with a spectrophotometer. In this analysis the phototoxic ranking of seven members of the tetracycline family were in the order of DMCT > doxycycline >> methacycline > tetracycline = oxytetracycline = chlortetracycline > lymecycline > minocycline (which had no photohemolytic effect).

To replicate the dynamics associated with the milieu of the natural target tissue, several groups have developed three-dimensional skin models which can be used to determine tetracycline phototoxicity. A dermal equivalent system developed by Augustin et al. [12] uses a collagen/glycosaminoglycan/chitosan porous matrix containing normal human fibroblasts. Seeding human keratinocytes onto this system then leads to a fully differentiated epidermis referred to as the skin equivalent model. The investigators observed that the dermal equivalent model was more sensitive to tetracycline-induced phototoxicity than the skin equivalent model. Both systems were dependent on UVA dose and time of exposure. A deficiency of the model is the inability to subtract out cytotoxicity from phototoxicity at high tetracycline concentrations.

Edwards et al. [13] have also developed a fully differentiated 3-dimensional dermal support co-culture system consisting of dermal fibroblasts and a multilayered epidermis comprising differentiated keratinocytes. In this system a 1% tetracycline solution exhibited a time-dependent decrease in cell viability in the presence of UVA radiation.

All the assays described above provide a reasonable evaluation of the comparative phototoxic potential of tetracyclines. However, there are shortcomings to the assays. For example, a tetracycline derivative may not be phototoxic *in vitro*, but it is possible for a metabolite to be formed *in vivo* that may be very phototoxic. In addition, the absorbed energy in the *in vitro* photochemical reaction may result in the formation of new photoproducts and toxicities that may not be relevant *in vivo*. The inability of the assays to replicate the natural physiological process of tetracycline metabolism may result in misleading information concerning the phototoxic potential of a tetracycline derivative, but this deficiency is outweighed by the overall value of the accuracy of predicting the phototoxic potential of the starting product.

Mechanisms for tetracycline phototoxicity

The mechanisms responsible for tetracycline photosensitivity have not been entirely elucidated. The electronic absorption spectra of the tetracyclines change upon UVA and visible light irradiation, indicating the formation of different photoproducts. Interestingly, the absorption spectrum of minocycline does not change, indicating that minocycline is photostable [14]. In these studies, minocyline was also found to be non-phototoxic to proliferating lymphocytes in cell culture, whereas the other tetracycline derivatives each exhibited some degree of phototoxicity. The Hasan group also observed that tetracycline-induced photosensitization is at least partially dependent on the presence of oxygen and singlet oxygen may be involved. These conclusions are supported in studies by Miskoski et al. [15], in which dye-sensitized (rose bengal) photoreactions of the tetracyclines were detected only in the presence of molecular oxygen. In addition, Wiebe and Moore [16] demonstrated that photo-irradiated tetracyclines are capable of acting as photosensitizers for the oxygenation of suitable acceptor molecules such as dimethylfuran and limonene, and Glette and Sanberg [17], using tryptophan degradation as a measurement of singlet oxygen production, showed that the extent of singlet oxygen production by several photoirradiated tetracyclines generally correlated with their reported phototoxicities.

Although singlet oxygen production appears to play a role in tetracycline phototoxicity, additional mechanisms that dictate the extent of phototoxicity *in vivo* cannot be ruled out, including reactions mediated by the triplet state, tissue distribution, and cellular uptake. For example, Shea et al. [18] observed that lumidoxycycline, a doxycycline-derived photoproduct, was approximately 50-fold less phototoxic toward a human bladder carcinoma cell line than parental doxycycline. However, lumidoxycycline was approximately 2.6 times more effective than doxycycline in generating singlet oxygen and subsequent degradation of histidine. But lumidoxycycline was not taken up by the bladder cells as efficiently as doxycycline, which could partly explain the differences in cellular phototoxicity between the two tetracyclines. Finally, Nilsson et al. [19] argue against the singlet oxygen scenario because the extent of erythrocyte photolysis in the presence of DMTC was the same in H_2O and D_2O, and the photoreaction was not inhibited by histidine.

It has been suggested that binding of divalent cations can affect the therapeutic action of tetracyclines and that there may be a direct relationship between metal ion binding and reduction of photosensitization [16, 20]. However, eliminating one of the Ca^{+2} and Zn^{+2} binding sites in tetracycline results in only slightly decreased phototoxicity. A tetracycline derivative, CMT-5 (see Fig. 1), in which the carbonyl oxygen at C-11 and the hydroxyl group at C-12 were replaced with nitrogen atoms, thus producing the pyrazole analog, exhibits markedly reduced MMP inhibitory activity, supposedly due to the inability to competitively bind Zn^{+2} which is required for maximal MMP activity [1, 21]. In the NIH 3T3 cell-based neutral red uptake phototoxicity assay described in Zerler et al. [10], CMT-5 was approximately one-third less phototoxic than tetracycline. The mean photo effect (MPE), a measure of phototoxicity based on a comparison of cell viability curves generated in the presence and absence of UV light [22], was 0.455 for CMT-5 compared to 0.679 for tetracycline.

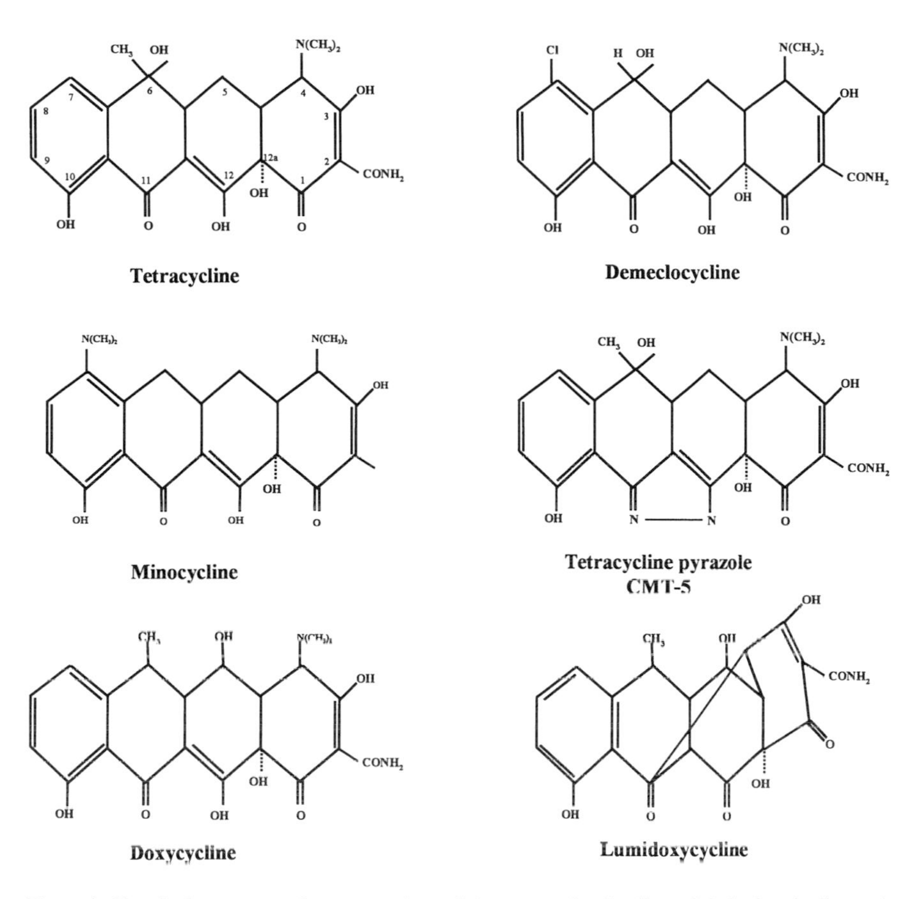

Figure 1. Chemical structures of representatives of the tetracycline family and their chemically modified derivatives.

Intracellular phototargets

The cell membrane is one obvious target for the tetracyclines and it has been suggested that lipophilicity may be directly correlated with the extent of cell membrane binding and subsequent phototoxicity. Although lipophilicity may contribute to the overall phototoxicity of the tetracyclines, other mechanisms appear to play a significant role. For example, Shea et al. [18] observed that lumidoxycycline, which is more hydrophobic than doxycycline, was less phototoxic than doxycycline in an *in vitro* assay. Interestingly, doxycycline was found to localize to the mitochondria, whereas lumidoxycycline exhibited a more diffuse cellular membrane localization pattern that included the mitochondria. Although the cellular staining patterns of the two derivatives differed, both compounds induced phototoxic injury to mitochondria. In a study by Pato using bacteria, tetracycline administration resulted in leakage of intracellular pools of nucleotides, amino acids and a sugar analog, leading to the cessation of DNA replication [23]. Pato suggests that the leakage may be due to tetracycline-induced perturbations to the cell membrane.

In addition to cell membranes, the tetracyclines have been reported to photoincorporate into eucaryotic ribosomes [24]. Tritiated tetracycline was observed to incorporate into both the 40-S and 60-S subunits upon photoirradiation. Furthermore, the tetracycline ribosome complex was defective in poly(U)-mediated protein synthesis. If photoirradiated intracellular tetracycline inhibits protein synthesis in eucaryotes by binding to ribosomes, then this may result in the secondary inhibition of DNA synthesis, ultimately triggering cell death pathways that culminate in a phototoxic response. However, in a purified system, tetracyclines have been observed to bind to DNA in the presence of divalent cations even in the absence of photoirradiation [25]. Therefore, the contributions of individual cellular targets to the phototoxic effects of tetracycline still need further characterization.

Minocycline

Interestingly, minocycline, although rarely reported to be phototoxic, is associated with other dermal disorders, including hyperpigmentation and various skin eruptions. Peyriere et al. [26] compiled data reported to the French Regional Centre of Pharmacovigilance between 1984 and 1996 for minocycline-induced photoreactions. There were 77 case reports documenting 13 photosensitive rashes, 52 skin pigmentations and 12 nail and/or intraoral pigmentations. The mechanism of minocycline-associated pigmentation is not fully understood. Iron and melanin have been observed in combination with minocyline in pigmented skin. In addition, lysosomal enzymes can oxidize minocycline into a quinone by-product which results in pigmentation.

Black pigmentation in the thyroid gland has also been observed in patients receiving long-term minocycline therapy. Unlike other tetracyclines, minocycline appears to accumulate in the thyroid [27]. One explanation for the pigmentation is that minocycline's strongly electron-donating dimethlylamino group increases the likelihood of oxidation [28]. Taurog et al. [29] suggest that thyroid peroxidase interacts with minocycline, resulting in the formation of a black product. Other tetracycline family members are not oxidized to dark pigmented products by thyroid peroxidase.

Summary

Phototoxic reactions caused by tetracyclines may present serious obstacles to the long-term use of these compounds for treatment of infectious diseases and various inflammatory disorders and cancer. Although the mechanisms responsible for tetracycline-mediated phototoxicity are still being characterized, there are several tissue culture assays available to identify and rank potential phototoxic tetracyclines and by-products. Several studies suggest that singlet oxygen formation contributes to tetracycline phototoxicity and cellular phototargets include membranes, mitochondria and ribosomes. Tetracycline phototoxicity is dose-dependent and there are subsets of tetracyclines that are not phototoxic even at high doses; however, they may cause other skin disorders such as the pigmentations observed with minocycline.

References

1 Golub LM, Lee H-M, Ryan ME (1998) Tetracyclines inhibit connective tissue breakdown by multiple non-antimicrobial mechanisms. *Adv Dent Res* 12: 12–26

2 Layton AM, Cunliffe WJ (1993) Phototoxic eruptions due to doxycycline – a dose-related phenomenon. *Clin Exp Dermatol* 18: 425–427

3 Frost P, Weinstein GD, Gomez EC (1972) Phototoxic potential of minocycline and doxycycline. *Arch Dermatol* 105: 681–683

4 Bjellerup M, Ljunggren B (1987) Double blind cross-over studies on phototoxicity to three tetracycline derivatives in human volunteers. *Photodermatology* 4: 281–287

5 Blank H, Stanley CI, Catalano PM (1968) Photosensitivity studies with demethylchlortetracycline and doxycycline. *Arch Dermatol* 97: 1–2

6 Bjellerup M (1986) Medium-wave ultraviolet radiation (UVB) is important in doxycycline phototoxicity. *Acta Dermato-Venereol (Stockh)* 66: 510–514

7 Spielmann H, Balls M, Dupuis J, Pape WJ, Pechovitch G, de Silva O, Holzhutter HG, Clothier R, Desolle P, Gerberick F et al (1998) The international EU/COLIPA *in vitro* phototoxicity validation study; results of phase II (blind trial). Part 1: The 3 T3 NRU phototoxicity test. *Toxicol Vitro* 12: 305–327

8 Spielmann H, Balls M, Dupuis J, Pape WJ, de Silva O, Holzhutter HG, Gerberick F, Liebsch M, Lovell WW, Pfannenbecker U (1997) A study on UV filter chemicals from Annex VII of European Union Directive 76/768/EEC, in the *In vitro* 3 T3 phototoxicity test. *ATLA* 26: 679–708

9 Lazarow RM, Isseroff R, Gomez EC (1992) Quantitative *in vitro* assessment of phototoxicity by a fibroblast-neutral red assay. *J Invest Dermatol* 98: 725–729

10 Zerler B, Roemer E, Raabe H, Sizemore A, Reeves A, Harbell J (2000) Evaluation of the phototoxic potential of chemically modified tetracyclines with the 3 T3 neutral red uptake phototoxicity test. *In*: M Balls, A-M van Zeller, ME Halder (eds): *Progress in the Reduction, Refinement, and Replacement of Animal Experimentation.* Elsevier Science B.V., Amsterdam, 545–554

11 Bjellerup M, Ljunggren B (1985) Photohemolytic potency of tetracyclines. *J Invest Dermatol* 84: 262–264

12 Augustin C, Collombel C, Damour O (1997) Use of dermal equivalent and skin equivalent models for identifying phototoxic compounds *in vitro*. *Photodermatol Photomed* 13: 27–36

13 Edwards SM, Donnelly TA, Sayre RM, Rheins LA (1994) Quantitative *in vitro* assessment of phototoxicity using a human skin model, skin^{2TM}. *Photodermatol Photoimmunol Photomed* 10: 111–117

14 Hasan T, Kochevar IE, McAuliffe Cooperman BS, Abdulah D (1984) Mechanism of tetracycline phototoxicity. *J Invest Dermatol* 83: 179–183

15 Miskoski S, Sanchez E, Garavano M, Lopez M, Soltermann AT, Garcia NA (1998) Singlet molevcular oxygen-mediated photo-oxidation of tetracyclines: kinetics, mechanism and microbiological implications. *J Photochem Photobiol B-Biol* 43: 164–171

16 Wiebe JA, Moore DE (1977) Oxidation photosensitized by tetracyclines. *J Pharm Sci* 66: 186–189

17 Glette J, Sandberg S (1986) Phototoxicity of tetracyclines as related to singlet oxygen production and uptake by polymorphonuclear leukocytes. *Biochem Pharmacol* 35: 2883–2885

18 Shea CR, Olack GA, Morrison H, Chen N, Hasan T (1983) Phototoxicity of lumidoxycycline. *J Invest Dermatol* 101: 329–333

19 Nilsson R, Swambeck G, Wennersten G (1975) Primary mechanisms of erythrocyte photolysis

induced by biological sensitizers and phototoxic drugs. *Photochem Photobiol* 22: 183–186
20 Riaz M, Pilpel N (1984) Complexation of tetracyclines with metal ions in relation to photosensitization. *J Pharm Pharmacol* 36: 153–156
21 Blackwood RK (1969) Tetracyclines. *Encyclop Chem Technol* 20: 1–33
22 Holzhutter HG (1997) A general measure of the *in vitro* phototoxicity derived from pairs of dose response curves and its use for predicting *in vivo* phototoxicity of chemicals. *ATLA* 25: 445–462
23 Pato ML (1977) Tetracycline inhibits propagation of deoxyribonucleic acid replication and alters membrane properties. *Antimicrob Agents Chemother* 11: 318–323
24 Reboud A-M, Dubost S, Reboud J-P (1982) Photoincorporation of tetracycline into rat-liver ribosomes and subunits. *Eur J Biochem* 124: 389–396
25 Kohn KW (1961) Mediation of divalent metal ions in the binding of tetracycline to macromolecules. *Nature* 191: 1156–1158
26 Peyriere H, Hillaire-Buys D, Dereure O, Meunier L, Blayac JP (1999) Muco-cutaneous pigmentation and photosensitization induced by minocycline hydrochloride. *J Dermatol Treatment* 10: 105–112
27 Kelly RG, Kanegis LA (1967) Metabolism and tissue distribution of radioisotopically labeled minocycline. *Toxicol Appl Pharmacol* 11: 171–183
28 Shapiro LE, Knowles SR, Shear NH (1997) Comparative safety of tetracycline, minocycline, and doxycycline. *Arch Dermatol* 133: 1224–1230
29 Taurog A, Dorris M, Doerge D (1996) Minocycline and the thyroid: antithyroid effects of the drug, and the role of thyroid peroxidase in minocycline-induced black pigemntation of the gland. *Thyroid* 6: 211–219

Subject index